T0281440

Graduate Texts in Physics

Graduate Texts in Physics

Graduate Texts in Physics publishes core learning/teaching material for graduate- and advanced-level undergraduate courses on topics of current and emerging fields within physics, both pure and applied. These textbooks serve students at the MS- or PhD-level and their instructors as comprehensive sources of principles, definitions, derivations, experiments and applications (as relevant) for their mastery and teaching, respectively. International in scope and relevance, the textbooks correspond to course syllabi sufficiently to serve as required reading. Their didactic style, comprehensiveness and coverage of fundamental material also make them suitable as introductions or references for scientists entering, or requiring timely knowledge of, a research field.

More information about this series at http://www.springer.com/series/8431

Gerd Keiser

Biophotonics

Concepts to Applications

 Springer

Gerd Keiser
Department of Electrical and Computer
   Engineering
Boston University
Newton, MA
USA

ISSN 1868-4513   ISSN 1868-4521   (electronic)
Graduate Texts in Physics
ISBN 978-981-10-9289-3   ISBN 978-981-10-0945-7   (eBook)
DOI 10.1007/978-981-10-0945-7

This Springer imprint is published by Springer Nature
The registered company is Springer Science+Business Media Singapore Pte Ltd.

The original version of the bookfrontmatter was revised: For detailed information please see Erratum. The Erratum to this chapter is available at DOI 10.1007/978-981-10-0945-7_12

To
Ching-yun, Nishla, Kai, and Neyla
for their loving patience and encouragement

Preface

The discipline of biophotonics or biomedical optics has undergone a fascinating journey in the past several decades and it is still growing rapidly. As the name *biophotonics* implies, the application of this field to biological disciplines is based on the wide range of photonics technologies, which involve the generation, detection, and control of photons for enabling functions such as transferring or processing information, analyzing material characteristics, sensing or measuring changes in physical parameters, and modifying material characteristics. Basically biophotonics deals with the interaction between light and biological material. The resulting interactions can be used in almost all biomedical areas for basic life science research and for biomedical diagnosis, therapy, monitoring, imaging, and surgery.

Owing to the importance of biophotonics to all aspects of human health, it is essential that a wide range of biomedical researchers, healthcare professionals, clinical technicians, and biomedical engineers have a good understanding of biophotonics and its applications. To address the attainment and implementation of these skills, this book provides the basic material for a one-semester entry-level course in the fundamentals and applications of biophotonics technology for senior or postgraduate students. It also will serve well as a working reference or as a short-course textbook for biomedical researchers, practicing physicians, healthcare professionals, clinical technicians, and biomedical engineers and technicians dealing with the design, development, and application of photonics components and instrumentation to biophotonics issues.

In Chap. 1–5 the sequence of topics takes the reader systematically from the underlying principles of light and biology, through the fundamentals of optical fiber light guiding, and then through optical sources and photodetection methods. Next, the topics in Chap. 6–10 address the concepts of light–tissue interactions, various optical probes and photonics sensing techniques, the principles of microscopy and spectroscopy, and biophotonics imaging modalities. The final chapter discusses advanced techniques and developments such as optical trapping, miniaturized instruments, single nanoparticles detection, and optogenetics procedures. By mastering these fundamental topics the reader will be prepared not only to contribute to current biomedical photonics disciplines, but also to understand quickly any further technology developments for future enhanced biophotonics developments.

The background required to study the book is that of typical senior-level science and engineering students. This background includes introductory biology and chemistry, calculus, and basic concepts of electromagnetism and optics as presented in a freshman physics course. To assist readers in learning the material and applying it to practical situations, 104 worked examples are given throughout the text. A collection of 129 homework problems is included to test the readers' comprehension of the material covered, and to extend and elucidate the text.

The articles and books cited as references in the text were selected from among numerous texts and thousands of papers in the literature relating to the material covered in each chapter. Because biophotonics brings together research, development, and application efforts from many different scientific, medical, and engineering disciplines, these references are a small sample of the major contributions to biophotonics. A number of these references are review papers and provide a starting point for delving deeper into any given topic.

Newton, USA   Gerd Keiser

Acknowledgments

For preparing this biophotonics book, I am extremely grateful to the many people worldwide with whom I had numerous beneficial discussions, who helped and inspired me in many different ways, who suggested modifications, and who supplied me with material. These individuals include Selim Ünlü and Irving Bigio, Boston University; Arthur Chiou and Fu-Jen Kao, National Yang Ming University; San-Liang Lee, Shih-Hsiang Hsu, Olivia Haobijam, and Puspa Pukhrambam, National Taiwan University of Science and Technology; Perry Ping Shum, Fei Xiong, Nan Zhang, and Ying Cui, Nanyang Technological University; Dora Juan Juan Hu, U.S. Dinish, and Jianzhong Hao, A*STAR, Singapore; Jürgen Popp, Friedrich-Schiller University Jena; David Sampson and Brendan Kennedy, the University of Western Australia; Sheng-Lung Huang, Lon Wang, and Chih-Chung Yang, National Taiwan University; Aaron H.P. Ho, the Chinese University of Hong Kong; Christina Lim, Elaine Wong, and Thas Nirmalathas, the University of Melbourne; Heidi Abrahamse, University of Johannesburg; Patricia S.P. Thong, National Cancer Center Singapore; Rozhin Penjweini, University of Pennsylvania, and Praveen Arany, University at Buffalo.

In addition, I would like to thank Loyola D'Silva and the other editorial and production team members of Springer. Their assistance and attentiveness during the writing, editing, and production cycles were much appreciated. As a final personal note, I am grateful to my wife Ching-yun and my family members Nishla, Kai, and Neyla for their patience and encouragement during the time I devoted to writing this book.

Newton, USA   Gerd Keiser

Contents

About the Author

Gerd Keiser is Research Professor at Boston University and Professor and consultant at PhotonicsComm Solutions, a firm specializing in education and consulting for the optical communications and biophotonics industries. Previously he was involved with telecom technologies at Honeywell, GTE, and General Dynamics. His technical achievements at GTE earned him the prestigious Leslie Warner Award. In addition, he has served as Adjunct Professor of Electrical Engineering at Northeastern University, Boston University, and Tufts University, and was an industrial advisor to the Wentworth Institute of Technology. Formerly he was a chair professor in the Electronics Engineering Department at the National Taiwan University of Science and Technology. He also was a visiting researcher at the Agency for Science, Technology, and Research (A*STAR) in Singapore and at the University of Melbourne, Australia. He is a life fellow of the IEEE, a fellow of OSA and SPIE, an associate editor and reviewer of several technical journals, and the author of five books. He received his B.A. and M.S. degrees in Mathematics and Physics from the University of Wisconsin and a Ph.D. in Physics from Northeastern University. His professional experience and research interests are in the general areas of optical networking and biophotonics.

Abbreviations

ADP	Adenosine diphosphate
AMP	Adenosine monophosphate
APD	Avalanche photodiode
ATP	Adenosine triphosphate
BFP	Blue fluorescent protein
CARS	Coherent anti-Stokes Raman spectroscopy
CCD	Charge-coupled device
CFP	Cyan fluorescent protein
CMOS	Complementary metal-oxide-semiconductor
CW	Continuous wave
DCF	Double-clad fiber
DCS	Diffuse correlation spectroscopy
DFB	Distributed feedback (laser)
DIC	Differential interference contrast
DNA	Deoxyribonucleic acid
DOF	Depth of field
DPSS	Diode-pumped solid-state
DRS	Diffuse reflectance spectroscopy
ELISA	Enzyme-linked immunosorbent assay
ESS	Elastic scattering spectroscopy
eV	Electron volt
FBG	Fiber Bragg grating
FCS	Fluorescence correlation spectroscopy
FD-OCT	Fourier domain OCT
FEL	Free-electron laser
FIDA	Fluorescence intensity distribution analysis
FLIM	Fluorescence lifetime imaging microscopy
FN	Field number
FOV	Field of view
FRET	Fluorescence resonance energy transfer
FRET	Förster resonance energy transfer

FTIR	Fourier transform infrared
FWHM	Full-width half-maximum
FWM	Four-wave mixing
GFP	Green fluorescent protein
GRIN	Graded-index
HCPCF	Hollow-core photonic crystal fiber
HCS	Hard-clad silica
HMFG	Heavy metal fluoride glasses
HRCT	High-resolution computed tomography
HSI	Hyperspectral imaging
IR	Infrared
IRIS	Interferometric reflectance imaging sensor
LED	Light-emitting diode
LITT	Laser-induced interstitial thermotherapy
LLLT	Low-level light therapy (photobiomodulation)
LOC	Lab-on-a-chip
LP	Linearly polarized
LPG	Long-period grating
LSFM	Light sheet fluorescence microscopy
LSPR	Localized surface plasmon resonance
LSS	Light scattering spectroscopy
MFD	Mode-field diameter
MI	Michelson interferometer
MMF	Multimode fiber
MOSFET	Metal-oxide-semiconductor field-effect transistor
MRI	Magnetic resonance imaging
MZI	Mach–Zehnder interferometer
NA	Numerical aperture
Nd:YAG	Neodymium:yttrium aluminum garnet
NEP	Noise equivalent power
NIR	Near-infrared
NO	Nitric oxide
OCE	Optical coherence elastography
OCT	Optical coherence tomography
OSA	Optical spectrum analyzer
PALM	Photo activated localization microscopy
PAT	Photoacoustic tomography
PCF	Photonic crystal fiber
PCH	Photon counting histograms
PCS	Photon correlation spectroscopy
PDL	Polarization dependent loss
PDT	Photodynamic therapy
PLA	Percutaneous laser ablation
PMMA	Polymethylmethacrylate
PMT	Photomultiplier tube

POC	Point of care
POF	Plastic optical fiber
POF	Polymer optical fiber
PSF	Point-spread function
QCL	Quantum cascade laser
RET	Resonance energy transfer
RI	Refractive index
RNA	Ribonucleic acid
ROS	Reactive oxygen species
SD-OCT	Spectral domain OCT
SERS	Surface-enhanced Raman scattering
SFG	Sum-frequency generation
SHG	Second-harmonic generation
SM	Single-mode
SMF	Single-mode fiber
SNR	Signal-to-noise ratio
SPR	Surface plasmon resonance
SRG	Stimulated Raman gain
SRI	Surrounding refractive index
SRL	Stimulated Raman loss
SRS	Stimulated Raman scattering
SSIM	Saturated structured illumination microscopy
SS-OCT	Swept source OCT
STED	Stimulated emission depletion microscopy
STORM	Stochastic optical reconstruction microscopy
TD-OCT	Time domain OCT
TE	Transverse electric
TFF	Thin film filter
THG	Third-harmonic generation
TPEF	Two-photon excitation fluorescence
TRT	Thermal relaxation time
UV	Ultraviolet
VCSEL	Vertical cavity surface-emitting laser
WDL	Wavelength dependent loss
YAG	Yttrium aluminum garnet
YFP	Yellow fluorescent protein

Chapter 1
Overview of Biophotonics

Abstract Biophotonics or biomedical optics has become an indispensible tool for basic life sciences research and for biomedical diagnosis, therapy, monitoring, imaging, and surgery. This chapter first describes what biophotonics is and what its benefits and applications are. Then some basic concepts of light and of light-tissue interactions are described, including what specific lightwave windows are needed to carry out biophotonics processes. The final section gives a brief tutorial of biological cell and molecular structures, cellular and molecular functions, and the vocabulary used to describe these structures and functions. This topic is essential for understanding the biophotonics tools and light-tissue interaction processes described in this book.

Throughout the history of the world, humans have always been fascinated by the power and importance of light. This fascination appears in religious writings and rituals, in art and literary works ranging from ancient to modern, in lights for buildings and vehicles, in a wide variety of displays for computing and communication devices, and in the dramatic visualization effects of cinematography and other entertainment settings. Analogous to the use of electrons in the electronics world, photons are the key enabling entities in the world of light-based technology or photonics. *Photonics* is the discipline that involves the generation, detection, and control of photons for enabling functions such as transferring or processing information, sensing or measuring changes in physical parameters, and physically modifying material characteristics. The applications of photonics are found in almost every technical discipline including illumination, imaging, manufacturing, material processing, telecommunications, data storage, displays, photography, forensics, power generation, bar codes, and quick response (QR) codes for smart phones.

Photonics technology also has become an indispensible tool for basic life sciences research and for biomedical diagnosis, therapy, monitoring, imaging, and surgery. This category of photonics deals with the interaction between light and biological material and is referred to as *biophotonics* or *biomedical optics*. As indicated in Fig. 1.1, the resulting reflection, absorption, fluorescence, and scattering manifestations of this interaction can be used to analyze the characteristics of

1

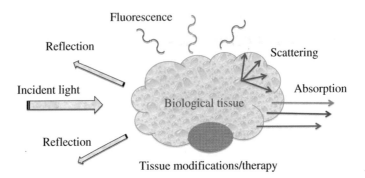

Fig. 1.1 Biophotonics involves all aspects of light-tissue interactions

biological tissues and to carry out tissue modifications and therapy. The purpose of this chapter is to give an overview of this rapidly growing multidisciplinary field and to note its dramatically increasing uses, which are addressed in the following chapters.

In this chapter, first Sect. 1.1 describes what biophotonics is and what its benefits are. Next Sect. 1.2 gives an overview of the areas where biophotonics is used. Of particular interest in this book are biophotonics implementations in biomedical research and clinical practices. Among the numerous diverse applications are various biomedical imaging techniques, microscopic and spectroscopic procedures, endoscopy, tissue pathology, blood flow monitoring, light therapy, biosensing, laser surgery, dentistry, and health status monitoring.

Then Sect. 1.3 gives some basic concepts of light and of light-tissue interactions. This section also describes the fundamental background as to why specific light-wave windows are needed to carry out most biophotonics processes. The light-tissue interaction topic is expanded on in Chap. 6. Finally, Sect. 1.4 gives a brief tutorial of biological cell and molecular structures, cellular and molecular functions, and the vocabulary used to describe these structures and functions. This topic is essential to understanding the biophotonics tools and light-tissue interaction processes described in this book.

1.1 What Is Biophotonics?

Biophotonics is a multidisciplinary field that deals with all aspects of the interactions between light and biological material [1–12]. As Fig. 1.2 illustrates, biophotonics draws from the resources of many technical fields, including biology, biotechnology, chemistry, physics, various engineering fields, and basically every medical discipline. From a global viewpoint, biophotonics refers to the detection, reflection, emission, modification, absorption, creation, and manipulation of

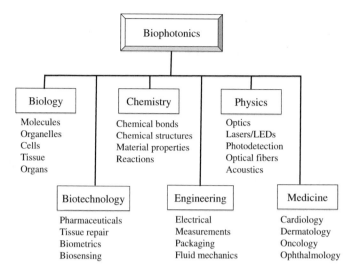

Fig. 1.2 Biophotonics draws from the resources of many technical fields

photons as they interact with biological cells, organisms, molecules, tissues, and substances. The applications of biophotonics processes include (a) 2D and 3D imaging of cells, tissues, and organs, (b) noninvasive measurements of biometric parameters such as blood oxygen and glucose levels, (c) therapeutic photonic treatment of injured, diseased, or unwanted cells, (d) detection of injured or diseased cells and tissue, (e) monitoring of wound healing and progress of therapeutic treatments, and (f) surgical procedures such as laser cutting, tissue ablation, and removal of cells and tissue.

The technologies supporting biophotonics include optical fibers, optical sources and photodetectors, test and measurement instrumentation, nanotechnology, microscopy, spectroscopy, and miniaturization methodologies. Therefore, biophotonics combines a wide variety of optical methods to investigate the structural, functional, mechanical, biological, and chemical properties of biological material and systems. In addition, biophotonics methodologies are being used extensively to investigate and monitor the health and wellbeing of humans. The wavelengths used for biophotonics typically range from 190 nm in the ultraviolet to 10.6 μm in the infrared region, with numerous applications being in the visible 400–700 nm spectrum. Thus a broad range of diverse tools and techniques are employed in biophotonics.

Several terms are commonly used when studying the characteristics of biological cells, molecules, or tissues or when determining the behavior of such biological samples when they are exposed to various external stimuli. These terms include the following expressions.

- In vivo: The term in vivo (Latin for "in the living") refers to tests or procedures on isolated biological components within whole, living organisms such as animals, humans, or plants. Such tests allow the observation of the condition, the temporal changes, or the effects of an experiment on biological components within a living entity in its natural surroundings.

- In vitro: The term in vitro (Latin for "within the glass") refers to tests done in a laboratory setting on biological components that have been extracted from a living organism. These tests often are made using containers such as glass test tubes, flasks, and petri dishes. Thus the evaluations are done on microorganisms, cells, molecules, or tissue samples outside of their normal biological environment.

- Ex vivo: The term ex vivo (Latin for "from the living") refers to procedures that typically involve taking living cells, tissues, or an organ from an organism and examining, modifying, or repairing these biological components in a controlled environment under sterile conditions with minimal alteration of the natural conditions from which the samples originated.

There are several aspects of light that make it a powerful tool in the life sciences and medicine.

- The use of photonics techniques in biophotonics research allows contactless measurements to be made on a tissue sample or within a cell or molecule with no or minimal disturbance of the biological activity. In addition, many either contactless or minimally invasive biomedical procedures can be carried out in a clinical environment. For example, a light source and a camera can be positioned close to the biological tissue of an evaluation or treatment site for in vivo contactless procedures. In addition, a thin optical fiber probe and associated miniaturized surgical instruments can be inserted through a natural body opening or through one or more minor incisions in the body for minimally invasive procedures. A generic example of a minimally invasive procedure is shown in Fig. 1.3.

- A large selection of ultrasensitive photonics and biophotonics detection instruments can be used over spatial scales covering six orders of magnitude from fractions of a nanometer to centimeters. Table 1.1 shows examples of the sizes of some biological components and measurement aids. The biological components that can be observed, treated, or manipulated vary from microscopic nanometer-sized molecules to macroscopic tissue samples.

- As Table 1.2 shows, the measurement time scales for life sciences research techniques and for biomedical processes can vary from femtoseconds (10^{-15} s) to hours (10^3 s). Such a broad temporal range can be satisfied by a variety of photonics devices. For example, as Chap. 4 describes, ultrafast lasers that can emit short pulse durations (e.g., a few femtoseconds) are available for use in applications such as fluorescence spectroscopy where the pulse width needs to be shorter than the desired time-resolution measurement. At the other time extreme, highly stable lasers are available for processes in which a relatively constant light output is required for measurements and monitoring of processes that take place over periods of several hours.

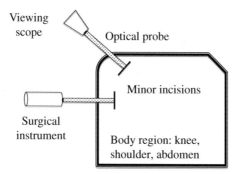

Fig. 1.3 Example of a minimally invasive surgery procedure with an optical fiber probe and a surgical tool

Table 1.1 Example sizes of some biological components and measurement aids

Component	Generic size
Molecule	Water molecule: 0.29 nm Glucose: 0.70 nm Amino acid: 0.80 nm
Virus	30–100 nm
Bacteria	Typical: 0.2 μm in diameter and 2–8 μm in length
Cells	Human red blood cell: 9 μm Eukaryotic cell: 10–20 μm
Tissue sample	1 mm–1 cm (nominal)
Nanoparticle	An example probe consists of an array of 450 nm diameter nanoparticles with center-to-center spacing of 900 nm
Optical fiber	Used for delivering and collecting light; core sizes can range from 9 μm to several mm
Needle probe	30-gauge needle used in conjunction with an internal optical fiber probe; 310 μm outer diameter (nominal)

Table 1.2 Photonics techniques used in biophotonics cover time scales over 18 orders of magnitude

Technique	Generic time and application
Spectroscopy	Fluorescent decay processes occur in 10^{-15} s
Plasma-induced ablation	100 fs–500 ps exposures for dentistry and ophthalmology
Photoablation	10–100 ns treatments in ophthalmology
Thermal irradiation	1 μs–1 min exposures for coagulation and tissue vaporization
Photodynamic therapy	5–40 min of light exposure for cancer treatments
Photobiomodulation	Minutes to hours for therapeutic or cosmetic effects

1.2   Diverse Applications

Among the disciplines that use biophotonics tools and techniques are biomedical research, clinical procedures, dentistry, drug development, healthcare, environmental monitoring, food safety, and manufacturing process control, as Fig. 1.4 illustrates [2, 13–24].

- **Biomedical research**: Biophotonics tools and techniques are being used to understand the basic functions of cells and molecules in life sciences, to discover the genesis of diseases in order to devise processes for their early detection, to develop targeted therapies and minimally invasive treatments, and to monitor health conditions for the prevention or control of diseases.
- **Clinical procedures**: Almost every medical specialty makes use of some aspect of biophotonics tools and techniques. Table 1.3 lists some examples from a range of clinical procedures. In many of these disciplines the use of miniaturized optical components together with fast photonics devices have made dramatic improvements in response times and measurement accuracies.
- **Dentistry**: The biophotonics concepts of laser-tissue interactions are having widespread uses in dentistry for precisely focusing lasers at specific points in hard dental enamel and bone without damaging the surrounding tissue to remove infected tissue and to prepare teeth for fillings or crowns. Other laser applications in dentistry include gum surgery, root canal sterilization, treatment of oral cavity ulcers, and whitening of teeth.
- **Drug development**: Processes such as flow cytometry, fluorescence detection methods to find out whether biological reactions are taking place in drug candidates, the use of photo-switchable inhibitory peptides that can be used to manipulate protein-protein interactions inside cells by applying light, and surface plasmon resonance techniques for rapidly assessing biological reactions in

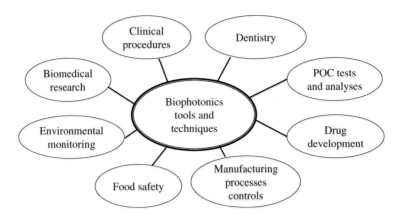

Fig. 1.4 Applications of biophotonics tools and techniques

Table 1.3 Uses of biophotonics tools and techniques in selected medical specialties

Medical specialty	Applications of biophotonics methodologies
Photobiomodulation: uses low irradiance levels to aid in pain relief and tissue healing	Photonic stimulation of tissue to alleviate acute and chronic pain, treat sprains and strains from sports injuries, speed up wound healing, treat nerve and brain injuries, and promote tissue and bone healing
Cardiology: deals with diseases and abnormalities of the heart	Use of lasers and optical fibers for imaging atherosclerotic plaques, diagnosing heart functions, monitoring laser surgery of atrial fibrillation, reducing angina or cardiac pain
Dentistry: deals with tooth repair and the prevention, diagnosis, and treatment of tooth and gum disorders	Detection of caries and cracks in teeth, soft and hard tissue ablation, endodontic therapy, tooth restoration treatments, and detection and treatment of diseases of the teeth, gums, and mouth structures
Dermatology: deals with the skin and its disorders and diseases	Treatment of skin atrophy, skin thickening, varicose veins, vascular lesions, unwanted hair, age spots, surgical and acne scars, and pigmented lesions
Gastroenterology: focuses on the digestive system and its disorders and diseases	Use of endoscopes for imaging the lower and upper gastrointestinal tract to detect abnormalities; use of lasers to destroy esophageal and gastric cancers; photocoagulation of hemorrhaging peptic ulcers
Oncology: deals with the study and treatment of tumors	Early and accurate non-invasive in vivo cancer detection and diagnosis through a variety of optical methods for tissue diagnosis, treatment of cancer through photodynamic therapy, identification of cancer site boundaries
Ophthalmology: deals with eye structures, functions, and diseases	Retinal surgery (treatment of cataracts, glaucoma, and age-related macular degeneracy), refractive corneal surgery, imaging of retinal nerve fiber layers, and sensing changes in eye anatomy and physiology
Neurophotonics/Optogenetics: deals with the structure or function of the nervous system and brain	The neuroscience discipline of optogenetics uses light to activate or inhibit signaling among neurons to study the link between neural network operation and behavioral or sensory functions to treat neuropsychiatric diseases
Vascular medicine or **angiology**: deals with preventing, diagnosing, and treating vascular and blood vessel related diseases	Use of lasers and optical fibers for imaging atherosclerotic plaques, diagnosing microcirculation, treating varicose veins, and treating diseases of the circulatory and lymphatic systems, i.e., arteries, veins and lymphatic vases

candidate drug compounds have become effective pharmaceutical research and development tools.

- **Point of care (POC) tests and analyses**: *Point of care* (POC) testing or diagnostics is a broad term that refers to the delivery of healthcare evaluations, products, and services by clinicians to patients at the time of care. Photonics-based devices include ophthalmological instruments, optical oximeters, compact portable microscopes, microscopes running on smartphones, fluorescent imaging flow cytometers on a smartphone, and webcam-based biosensors.
- **Environmental monitoring**: Biophotonics spectroscopic techniques and tools (including the use of spectrometric add-ons to smartphones) are being used to measure and monitor concentrations of airborne particles in the atmosphere (e.g., fine dust, pollens, and chemical pollutants) and to detect pathogens (e.g., bacteria, viruses, and protozoa) in bodies of water.
- **Food safety**: Concerns about food safety (e.g., the presence of contaminants and natural pathogens, food fraud, and food adulteration) are being addressed through biophotonics techniques and tools such as high-performance chromatography, Fourier transform infrared spectroscopy, surface-enhanced Raman spectroscopy, and biosensors.
- **Manufacturing process control**: Spectroscopic techniques are being used and developed further for the non-contact, non-destructive compositional analyses and quality control of complex biological material used in the manufacturing of pharmaceutical, nutritional, and cosmetic products.

1.3   Biophotonics Spectral Windows

As described in Sect. 2.1, light propagates in the form of electromagnetic waves [25–28]. In free space, light travels at a constant speed c = 299,792,458 m/s (or approximately 3×10^8 m/s). In biophotonics the free-space speed of light could be expressed in units such as 30 cm/ns, 0.3 mm/ps, or 0.3 μm/fs. The speed of light in free space is related to the wavelength λ and the wave frequency ν through the equation c = νλ. Upon entering a transparent or translucent dielectric medium or a biological tissue the lightwave will travel at a slower speed s. The ratio between *c* and the speed s at which light travels in a material is called the *refractive index* (or *index of refraction*) n of the material (with n ≥ 1), so that

$$s = c/n \qquad (1.1)$$

Table 1.4 lists the indices of refraction for a variety of substances. Note that in many cases the refractive index changes with wavelength.

Table 1.4 Indices of refraction for various substances

Material	Refractive index
Air	1.000
Water	1.333
Cornea	1.376
Cytoplasm	1.350–1.375
Epidermis	1.34–1.43
Extracellular fluids	1.35–1.36
Human liver	1.367–1.380
Melanin	1.60–1.70
Mitochondria	1.38–1.41
Tooth enamel	1.62–1.73
Whole blood	1.355–1.398

Example 1.1 If a biological tissue sample has a refractive index n = 1.36, what is the speed of light in this medium?

Solution: From Eq. (1.1) the speed of light in this tissue sample is

$$s = \frac{c}{n} = \frac{3 \times 10^8 \, \text{m/s}}{1.36} = 2.21 \times 10^8 \, \text{m/s} = 22.1 \, \text{cm/ns}$$

Figure 1.5 shows the spectral range covered by optical wavelengths. The ultraviolet (UV) region ranges from 10 to 400 nm, the visible spectrum runs from the 400 nm violet to the 700 nm red wavelengths, and the infrared (IR) region ranges from 700 nm to 300 μm. Biophotonics disciplines are concerned mainly with wavelengths falling into the spectrum ranging from the mid-UV (about 190 nm) to the mid-IR (about 10 μm).

Also shown in Fig. 1.5 is the relationship between the energy of a photon and its frequency (or wavelength), which is determined by the equation known as *Planck's Law*

$$E = h\nu = \frac{hc}{\lambda} \tag{1.2}$$

where the parameter $h = 6.63 \times 10^{-34}$ J·s $= 4.14 \times 10^{-15}$ eV·s is *Planck's constant*. The unit J means *joules* and the unit eV stands for *electron volts*. In terms of wavelength (measured in units of μm), the energy in electron volts is given by

$$E(eV) = \frac{1.2406}{\lambda(\mu m)} \tag{1.3}$$

Table 1.5 lists the correspondence between the wavelength range and the range of photon energies for various biophotonics spectral bands. The spectral band

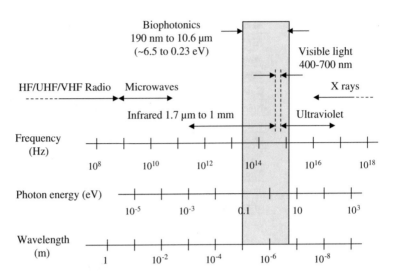

Fig. 1.5 Location of the biophotonics discipline in the electromagnetic spectrum

Table 1.5 Wavelength and photon energy ranges for various optical spectrum bands (Spectral band ranges are based on *Document ISO-21348* of the International Standards Organization)

Band designation	Wavelength range (nm)	Photon energy range (eV)
UV-C	100–280	12.4–4.43
UV-B	280–315	4.43–3.94
UV-A	315–400	3.94–3.10
Visible	400–700	3.10–1.63
Near infrared (IR-A)	700–1400	1.63–0.89
Mid infrared (IR-B)	1400–3000	0.89–0.41

ranges listed in the literature often vary slightly from those shown in Table 1.5, which are based on the designations given in *Document ISO-21348* of the International Standards Organization. Note that the UV spectral region is partitioned into three bands that are classified as UV-C, UV-B, and UV-A. The reason for these designations is that the UV light energy is stronger than visible light and its effects are different in each band. Natural sunlight at the surface of the earth contains mainly UV-A and some UV-B lightwaves. UV-A light is slightly more energetic than visible blue light. Many plants use UV-A for healthy growth and, in moderate doses, UV-A light is responsible for skin tanning. Health concerns for exposure to UV light are mostly in the UV-B range. The most effective biological wavelength for producing skin burns is 297 nm. The adverse biological effects increase logarithmically with decreasing wavelength in the UV range, with 330 nm being only 0.1 % as effective as 297 nm. Thus it is quite important to control human exposure to UV-B. UV-C light has the highest energy. Although the sun emits UV-C wavelengths, the upper atmosphere filters out this band, so there is no exposure to UV-C from sunlight on earth. However, because UV-C light is used for sterilizing instruments, great care must be exercised to avoid human exposure to this light.

Example 1.2 Show that photon energies decrease with increasing wavelength. Use wavelengths at 850, 1310, and 1550 nm.

Solution: From Eq. (1.3) it follows that E(850 nm) = 1.46 eV, E (1310 nm) = 0.95 eV, and E(1550 nm) = 0.80 eV. Thus, the energy decreases with increasing wavelength.

In relation to Fig. 1.5, biophotonics disciplines are concerned mainly with wavelengths falling into the spectrum ranging from the mid-UV (about 190 nm) to the mid-IR (about 10.6 μm). Generally, because optical properties vary from one tissue type to another, specific lightwave windows are needed to carry out most therapeutic and diagnostic biophotonics processes. Thus, knowing the details of these properties allows the specification and selection of photonics components that meet the criteria for carrying out a biological process in a selected optical wavelength band.

The interaction of light with biological tissues and fluids is a complex process because the constituent materials are optically inhomogeneous. More details on these effects are given in Chap. 6. Owing to the fact that diverse biological tissue components have different indices of refraction, the refractive index along some path through a given tissue volume can vary continuously or undergo abrupt changes at material boundaries, such as at flesh and blood vessel interfaces. This spatial index variation gives rise to scattering, reflection, and refraction effects in the tissue. Thus, although light can penetrate several centimeters into a tissue, strong scattering of light can prevent observers from getting a clear image of tissue characteristics beyond a few millimeters in depth.

1.4   Light Absorption

Light absorption is another important factor in the interaction of light with tissue, because the degree of absorption determines how far light can penetrate into a specific tissue. Figure 1.6 shows the absorption coefficients as a function of wavelength for several major tissue components. These components include water (about 70 % of the body), proteins, whole blood, melanin (pigments that give skin, hair, and eyes their color), and the epidermis (outer layer of the skin).

Most tissues exhibit comparatively low absorption in the spectral range that extends from 500 nm to about 1500 nm, that is, from the orange region in the visible spectrum to the near infrared (NIR). This wavelength band is popularly known as the *therapeutic window* or the *diagnostic window*, because it enables the best conditions for viewing or treating tissue regions within a living body by optical means. Note that although the absorption coefficients for tissue components such as

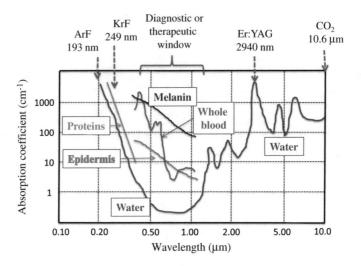

Fig. 1.6 Absorption coefficients of several major tissue components as a function of wavelength with example light source emission peaks

melanin, blood, and the epidermis are larger than that of water, they have relatively small concentrations compared to water and thus account for a comparatively small amount of absorption in the therapeutic window.

Light absorption characteristics of tissue for regions outside the therapeutic window are important for implementing functions that depend on high optical power absorption, such as drilling, cutting, bonding, and ablation of tissue. As Chap. 4 discusses, a wide variety of optical sources can be used to carry out these functions. For example, as is indicated in Fig. 1.6, ultraviolet light from ArF or KrF lasers emitting at wavelengths of 193 and 249 nm, respectively, is strongly absorbed in the surface of a tissue and thus can be used for many surgical applications. In the infrared region, 2940 nm light from an Er:YAG laser is strongly absorbed by osseous minerals, which makes optical sawing and drilling in bone and teeth possible. Also in the infrared region, strongly absorbed light from CO_2 lasers are used in applications such as surgery, ophthalmology, and cosmetic treatments.

1.5 Signal Attenuation

Some type of optical link configuration is used for many biophotonics imaging, therapeutic, monitoring, and tissue evaluation techniques. Such a link typically contains a light source, an optical fiber, a photodetector, and various intermediate optical components. As an optical signal passes through such a link, the signal loses a percent of its power within each constituent of the link. An understanding of the degree of signal attenuation is important for assuring that the correct optical power level reaches its destination at either the light delivery site or at the photodetector where the signal is evaluated and interpreted.

A standard and convenient method for measuring attenuation through an optical link or an optical device is to reference the output signal level to the input level. For convenience in calculation, signal attenuation or amplification can be designated in terms of a logarithmic power ratio measured in *decibels* (dB). The dB unit is defined by

$$\text{Power ratio in dB} = 10 \log \frac{P_{out}}{P_{in}} \tag{1.4}$$

where P_{in} and P_{out} are the input and output optical powers, respectively, to an optical fiber or an optical component and *log* is the base-10 logarithm. The logarithmic nature of the decibel allows a large ratio to be expressed in a fairly simple manner. Power levels differing by many orders of magnitude can be compared easily through an addition and subtraction process when they are in decibel form. For example, a power reduction by a factor of 1000 is a −30 dB loss, an attenuation of 50 % is a −3 dB loss, and a 10-fold amplification of the power is a +10 dB gain. Thus an attractive feature of the decibel is that to measure the changes in the strength of a signal, the decibel loss or gain numbers in a series of connected optical link elements between two different points are merely added. Table 1.6 shows some sample values of power loss given in decibels and the percent of power remaining after this loss.

Example 1.3 Assume that after traveling a certain distance in some transmission link, the power of an optical signal is reduced to half, that is, $P_{out} = 0.5 \, P_{in}$. What is the loss in dB of the link?

Solution: From Eq. (1.4) it follows that the attenuation or loss of power is

$$10 \log \frac{P_{out}}{P_{in}} = 10 \log \frac{0.5 P_{in}}{P_{in}} = 10 \log 0.5 = 10(-0.3) = -3 \, \text{dB}$$

Thus, the value −3 dB (or a 3 dB loss or attenuation) means that the signal has lost half its power.

Table 1.6 Representative values of decibel power loss and the remaining percentages

Power loss (in dB)	Percent of power left
0.1	98
0.5	89
1	79
2	63
3	50
6	25
10	10
20	1

Example 1.4 Consider an optical link that has four optical components hooked together sequentially. If the losses of these four components are 0.5, 0.1, 1.5, and 0.8 dB, what is the loss in dB of the link?

Solution: The total power loss is found by adding the losses between the two link end points, that is,

$$\text{Link loss} = 0.5\,\text{dB} + 0.1\,\text{dB} + 1.5\,\text{dB} + 0.8\,\text{dB} = 2.9\,\text{dB}$$

Because the decibel is used to refer to ratios or relative units, it gives no indication of the absolute power level. However, a derived unit can be used for this purpose. Such a unit that is particularly common is the *dBm*. This unit expresses the power level P as a logarithmic ratio of P referred to 1 mW. In this case, the power in dBm is an absolute value defined by

$$\text{Power level in dBm} = 10\log\frac{P(\text{in mW})}{1\,\text{mW}} \qquad (1.5)$$

Example 1.5 Consider three different light sources that have the following optical output powers: 50 µW, 1 mW, and 50 mW. What are the power levels in dBm units?

Solution: Using Eq. (1.5) the power levels in dBm units are −13, 0, and +17 dBm.

Example 1.6 A product data sheet for a certain photodetector states that an optical power level of −28 dBm is needed at the photodetector to satisfy a specific performance requirement. What is the required power level in µW (microwatt) units?

Solution: Eq. (1.5) shows that −28 dBm corresponds to a power in µW of

$$P = 10\exp\left(-28/10\right)\text{mW} = 0.001585\,\text{mW} = 1.585\,\mu\text{W}$$

1.6 Structures of Biological Cells and Tissues

A basic knowledge of biological cell and molecular structures, cellular and molecular functions, and the vocabulary used to describe these structures and functions is essential to understanding the biophotonics tools and light-tissue interaction processes described in this book. This section gives a brief tutorial of these concepts [29–32]. The basis of this discussion is the observation that a living cell consists of about 70 % water. The rest of the cell is composed mostly of macromolecules that are composed of thousands of atoms. First, Sect. 1.6.1 describes the basic macromolecules, which are the chemical building blocks of the various components of a cell. Next Sect. 1.6.2 identifies the various components that make up a cell and describes their different functions depending on where they are found in a body. Then Sect. 1.6.3 describes the major categories and functions of biological tissues, which are ensembles of similar cells.

1.6.1 Macromolecules

Macromolecules are large molecules that are made up of selections from a set of smaller molecular chains such as monomers, glycerin, and fatty acids. The four basic groups of macromolecules found in living organisms are proteins, carbohydrates, nucleic acids, and lipids. Proteins, carbohydrates, and nucleic acids are known as *polymers* because they are made up of many monomers. Monomers include amino acids for forming proteins, nucleotides that form nucleic acids, and simple sugars such as glucose and fructose that are the building blocks for carbohydrates. Lipids are not polymers.

After water, proteins make up the second largest component in human cells, tissues, and muscles. *Proteins* consist of one or more long chains of interconnected monomers called amino acids. *Amino acids* are small organic molecules composed of amine ($-NH_2$) and carboxylic acid ($-COOH$) *functional groups* (groups of atoms or bonds that are responsible for the characteristic chemical reactions of a molecule) plus *molecular side chains* (called R groups) that are specific to each amino acid. The basic elements of an amino acid are carbon, hydrogen, oxygen, and nitrogen. Amino acids are connected by covalent bonds (referred to as *peptide bonds*) to form chains of molecules called *polypeptide chains*. A protein consists of one or more of these polypeptide chains that are twisted, wound, and folded upon themselves to form a macromolecule. Although about 500 amino acids are known, there are only 20 naturally occurring amino acids.

Proteins are an essential substance of living organisms and participate in almost every process within cells. Common types of proteins include the following:

- *Enzymes* for catalyzing chemical reactions that are vital to metabolism
- *Structural proteins* (such as keratin found in hair and fingernails and collagen found in skin) that provide support for maintaining cell shape. Note that *collagen* is the most abundant protein in the human body and is found in the bones, muscles, skin and tendons.
- *Antibodies* that bind to foreign particles or cells and assist in their elimination from the body
- *Transport proteins* that carry compounds or chemicals from one location to another throughout living systems. An example of a transport protein is *hemoglobin* that carries oxygen through the blood.

Fig. 1.7 Example of a
hypothetical protein showing
twisted and folded
interconnected polypeptide
chains

An example of a hypothetical protein macromolecule is illustrated in Fig. 1.7, which shows the twisting and folding of interconnected polypeptide chains.

Carbohydrates are important in defining the structure of cells and are a source of immediate energy needs in living systems. These macromolecules consist of interconnected chains or rings of carbon (C), hydrogen (H) and oxygen (O) atoms with a ratio of H twice that of C and O. Carbohydrates include compounds such as sugars, starches, and cellulose. The most elementary monomers of carbohydrates are simple sugars or *monosaccharides*. Two fundamental monosaccharides are *glucose* and *fructose*, which both have the chemical formula $C_6H_{12}O_6$ but have different chemical structures as shown in Fig. 1.8. These simple sugars can combine with each other to form more complex carbohydrates. Combinations of two simple sugars are called *disaccharides* (for example, maltose = glucose + glucose and sucrose = glucose + fructose). *Polysaccharides* consist of a larger number of simple sugars.

Nucleic acids are the macromolecules inside of a cell that store and process genetic or hereditary information by means of protein synthesis. The two principal forms of nucleic acids that are essential for all forms of life are *deoxyribonucleic acid* (DNA) and *ribonucleic acid* (RNA). The monomers that make up nucleic acids are called *nucleotides*. The chemical building blocks of each nucleotide include a sugar molecule, a nitrogenous base, and a phosphate group, as Fig. 1.9 illustrates.

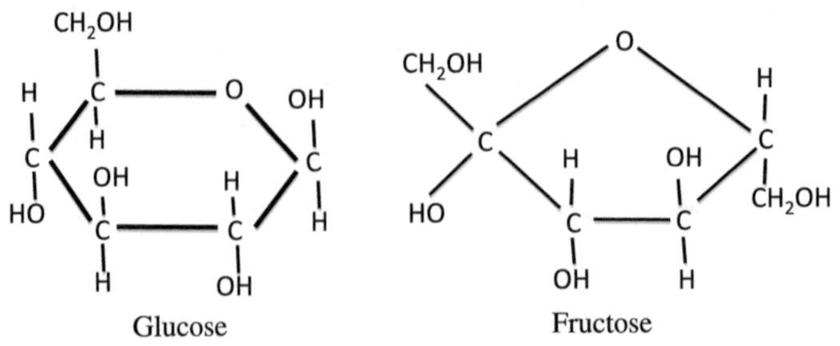

Glucose   Fructose

Fig. 1.8 Glucose and fructose are two basic monomers for creating carbohydrates

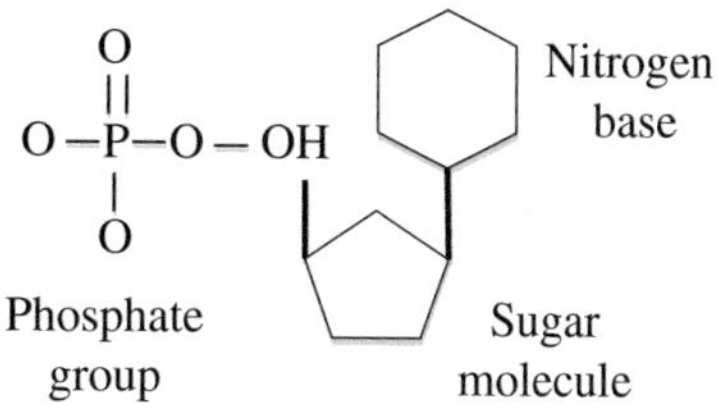

Fig. 1.9 Basic nucleotide monomer for creating nucleic acids

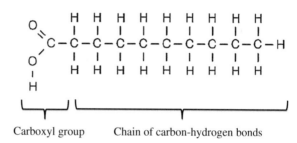

Fig. 1.10 Example of a lipid molecular structure

An important nucleotide is *adenosine triphosphate* (ATP). One molecule of ATP contains three phosphate groups. A large amount of chemical energy is stored in the phosphate bonds. When ATP is broken down (hydrolyzed) the energy that is released is used for many metabolic processes. Because of this function, ATP is considered to be a universal energy currency for metabolism.

Lipids are non-polymer macromolecules that are all insoluble in water. Simple lipids include fats, phospholipids (e.g., lecithin), and steroids. Fats provide long-term energy storage and insulation in humans and animals. Fats are composed of glycerin and fatty acids. As shown in Fig. 1.10, a fatty acid is a long chain of carbon-hydrogen (CH) bonds, with a carboxyl group (COOH) at one end. Phospholipids are found mainly in the cell membranes of living systems and are used for cell-to-cell signaling. Examples of steroids include cholesterol, estrogen, and testosterone.

1.6.2   Biological Cells

All organisms consist of two structurally different kinds of cells called prokaryotic cells and eukaryotic cells. *Prokaryotic cells* have few internal structures and no true cell nucleus. These cells are found only in bacteria, with the smallest cells being 0.1–1 µm in diameter. Most bacteria are 1–10 µm in diameter. *Eukaryotic cells* are found in all animals, plants, fungi, and protists (one-celled eukaryotic organisms that are not fungi, plants, or animals). As is shown in Fig. 1.11, eukaryotic cells are typically 10–100 µm in diameter and have a true nucleus that is surrounded by a membranous envelope. The nucleus averages 5 µm in diameter and contains most

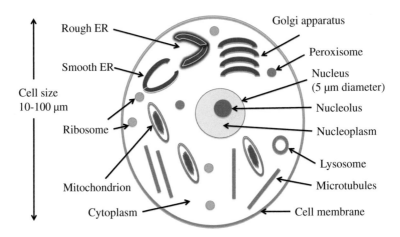

Fig. 1.11 Major constituents of a eukaryotic cell

Table 1.7 Summary of the various key organelles and their functions in a eukaryotic cell

General function	Organelle and its functions
Genetic control	• *Chromosomes* in the nucleus: serve as the location for inherited DNA, which program the synthesis of proteins • *Nucleolus* in the nucleus: produce ribosomes
Production of macromolecules	• *Ribosomes*: responsible for synthesis of polypeptides • *Rough endoplasmic reticulum* (ER): assist in protein synthesis and membrane production • *Smooth ER*: synthesize lipids; store and release calcium ions in muscle cells; handle liver cell functions • *Golgi apparatus*: handle chemical products of the cell
Maintenance of the cell	• *Lysosomes*: digest food and other materials • *Peroxisomes*: metabolize lipids and carbohydrates; handle enzymatic disposal of hydrogen peroxide
Energy processing	• *Mitochondria*: responsible for cellular respiration and synthesis of adenosine triphosphate (ATP)
Support, movement, and communication between cells	• *Cytoskeleton*: maintains cell shape, anchors organelles, moves cells and organelles within cells • *Glycocalyx*: protects surfaces; binds cells in tissues • *Intercellular junctions*: handle communication between cells; bind cells in tissues

of the genes that control the cell. The entire region between the nucleus and the surrounding membrane is called the *cytoplasm*. The cytoplasm contains a semifluid medium in which are suspended various subcellular structures called *organelles*. Table 1.7 gives a summary of the various key organelles and their functions in a eukaryotic cell.

Table 1.8 Characteristics and functions of basic biological tissues

Biological tissue	Characteristic	Functions
Epithelial tissue	Cover outside skin surfaces; line inner surfaces of organs and cavities in the body	• Protect against invading microorganisms, fluid loss, and mechanical injury • Absorb or secret special chemical solutions
Connective tissue	Examples: loose and fibrous connective tissue, adipose tissue, cartilage, bone, blood	Bind together various tissues, give shape to organs, and hold organs in place
Muscle tissue	Long, excitable cells capable of considerable contraction	The three main types are skeletal, cardiac, and visceral muscles
Nervous tissue	Cells that make up the central and peripheral nervous systems	• Central nervous system: brain and spinal cord • Peripheral nervous system: cranial and spinal nerves

1.6.3 Biological Tissues and Organs

A *biological tissue* (or simply a *tissue*) is an ensemble of similar cells that collectively carry out a specific function. The four basic categories are epithelial, connective, muscle, and nervous tissues. The characteristics of these tissues are described in the following paragraphs and are summarized in Table 1.8.

Epithelial tissues cover the outside surface of the skin and line the inner surfaces of organs and cavities within the body. The cells of an epithelium are connected by tight junctions and thus serve as a barrier to protect against mechanical injury, invading microorganisms, and fluid loss. In addition to protecting the organs that they cover or line, some epithelial tissues are specialized for absorbing or secreting chemical solutions. One example is the mucous membrane, which secretes a slippery substance called the mucous that lubricates and moistens a specific tissue surface.

The functions of *connective tissue* are to bind together other tissues, to give shape to organs, and to hold them in place. Examples of connective tissue are loose connective tissue, adipose tissue, fibrous connective tissue, cartilage, bone, and blood. *Loose connective tissue* consists of a loose weave of protein fibers, which are categorized as collagenous fibers (a rope-like coil of three collagen molecules that resist stretching), elastic fibers (rubbery stretchable fibers made of elastin that quickly restore skin shape if it is stretched), fibroblasts (cells that maintain the structural integrity of connective tissue), and microphages (travelling cells that swallow bacteria and the debris of dead cells and eliminate them from the body). *Adipose tissue* is located beneath the skin, around internal organs, and in bone marrow. Its functions include providing insulation from heat and cold, providing a protective padding around organs, and storing lipids that can be used to meet the energy needs of the body. *Fibrous connective tissue* are dense collagen-enriched fiber bundles that have a high tensile strength in order to attach muscles to bones (tendons) and to hold bones together at their joints (ligaments). *Cartilage* is a strong but yet flexible support material consisting of collagenous fibers and a protein-carbohydrate complex called chondrin. Cartilage provides support in locations such as the nose, the ears, the rings that reinforce the windpipe, and the cushioning discs between vertebrae. *Bone* is a combination of mineralized connective tissue and flexible collagen that supports the body of most vertebrates. The interior of bones contains a spongy tissue called *marrow*. *Blood* is categorized as a connective tissue because, like other connective tissues, it has an extensive extracellular matrix. This matrix is in the form of a liquid called *plasma*. This liquid has red blood cells, white blood cells, and platelets suspended in it. Red blood cells carry oxygen throughout the body, white blood cells act against invaders such as bacteria and viruses, and platelets are involved in the clotting of blood.

Muscle tissue is composed of long, excitable cells that are capable of considerable contraction. The three main types are skeletal muscles (which are responsible for voluntary movements of the body), cardiac muscles (which form the contractile walls of the heart), and visceral muscles (which are in the walls of the digestive tract, bladder, arteries, and other internal organs).

Nervous tissues are the cells that make up the central and the peripheral nervous systems. The neural tissue in the central nervous system forms the brain and spinal cord. In the peripheral nervous system, the neural tissue forms the cranial nerves and spinal nerves.

In humans and most animals, a number of specialized functional units called *organs* are composed of multiple layers of different types of biological tissues. In the human body there are over 70 organs that have various sizes and functions. Some of the major organs include the bladder, brain, eyes, heart, kidneys, liver, lungs, skin, spleen, stomach, and thyroid. As an example of layers and their functions, the human stomach consists of an arrangement of four tissue layers. From the inner to the outer layer these are a thick epithelium that secrets mucous and digestive juices, a layer of connective tissue followed by a layer of smooth muscle, and finally another layer of connective tissue. A collection of certain organs that carry out a specific function is called an *organ system*. These include the digestive, cardiovascular, muscular, skeletal, nervous, respiratory, endocrine, and lymphatic systems.

1.7 Summary

Biophotonics technology deals with the interaction between light and biological matter. The techniques and tools used in this discipline have become indispensable for basic life sciences research and for biomedical diagnosis, therapy, monitoring, and surgery. Among the diverse applications are biomedical imaging techniques, microscopic and spectroscopic procedures, endoscopy, tissue pathology, blood flow monitoring, light therapy, biosensing, laser surgery, dentistry, and health status monitoring.

This chapter first describes the concepts and benefits of biophotonics, gives illustrations of the disciplines where biophotonics techniques are used, and gives some basic concepts of light-tissue interactions. Further details of light-tissue interactions are given in Chap. 6. In order to understand the various biophotonics concepts and their application to medical issues, the final section of this chapter gives a brief tutorial of biological cell and molecular structures, cellular and molecular functions, and the vocabulary used to describe these structures and functions.

1.8 Problems

1.1 If a biological tissue sample has a refractive index n = 1.34, show that the speed of light in this medium is s = 2.24×10^8 m/s = 22.4 cm/ns.

1.2 Show that the photon energies for UV-C light at 190 nm and UV-A light at 315 nm are 6.53 and 3.94 eV, respectively.

1.3 Collagen is the most abundant protein in humans and animals. It is organized into long chains of insoluble fibers that have great tensile strength. Collagen is found in tissues such as bone, teeth, cartilage, tendon, ligament, and the fibrous matrices of skin and blood vessels. (a) Using literature or Web resources, describe the basic structural unit of collagen, which is a triple-stranded helical molecule. (b) What are the three principal amino acids present in collagen?

1.4 Elastin is a protein consisting of tissue fibers that can stretch to several times their normal length and then return to their original configuration when the stretching tension is released. (a) Using Web resources, what are the four basic amino acids found in elastin? (b) Give at least four organs in the human body that contain elastin and briefly describe its function in these organs.

1.5 When ATP is broken down (hydrolyzed) the energy that is released is used for many metabolic processes. (a) Using Web resources, draw the chemical structure of the nucleotide ATP to show the triple phosphate group, the sugar molecule, and the nitrogen base. (b) Describe how ATP releases energy by producing adenosine diphosphate (ADP) or adenosine monophosphate (AMP). (c) What happens to these molecules when an orgasm is resting and does not need energy immediately?

1.6 For certain biophotonics procedures, an important factor is the ratio of the surface area of a cell to its volume. Consider two different spherical cells called Cell #1 and Cell #2. Suppose Cell #1 has a membrane surface area of 1 μm^2 for every microliter (μL) of volume. If the diameter of Cell #2 is five times smaller than the diameter of Cell #1, show that the ratio of the surface area to the volume for Cell #2 is 5 $\mu m^2/\mu L$.

1.7 Suppose the body of a 70 kg adult human male contains about 44 billion cells. If the average mass density of humans is about the same as that of water (1 gm/cm^3), show that the average volume of a human cell is about 1.59×10^6 μm^3.

1.8   A product data sheet for a certain photodetector states that a −32 dBm optical
      power level is needed at the photodetector to satisfy a specific performance
      requirement. Show that the required power level in nW (nanowatt) units is
      631 nW.

References

1. N.H. Niemz, *Laser-Tissue Interaction*, 3rd edn. (Springer, 2007)
2. L.V. Wang, H.I. Wu, *Biomedical Optics: Principles and Imaging* (Wiley, Hoboken, NJ, 2007)
3. L. Pavesi, P.M. Fauchet (eds.), *Biophotonics* (Springer, 2008)
4. J. Popp, V.V. Tuchin, A. Chiou, S.H. Heinemann (eds.), *Handbook of Biophotonics: Vol. 1: Basics and Techniques* (Wiley, 2011)
5. J. Popp, V.V. Tuchin, A. Chiou, S.H. Heinemann (eds.), *Handbook of Biophotonics: Vol. 2: Photonics for Healthcare* (Wiley, 2011)
6. A.J. Welch, M.J.C. van Gemert (eds.), *Optical-Thermal Response of Laser-Irradiated Tissue*, 2nd edn. (Springer, 2011)
7. S.L. Jacques, Optical properties of biological tissues: a review. Phys. Med. Biol. **58**, R37–R61 (2013)
8. T. Vo-Dinh (ed.), *Biomedical Photonics Handbook*, Vol. I-III, 2nd edn. (CRC Press, Boca Raton, FL, 2014)
9. K. Kulikov, *Laser Interaction with Biological Material* (Springer, 2014)
10. K. Tsia, *Understanding Biophotonics* (Pan Stanford Publishing, Singapore, 2015)
11. M. Olivo, U.S. Dinish (eds.), *Frontiers in Biophotonics for Translational Medicine* (Springer, Singapore, 2016)
12. A.H.-P. Ho, D. Kim, M.G. Somekh (eds.), *Handbook of Photonics for Biomedical Engineering* (Springer, 2016)
13. A. Kishen, A. Asundi, *Fundamentals and Applications of Biophotonics in Dentistry* (Imperial College Press, 2008)
14. H. Zhu, S.O. Isikman, O. Mudanyali, A. Greenbaum, A. Ozcan, Optical imaging techniques for point-of-care diagnostics. Lab Chip **13**(1), 51–67 (7 Jan 2013)
15. M.W. Collins, C.S. König (eds.), *Micro and Nano Flow Systems for Bioanalysis* (Springer, New York, 2013)
16. A. Ricciardi, M. Consales, G. Quero, A. Crescitelli, E. Esposito, A. Cusano, Lab-on-fiber devices as an all around platform for sensing. Opt. Fiber Technol. **19**(6), 772–784 (2013)
17. P. Vitruk, Oral soft tissue laser ablative and coagulative efficiencies spectra. Implant Practices **7**(6), 23–27 (2014)
18. X. Yang, D. Lorenser, R.A. McLaughlin, R.W. Kirk, M. Edmond, M.C. Simpson, M.D. Grounds, D.D. Sampson, Imaging deep skeletal muscle structure using a high-sensitivity ultrathin side-viewing optical coherence tomography needle probe. Biomed. Opt. Express **5** (1), 136–148 (2014)
19. J. Albert, A lab on fiber. IEEE Spectr. **51**, 48–53 (2014)
20. S.P. Morgan, B.C. Wilson, A. Vitkin, F.R.A.J. Rose, *Optical Techniques in Regenerative Medicine* (CRC Press, Boca Raton, FL, 2014)
21. S.A. Boppart, R. Richards-Kortum, Point-of-care and point-of-procedure optical imaging technologies for primary care and global health. Sci. Transl. Med. **6**(253), article 253rv2 (Sept. 2014) [Review article]
22. F. Guarnieri (ed.), *Corneal Biomechanics and Refractive Surgery* (Springer, New York, 2015)

23. J.M. Cayce, J.D. Wells, J.D. Malphrus, C. Kao, S. Thomsen, N.B. Tulipan, P.E. Konrad, E.D. Jansen, A. Mahadevan-Jansen, Infrared neural stimulation of human spinal nerve roots in vivo. Neurophotonics **2**, article 015007 (Jan–Mar 2015)
24. B.J.-F. Wong, J. Ligner, *Biomedical Optics in Otorhinolaryngology* (Springer, 2016)
25. B.E.A. Saleh, M.C. Teich, *Fundamentals of Photonics*, 2nd edn. (Wiley, Hoboken, NJ, 2007)
26. C.A, Diarzio, *Optics for Engineers* (CRC Press, Boca Raton, FL, 2012)
27. G. Keiser, *Optical Fiber Communications* (McGraw-Hill, 4th US edn., 2011; 5th international edn., 2015)
28. E. Hecht, *Optics*, 5th edn. (Addison-Wesley, 2016)
29. A. Tözeren, S.W. Byers, *New Biology for Engineers and Computer Scientists* (Prentice Hall, Upper Saddle River, NJ, 2004)
30. H. Lodish, A. Berk, C.A. Kaiser, M. Krieger, A. Bretscher, H. Ploegh, A. Amon, M.P. Scott, W.H. Freeman, *Molecular Cell Biology*, 7th edn. (2013)
31. J.B. Reece, L.A. Urry, M.L. Cain, S.A. Wasserman, P.V. Minorsky, R.B. Jackson, *Campbell Biology*, 10th edn. (Benjamin Cummings, Redwood City, CA, 2014)
32. B. Alberts, A. Johnson, J. Lewis, D. Morgan, M. Raff, K. Roberts, P. Walter, *Molecular Biology of the Cell*, 6th edn. (Garland Science, 2015)

Chapter 2
Basic Principles of Light

Abstract The purpose of this chapter is to present an overview of the fundamental behavior of light. Having a good grasp of the basic principles of light is important for understanding how light interacts with biological matter, which is the basis of biophotonics. A challenging aspect of applying light to biological materials is that the optical properties of the materials generally vary with the light wavelength and can depend on factors such as the optical power per area irradiated, the temperature, the light exposure time, and light polarization. The following topics are included in this chapter: the characteristics of lightwaves, polarization, quantization and photon energy, reflection and refraction, and the concepts of interference and coherence.

Having a good grasp of the basic principles of light is important for understanding how light interacts with biological matter, which is the basis of biophotonics. As Chap. 6 describes, light impinging on a biological tissue can pass through or be absorbed, reflected, or scattered in the material. The degree of these interactions of light with tissue depends significantly on the characteristics of the light and on the optical properties of the tissue. A challenging aspect of applying light to biological materials is the fact that the optical properties of the material generally vary with the light wavelength. In addition, the optical properties of tissue can depend on factors such as the optical power per area irradiated, the temperature, the light exposure time, and light polarization. Chapter 6 and subsequent chapters present further details on these light-tissue interaction effects.

The purpose of this chapter is to present an overview of the fundamental behavior of light [1–11]. The theory of quantum optics describes light as having properties of both waves and particles. This phenomenon is called the *wave-particle duality*. The dual wave-particle picture describes the transportation of optical energy either by means of the classical electromagnetic wave concept of light or by means of quantized energy units or *photons*. The wave theory can be employed to understand fundamental optical concepts such as reflection, refraction, dispersion, absorption, luminescence, diffraction, birefringence, and scattering. However, the particle theory is needed to explain the processes of photon generation and absorption.

© Springer Science+Business Media Singapore 2016
G. Keiser, *Biophotonics*, Graduate Texts in Physics,
DOI 10.1007/978-981-10-0945-7_2

The following topics are included in this chapter: the characteristics of light-waves, polarization, quantization and photon energy, reflection and refraction, and the concepts of interference and coherence.

2.1   Lightwave Characteristics

The theory of classical electrodynamics describes light as electromagnetic waves that are transverse, that is, the wave motion is perpendicular to the direction in which the wave travels. In this *wave optics* or *physical optics* viewpoint, a series of successive spherical wave fronts (referred to as a *train of waves*) spaced at regular intervals called a *wavelength* can represent the electromagnetic waves radiated by a small optical source with the source at the center as shown in Fig. 2.1a. A *wave front* is defined as the locus of all points in the wave train that have the same phase (the term *phase* indicates the current position of a wave relative to a reference point). Generally, one draws wave fronts passing through either the maxima or the minima of the wave, such as the peak or trough of a sine wave, for example. Thus the wave fronts (also called *phase fronts*) are separated by one wavelength.

When the wavelength of the light is much smaller than the object (or opening) that it encounters, the wave fronts appear as straight lines to this object or opening. In this case, the lightwave can be represented as a plane wave and a light ray can indicate its direction of travel, that is, the ray is drawn perpendicular to the phase front, as shown in Fig. 2.1b. The light-ray concept allows large-scale optical effects

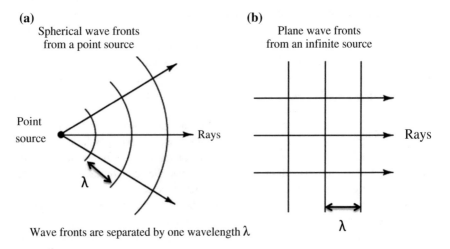

Fig. 2.1 Representations of **a** spherical waves radiating from a point source and **b** plane waves and their associated rays

Fig. 2.2 A monochromatic wave consists of a sine wave of infinite extent with amplitude A, wavelength λ, and period T

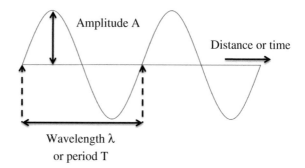

such as reflection and refraction to be analyzed by the simple geometrical process of ray tracing. This view of optics is referred to as *ray optics* or *geometrical optics*. The concept of light rays is very useful because the rays show the direction of energy flow in the light beam.

To help understand wave optics, this section first discusses some fundamental characteristics of waveforms. For mathematical convenience in describing the characteristics of light, first it is assumed that the light comes from an ideal monochromatic (single-wavelength) source with the emitted lightwave being represented in the time domain by an infinitely long, single-frequency sinusoidal wave, as Fig. 2.2 shows. In a real lightwave or in an optical pulse of finite time duration, the waveform has an arbitrary time dependence and thus is polychromatic. However, although the waveform describing a polychromatic wave is nonharmonic in time, it may be represented by a superposition of monochromatic harmonic functions.

2.1.1 Monochromatic Waves

A monochromatic lightwave can be represented by a real waveform u(**r**, t) that varies harmonically in time

$$u(\mathbf{r},t) = A(\mathbf{r})\cos[\omega t + \varphi(\mathbf{r})] \qquad (2.1)$$

where **r** is a position vector so that

A(**r**) is the wave amplitude at position **r**

$\varphi(\mathbf{r})$ is the phase of the wave at position **r**

$\omega = 2\pi\nu$ is the angular frequency measured in radians/s or s^{-1}

λ is the wavelength of the wave

$\nu = c/\lambda$ is the lightwave frequency measured in cycles/s or Hz

$T = 1/\nu = 2\pi/\omega$ is the wave period in units of seconds

As noted in Fig. 1.5, the frequencies of optical waves run from 1×10^{12} to 3×10^{16} Hz, which corresponds to maximum and minimum wavelengths of 300 μm and 10 nm, respectively, in the optical spectrum.

The waveform also can be expressed by the following complex function

$$U(\mathbf{r}, t) = A(\mathbf{r}) \exp\{i[\omega t + \varphi(\mathbf{r})]\} = U(\mathbf{r}) \exp(i\omega t) \qquad (2.2)$$

where the time-independent factor $U(\mathbf{r}) = A(\mathbf{r}) \exp[i\varphi(\mathbf{r})]$ is the complex amplitude of the wave. The waveform $u(\mathbf{r}, t)$ is found by taking the real part of $U(\mathbf{r}, t)$

$$u(\mathbf{r}, t) = \text{Re}\{U(\mathbf{r}, t)\} = \frac{1}{2}[U(\mathbf{r}, t) + U^*(\mathbf{r}, t)] \qquad (2.3)$$

where the symbol * denotes the complex conjugate.

The *optical intensity* $I(\mathbf{r}, t)$, which is defined as the optical power per unit area (for example, in units such as W/cm^2 or mW/mm^2), is equal to the average of the squared wavefunction. That is, by letting the operation $\langle x \rangle$ denote the average of a generic function x, then

$$I(\mathbf{r}, t) = \langle U^2(\mathbf{r}, t) \rangle = |U(\mathbf{r})|^2 \langle 1 + \cos(2[2\pi vt + \varphi(\mathbf{r})]) \rangle \qquad (2.4)$$

When this average is taken over a time longer than the optical wave period T, the average of the cosine function goes to zero so that the optical intensity of a monochromatic wave becomes

$$I(\mathbf{r}) = |U(\mathbf{r})|^2 \qquad (2.5)$$

Thus the intensity of a monochromatic wave is independent of time and is the square of the absolute value of its complex amplitude.

Example 2.1 Consider a spherical wave given by

$$U(r) = \frac{A_0}{r} \exp(-i2\pi r/\lambda)$$

where $A_0 = 1.5$ W$^{1/2}$ is a constant and r is measured in cm. What is the intensity of this wave?

Solution: From Eq. (2.5),

$$I(r) = |U(r)|^2 = (A_0/r)^2 = \left[\left(1.5\,W^{1/2}\right)/r\,(\text{in cm})\right]^2 = 2.25\,W/cm^2$$

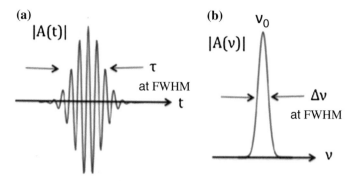

Fig. 2.3 The **a** temporal and **b** spectral characteristics of a pulsed plane wave

2.1.2 Pulsed Plane Waves

Many biophotonics procedures use pulses of light to measure or analyze some biological function. As noted above, these light pulses are polychromatic functions that can be represented by a pulsed plane wave, such as

$$U(\mathbf{r}, t) = A\left(t - \frac{z}{c}\right) \exp\left[i2\pi v_0\left(t - \frac{z}{c}\right)\right] \tag{2.6}$$

In this case, the complex envelope A is a time-varying function and the parameter v_0 is the central optical frequency of the pulse. Figure 2.3 shows the temporal and spectral characteristics of a pulsed plane wave. The complex envelope A(t) typically is of finite duration τ and varies slowly in time compared to an optical cycle. Thus its Fourier transform A(v) has a *spectral width* Δv, which is inversely proportional to the *temporal width* τ at the full-width half-maximum (FWHM) point and is much smaller than the central optical frequency v_0. The temporal and spectral widths usually are defined as the root-mean-square (rms) widths of the power distributions in the time and frequency domains, respectively.

2.2 Polarization

Light emitted by the sun or by an incandescent lamp is created by electromagnetic waves that vibrate in a variety of directions. This type of light is called *unpolarized light*. Lightwaves in which the vibrations occur in a single plane are known as *polarized light*. The process of *polarization* deals with the transformation of unpolarized light into polarized light. The polarization characteristics of lightwaves are important when describing the behavior of polarization-sensitive devices such as optical filters, light signal modulators, Faraday rotators, and light beam splitters.

These types of components typically incorporate polarization-sensitive birefringent crystalline materials such as calcite, lithium niobate, rutile, and yttrium vanadate.

Light consists of *transverse electromagnetic waves* that have both electric field (E field) and magnetic field (H field) components. The directions of the vibrating electric and magnetic fields in a transverse wave are perpendicular to each other and are orthogonal (at right angles) to the direction of propagation of the wave, as Fig. 2.4 shows. The waves are moving in the direction indicated by the *wave vector* **k**. The magnitude of the wave vector **k** is k = $2\pi/\lambda$, which is known as the *wave propagation constant* with λ being the wavelength of the light. Based on Maxwell's equations, it can be shown that E and H are both perpendicular to the direction of propagation. This condition defines a *plane wave*; that is, the vibrations in the electric field are parallel to each other at all points in the wave. Thus, the electric field forms a plane called the *plane of vibration*. An identical situation holds for the magnetic field component, so that all vibrations in the magnetic field wave lie in another plane. Furthermore, **E** and **H** are mutually perpendicular, so that **E**, **H**, and **k** form a set of orthogonal vectors.

An ordinary lightwave is made up of many transverse waves that vibrate in a variety of directions (i.e., in more than one plane), which is called *unpolarized light*. However any arbitrary direction of vibration of a specific transverse wave can be represented as a combination of two orthogonal plane polarization components. As described in Sect. 2.4 and in Chap. 3, this concept is important when examining the reflection and refraction of lightwaves at the interface of two different media, and when examining the propagation of light along an optical fiber. In the case when all the electric field planes of the different transverse waves are aligned parallel to each other, then the lightwave is *linearly polarized*. This describes the simplest type of polarization.

Fig. 2.4 Electric and magnetic field distributions in a train of plane electromagnetic waves at a given instant in time

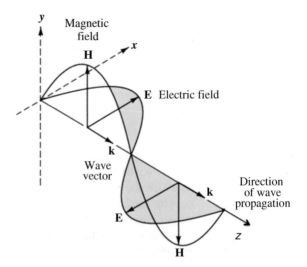

2.2.1 Linear Polarization

Using Eq. (2.2), a train of electric or magnetic field waves designated by $\mathbf{A}$ can be represented in the general form

$$\mathbf{A}(\mathbf{r},\, t) = \mathbf{e}_i\, A_0\, \exp[i(\omega t - \mathbf{k} \cdot \mathbf{r})] \qquad (2.7)$$

where $\mathbf{r} = x\mathbf{e}_x + y\mathbf{e}_y + z\mathbf{e}_z$ represents a general position vector, $\mathbf{k} = k_x\mathbf{e}_x + k_y\mathbf{e}_y + k_z\mathbf{e}_z$ is the wave propagation vector, $\mathbf{e}_j$ is a unit vector lying parallel to an axis designated by j (where j = x, y, or z), and k_j is the magnitude of the wave vector along the j axis. The parameter A_0 is the maximum amplitude of the wave and $\omega = 2\pi\nu$, where ν is the frequency of the light.

The components of the actual (measurable) electromagnetic field are obtained by taking the real part of Eq. (2.7). For example, if $\mathbf{k} = k\mathbf{e}_z$, and if $\mathbf{A}$ denotes the electric field $\mathbf{E}$ with the coordinate axes chosen such that $\mathbf{e}_i = \mathbf{e}_x$, then the measurable electric field is

$$\mathbf{E}_x(z,\, t) = \mathrm{Re}(\mathbf{E}) = \mathbf{e}_x\, E_{0x}\, \cos(\omega t - kz) = \mathbf{e}_x\, E_x(z) \qquad (2.8)$$

which represents a plane wave that varies harmonically as it travels in the z direction. Here E_{0x} is the maximum wave amplitude along the x axis and $E_x(z) = E_{0x}\cos(\omega t - kz)$ is the amplitude at a given value of z in the xz plane. The reason for using the exponential form is that it is more easily handled mathematically than equivalent expressions given in terms of sine and cosine. In addition, the rationale for using harmonic functions is that any waveform can be expressed in terms of sinusoidal waves using Fourier techniques.

The plane wave example given by Eq. (2.8) has its electric field vector always pointing in the $\mathbf{e}_x$ direction, so it is linearly polarized with polarization vector $\mathbf{e}_x$. A general state of polarization is described by considering another linearly polarized wave that is independent of the first wave and orthogonal to it. Let this wave be

$$\mathbf{E}_y(z,\, t) = \mathbf{e}_y\, E_{0y}\, \cos(\omega t - kz + \delta) = \mathbf{e}_y\, E_y(z) \qquad (2.9)$$

where δ is the relative phase difference between the waves. Similar to Eq. (2.8), E_{0y} is the maximum amplitude of the wave along the y axis and $E_y(z) = E_{0y}\cos(\omega t - kz + \delta)$ is the amplitude at a given value of z in the yz plane. The resultant wave is

$$\mathbf{E}(z,t) = \mathbf{E}_x(z,t) + \mathbf{E}_y(z,t) \qquad (2.10)$$

If δ is zero or an integer multiple of 2π, the waves are in phase. Equation (2.10) is then also a linearly polarized wave with a polarization vector making an angle

$$\theta = \arcsin\frac{E_{0y}}{E_{0x}} \qquad (2.11)$$

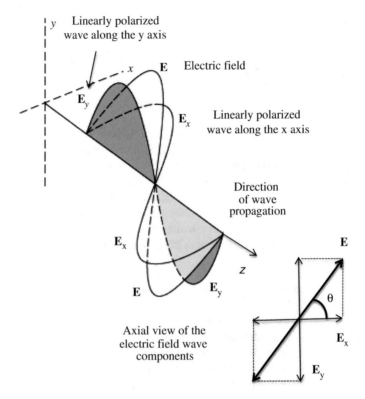

Fig. 2.5 Addition of two linearly polarized waves having a zero relative phase between them

with respect to $\mathbf{e}_x$ and having a magnitude

$$E = \left(E_{0x}^2 + E_{0y}^2 \right)^{1/2} \tag{2.12}$$

This case is shown schematically in Fig. 2.5. Conversely, just as any two orthogonal plane waves can be combined into a linearly polarized wave, an arbitrary linearly polarized wave can be resolved into two independent orthogonal plane waves that are in phase. For example, the wave $\mathbf{E}$ in Fig. 2.5 can be resolved into the two orthogonal plane waves $\mathbf{E}_x$ and $\mathbf{E}_y$.

Example 2.2 The general form of an electromagnetic wave is

$$y = (\text{amplitude in } \mu m) \times \cos(\omega t - kz) = A \cos[2\pi(\nu t - z/\lambda)]$$

Find the (a) amplitude, (b) the wavelength, (c) the angular frequency, and (d) the displacement at time $t = 0$ and $z = 4$ μm of a plane electromagnetic wave specified by the equation $y = 12\cos[2\pi(3t - 1.2z)]$.

Solution: From the above general electromagnetic wave equation

(a) Amplitude = 12 μm
(b) Wavelength: $1/\lambda = 1.2$ μm^{-1} so that $\lambda = 833$ nm
(c) The angular frequency is $\omega = 2\pi v = 2\pi$ (3) $= 6\pi$
(d) At time t = 0 and z = 4 μm, the displacement is

$$y = 12 \cos[2\pi(-1.2\,\mu\text{m}^{-1})(4\,\mu\text{m})] = 12 \cos[2\pi(-4.8)] = 10.38\,\mu\text{m}$$

2.2.2 Elliptical and Circular Polarization

For general values of δ the wave given by Eq. (2.10) is elliptically polarized. The resultant field vector **E** will both rotate and change its magnitude as a function of the angular frequency ω. Elimination of the $(\omega t - kz)$ dependence between Eqs. (2.8) and (2.9) for a general value of δ yields,

$$\left(\frac{E_x}{E_{0x}}\right)^2 + \left(\frac{E_y}{E_{0y}}\right)^2 - 2\left(\frac{E_x}{E_{0x}}\right)\left(\frac{E_y}{E_{0y}}\right)\cos\delta = \sin^2\delta \qquad (2.13)$$

which is the general equation of an ellipse. Thus as Fig. 2.6 shows, the endpoint of **E** will trace out an ellipse at a given point in space. The axis of the ellipse makes an angle α relative to the x axis given by

$$\tan 2\alpha = \frac{2E_{0x}E_{0y}\cos\delta}{E_{0x}^2 - E_{0y}^2} \qquad (2.14)$$

Aligning the principal axis of the ellipse with the x axis gives a better picture of Eq. (2.13). In that case, $\alpha = 0$, or, equivalently, $\delta = \pm\pi/2, \pm3\pi/2, \ldots$, so that Eq. (2.13) becomes

$$\left(\frac{E_x}{E_{0x}}\right)^2 + \left(\frac{E_y}{E_{0y}}\right)^2 = 1 \qquad (2.15)$$

This is the equation of an ellipse with the origin at the center and with amplitudes of the semi-axes equal to E_{0x} and E_{0y}.

When $E_{0x} = E_{0y} = E_0$ and the relative phase difference $\delta = \pm\pi/2 + 2m\pi$, where $m = 0, \pm1, \pm2, \ldots$, then the light is *circularly polarized*. In this case, Eq. (2.15) reduces to

$$E_x^2 + E_y^2 = E_0^2 \qquad (2.16)$$

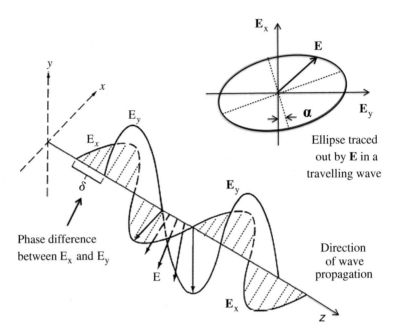

Fig. 2.6 Elliptically polarized light results from the addition of two linearly polarized waves of unequal amplitude having a nonzero phase difference δ between them

which defines a circle. Choosing the phase difference to be $\delta = +\pi/2$ and using the relationship $\cos(a+b) = (\cos a)(\cos b) + (\sin a)(\sin b)$ to expand Eq. (2.9), then Eqs. (2.8) and (2.9) become

$$\mathbf{E}_x(z, t) = \mathbf{e}_x E_0 \cos(\omega t - kz) \tag{2.17a}$$

$$\mathbf{E}_y(z, t) = -\mathbf{e}_y E_0 \sin(\omega t - kz) \tag{2.17b}$$

In this case, the endpoint of **E** will trace out a circle at a given point in space, as Fig. 2.7 illustrates. To see this, consider an observer located at some arbitrary point z_{ref} toward whom the wave is moving. For convenience, pick the reference point at $z_{ref} = \pi/k$ at $t = 0$. Then from Eq. (2.17a) it follows that

$$\mathbf{E}_x(z, t) = -\mathbf{e}_x E_0 \text{ and } \mathbf{E}_y(z, t) = 0 \tag{2.18}$$

so that **E** lies along the negative x axis as Fig. 2.7 shows. At a later time, say at $t = \pi/2\omega$, the electric field vector has rotated through 90° and now lies along the positive y axis. Thus as the wave moves toward the observer with increasing time, the resultant electric field vector **E** rotates clockwise at an angular frequency ω. It makes one complete rotation as the wave advances through one wavelength. Such a light wave is *right circularly polarized*.

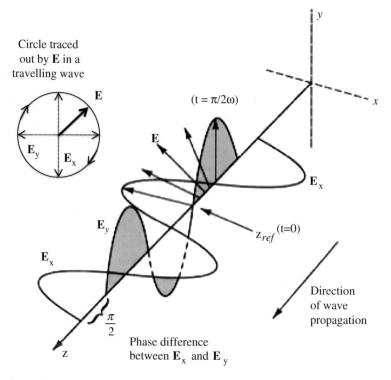

Fig. 2.7 Addition of two equal-amplitude linearly polarized waves with a relative phase difference $\delta = \pi/2 + 2m\pi$ results in a right circularly polarized wave

If the negative sign is selected for δ, then the electric field vector is given by

$$\mathbf{E} = E_0[\mathbf{e}_x \cos(\omega t - kz) + \mathbf{e}_y \sin(\omega t - kz)] \qquad (2.19)$$

Now $\mathbf{E}$ rotates counterclockwise and the wave is *left circularly polarized.*

2.3 Quantized Photon Energy and Momentum

The wave theory of light adequately accounts for all phenomena involving the transmission of light. However, in dealing with the interaction of light and matter, such as occurs in the emission and absorption of light, one needs to invoke the quantum theory of light, which indicates that optical radiation has particle as well as wave properties. The particle nature arises from the observation that light energy is always emitted or absorbed in discrete (quantized) units called *photons*. The photon energy depends only on the frequency v. This frequency, in turn, must be measured by observing a wave property of light.

The relationship between the wave theory and the particle theory is given by *Planck's Law*, which states that the energy E of a photon and its associated wave frequency v is given by the equation

$$E = hv = hc/\lambda \tag{2.20}$$

where $h = 6.625 \times 10^{-34}$ J·s is Planck's constant and λ is the wavelength. The most common measure of photon energy is the *electron volt* (eV), which is the energy an electron gains when moving through a 1-volt electric field. Note that $1 \text{ eV} = 1.60218 \times 10^{-19}$ J. As is noted in Eq. (1.3), for calculation simplicity, if λ is expressed in μm then the energy E is given in eV by using $E(\text{eV}) = 1.2405/\lambda$ (μm). The linear momentum p associated with a photon of energy E in a plane wave is given by

$$p = E/c = h/\lambda \tag{2.21}$$

The momentum is of particular importance when examining photon scattering by molecules (see Chap. 6).

When light is incident on an atom or molecule, a photon can transfer its energy to an electron within this atom or molecule, thereby exciting the electron to a higher electronic or vibrational energy levels or quantum states, as shown in Fig. 2.8a. In this process either all or none of the photon energy is imparted to the electron. For example, consider an incoming photon that has an energy hv_{12}. If an electron sits at an energy level E_1 in the ground state, then it can be boosted to a higher energy level E_2 if the incoming photon has an energy $hv_{12} = E_2 - E_1$. Note that the energy absorbed by the electron from the photon must be exactly equal to the energy required to excite the electron to a higher quantum state. Conversely, an electron in an excited state E_3 can drop to a lower energy level E_1 by emitting a photon of energy exactly equal to $hv_{13} = E_3 - E_1$, as shown in Fig. 2.8b.

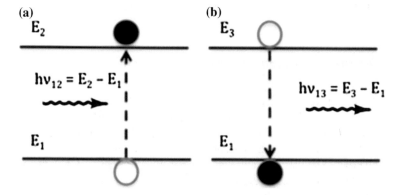

Fig. 2.8 Electron transitions between the ground state and higher electronic or vibrational energy levels

Example 2.3 Two commonly used sources in biomedical photonics are Er: YAG and CO_2 lasers, which have peak emission wavelengths of 2.94 and 10.6 μm, respectively. Compare the photon energies of these two sources.

Solution: Using the relationship $E = hc/\lambda$ from Eq. (2.20) yields

$$E(2.94\,\mu m) = (6.625 \times 10^{-34}\,J \cdot s)\,(3 \times 10^8\,m/s)/(2.94 \times 10^{-6}\,m)$$
$$= 6.76 \times 10^{-20}\,J\ = 0.423\,eV$$

Similarly, $E\,(10.6\,\mu m) = 0.117\,eV$

Example 2.4 Compare the photon energies for an ultraviolet wavelength of 300 nm and an infrared wavelength of 1550 nm.

Solution: From Eq. (2.20), $E(300\ nm) = 4.14$ eV and $E(1550\ nm) = 0.80$ eV.

Example 2.5 (a) Consider an incoming photon that boosts an electron from a ground state level E_1 to an excited level E_2. If $E_2 - E_1 = 1.512$ eV, what is the wavelength of the incoming photon? (b) Now suppose this excited electron loses some of its energy and moves to a slightly lower energy level E_3. If the electron then drops back to level E_1 and if $E_3 - E_1 = 1.450$ eV, what is the wavelength of the emitted photon?

Solution: (a)   From   Eq.   (2.20), $\lambda_{incident} = 1.2405/1.512$ eV = 0.820 μm = 820 nm. (b) From Eq. (2.20), $\lambda_{emitted} = 1.2405/1.450$ eV = 0.855 μm = 855 nm.

2.4   Reflection and Refraction

The concepts of reflection and refraction can be described by examining the behavior of the light rays that are associated with plane waves, as is shown in Fig. 2.1. When a light ray encounters a smooth interface that separates two different dielectric media, part of the light is reflected back into the first medium. The remainder of the light is bent (or refracted) as it enters the second material. This is shown in Fig. 2.9 for the interface between a glass material and air that have refractive indices n_1 and n_2, respectively, where $n_2 < n_1$.

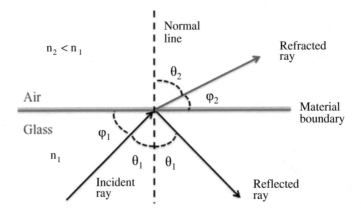

Fig. 2.9 Reflection and refraction of a light ray at a material boundary

2.4.1 Snells' Law

The bending or refraction of light rays at a material interface is a consequence of the difference in the speed of light in two materials with different refractive indices. The relationship at the interface is known as *Snell's law* and is given by

$$n_1 \sin\theta_1 = n_2 \sin\theta_2 \tag{2.22}$$

or, equivalently, as

$$n_1 \cos\varphi_1 = n_2 \cos\varphi_2 \tag{2.23}$$

where the angles are defined in Fig. 2.9. The angle θ_1 between the incident ray and the normal to the surface is known as the *angle of incidence*.

In accordance with the law of reflection, the angle θ_1 at which the incident ray strikes the interface is the same as the angle that the reflected ray makes with the interface. Furthermore, the incident ray, the reflected ray, and the normal to the interface all lie in a common plane, which is perpendicular to the interface plane between the two materials. This common plane is called the *plane of incidence*. When light traveling in a certain medium reflects off of a material that has a higher refractive index (called an optically denser material), the process is called *external reflection*. Conversely, the reflection of light off of a material that has a lower refractive index and thus is less optically dense (such as light traveling in glass being reflected at a glass–air interface) is called *internal reflection*.

As the angle of incidence θ_1 in an optically denser material increases, the refracted angle θ_2 approaches $\pi/2$. Beyond this point no refraction is possible as the incident angle increases and the light rays become totally internally reflected. The application of Snell's law yields the conditions required for *total internal reflection*. Consider Fig. 2.10, which shows the interface between a glass surface and air. As a

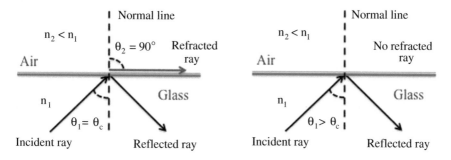

Fig. 2.10 Representation of the critical angle and total internal reflection at a glass-air interface, where n_1 is the refractive index of glass

light ray leaves the glass and enters the air medium, the ray gets bent toward the glass surface in accordance with Snell's law. If the angle of incidence θ_1 is increased, a point will eventually be reached where the light ray in air is parallel to the glass surface. This situation defines the *critical angle of incidence* θ_c. The condition for total internal reflection is satisfied when the angle of incidence θ_1 is greater than the critical angle, that is, all the light is reflected back into the glass with no light penetrating into the air.

To find the critical angle, consider Snell's law as given by Eq. (2.22). The critical angle is reached when $\theta_2 = 90°$ so that $\sin \theta_2 = 1$. Substituting this value of θ_2 into Eq. (2.22) shows that the critical angle is determined from the condition

$$\sin \theta_c = \frac{n_2}{n_1} \tag{2.24}$$

Example 2.6 Consider the interface between a smooth biological tissue with $n_1 = 1.45$ and air for which $n_2 = 1.00$. What is the critical angle for light traveling in the tissue?

Solution: From Eq. (2.24), for light traveling in the tissue the critical angle is

$$\theta_c = \sin^{-1} \frac{n_2}{n_1} = \sin^{-1} 0.690 = 43.6°$$

Thus any light ray traveling in the tissue that is incident on the tissue–air interface at an angle θ_1 with respect to the normal (as shown in Fig. 2.9) greater than $43.6°$ is totally reflected back into the tissue.

Example 2.7 A light ray traveling in air ($n_1 = 1.00$) is incident on a smooth, flat slab of crown glass, which has a refractive index $n_2 = 1.52$. If the incoming ray makes an angle of $\theta_1 = 30.0°$ with respect to the normal, what is the angle of refraction θ_2 in the glass?

Solution: From Snell's law given by Eq. (2.24),

$$\sin \theta_2 = \frac{n_1}{n_2} \sin \theta_1 = \frac{1.00}{1.52} \sin 30° = 0.658 \times 0.50 = 0.329$$

Solving for θ_2 then yields $\theta_2 = \sin^{-1} (0.329) = 19.2°$.

2.4.2 The Fresnel Equations

As noted in Sect. 2.2, one can consider unpolarized light as consisting of two orthogonal plane polarization components. For analyzing reflected and refracted light, one component can be chosen to lie in the plane of incidence (the plane containing the incident and reflected rays, which here is taken to be the yz-plane) and the other of which lies in a plane perpendicular to the plane of incidence (the xz-plane). For example, these can be the E_y and E_x components, respectively, of the electric field vector shown in Fig. 2.5. These then are designated as the *perpendicular polarization* (E_x) and the *parallel polarization* (E_y) components with maximum amplitudes E_{0x} and E_{0y}, respectively.

When an unpolarized light beam traveling in air impinges on a nonmetallic surface such as biological tissue, part of the beam (designated by E_{0r}) is reflected and part of the beam (designated by E_{0t}) is refracted and transmitted into the target material. The reflected beam is partially polarized and at a specific angle (known as *Brewster's angle*) the reflected light is completely perpendicularly polarized, so that $(E_{0r})_y = 0$. This condition holds when the angle of incidence is such that $\theta_1 + \theta_2 = 90°$. The parallel component of the refracted beam is transmitted entirely into the target material, whereas the perpendicular component is only partially refracted. How much of the refracted light is polarized depends on the angle at which the light approaches the surface and on the material composition.

The amount of light of each polarization type that is reflected and refracted at a material interface can be calculated using a set of equations known as the *Fresnel equations*. These field-amplitude ratio equations are given in terms of the perpendicular and parallel *reflection coefficients* r_x and r_y, respectively, and the perpendicular and parallel *transmission coefficients* t_x and t_y, respectively. Given that E_{0i}, E_{0r}, and E_{0t} are the amplitudes of the incident, reflected, and transmitted waves, respectively, then

$$r_\perp = r_x = \left(\frac{E_{0r}}{E_{0i}}\right)_x = \frac{n_1 \cos \theta_1 - n_2 \cos \theta_2}{n_1 \cos \theta_1 + n_2 \cos \theta_2} \tag{2.25}$$

$$r_{\parallel} = r_y = \left(\frac{E_{0r}}{E_{0i}}\right)_y = \frac{n_2 \cos\theta_1 - n_1 \cos\theta_2}{n_1 \cos\theta_2 + n_2 \cos\theta_1} \qquad (2.26)$$

$$t_{\perp} = t_x = \left(\frac{E_{0t}}{E_{0i}}\right)_x = \frac{2n_1 \cos\theta_1}{n_1 \cos\theta_1 + n_2 \cos\theta_2} \qquad (2.27)$$

$$t_{\parallel} = t_y = \left(\frac{E_{0t}}{E_{0i}}\right)_y = \frac{2n_1 \cos\theta_1}{n_1 \cos\theta_2 + n_2 \cos\theta_1} \qquad (2.28)$$

If light is incident perpendicularly on the material interface, then the angles are $\theta_1 = \theta_2 = 0$. From Eqs. (2.25) and (2.26) it follows that the reflection coefficients are

$$r_x(\theta_1 = 0) = -r_y(\theta_2 = 0) = \frac{n_1 - n_2}{n_1 + n_2} \qquad (2.29)$$

Similarly, for $\theta_1 = \theta_2 = 0$, the transmission coefficients are

$$t_x(\theta_1 = 0) = t_y(\theta_2 = 0) = \frac{2n_1}{n_1 + n_2} \qquad (2.30)$$

Example 2.8 Consider the case when light traveling in air ($n_{air} = 1.00$) is incident perpendicularly on a smooth tissue sample that has a refractive index $n_{tissue} = 1.35$. What are the reflection and transmission coefficients?

Solution: From Eq. (2.29) with $n_1 = n_{air}$ and $n_2 = n_{tissue}$ it follows that the reflection coefficient is

$$r_y = -r_x = (1.35 - 1.00)/(1.35 + 1.00) = 0.149$$

and from Eq. (2.30) the transmission coefficient is

$$t_x = t_y = 2(1.00)/(1.35 + 1.00) = 0.851$$

The change in sign of the reflection coefficient r_x means that the field of the perpendicular component shifts by 180° upon reflection.

The field amplitude ratios can be used to calculate the *reflectance* R (the ratio of the reflected to the incident flux or power) and the *transmittance* T (the ratio of the transmitted to the incident flux or power). For linearly polarized light in which the vibrational plane of the incident light is perpendicular to the interface plane, the total reflectance and transmittance are

$$R_\perp = \left(\frac{E_{0r}}{E_{0i}}\right)^2_x = R_x = r_x^2 \tag{2.31}$$

$$R_\parallel = \left(\frac{E_{0r}}{E_{0i}}\right)^2_y = R_y = r_y^2 \tag{2.32}$$

$$T_\perp = \frac{n_2 \cos\theta_2}{n_1 \cos\theta_1}\left(\frac{E_{0t}}{E_{0i}}\right)^2_x = T_x = \frac{n_2 \cos\theta_2}{n_1 \cos\theta_1}t_x^2 \tag{2.33}$$

$$T_\parallel = \frac{n_2 \cos\theta_2}{n_1 \cos\theta_1}\left(\frac{E_{0t}}{E_{0i}}\right)^2_y = T_y = \frac{n_2 \cos\theta_2}{n_1 \cos\theta_1}t_y^2 \tag{2.34}$$

The expression for T is a bit more complex compared to R because the shape of the incident light beam changes upon entering the second material and the speeds at which energy is transported into and out of the interface are different.

If light is incident perpendicularly on the material interface, then substituting Eq. (2.29) into Eqs. (2.31) and (2.32) yields the following expression for the reflectance R

$$R = R_\perp(\theta_1 = 0) = R_\parallel(\theta_1 = 0) = \left(\frac{n_1 - n_2}{n_1 + n_2}\right)^2 \tag{2.35}$$

and substituting Eq. (2.30) into Eqs. (2.33) and (2.34) yields the following expression for the transmittance T

$$T = T_\perp(\theta_2 = 0) = T_\parallel(\theta_2 = 0) = \frac{4n_1 n_2}{(n_1 + n_2)^2} \tag{2.36}$$

Example 2.9 Consider the case described in Example 2.8 in which light traveling in air ($n_{air} = 1.00$) is incident perpendicularly on a smooth tissue sample that has a refractive index $n_{tissue} = 1.35$. What are the reflectance and transmittance values?

Solution: From Eq. (2.35) and Example 2.8 the reflectance is

$$R = [(1.35 - 1.00)/(1.35 + 1.00)]^2 = (0.149)^2 = 0.022 \text{ or } 2.2\,\%.$$

From Eq. (2.36) the transmittance is

$$T = 4(1.00)(1.35)/(1.00 + 1.35)^2 = 0.978 \text{ or } 97.8\,\%$$

Note that R + T = 1.00.

Example 2.10 Consider a plane wave that lies in the plane of incidence of an air-glass interface. What are the values of the reflection coefficients if this lightwave is incident at 30° on the interface? Let $n_{air} = 1.00$ and $n_{glass} = 1.50$.

Solution: First from Snell's law as given by Eq. (2.22), it follows that $\theta_2 = 19.2°$ (see Example 2.7). Substituting the values of the refractive indices and the angles into Eqs. (2.25) and (2.26) then yield $r_x = -0.241$ and $r_y = 0.158$. As noted in Example 2.8, the change in sign of the reflection coefficient r_x means that the field of the perpendicular component shifts by 180° upon reflection.

2.4.3 Diffuse Reflection

The amount of light that is reflected by an object into a certain direction depends greatly on the texture of the reflecting surface. Reflection off of smooth surfaces such as mirrors, polished glass, or crystalline materials is known as *specular reflection*. In specular reflection the surface imperfections are smaller than the wavelength of the incident light, and basically all of the incident light is reflected at a definite angle following Snell's law. *Diffuse reflection* results from the reflection of light off of surfaces that are microscopically rough, uneven, granular, powdered, or porous. This type of reflection tends to send light in all directions, as Fig. 2.11 shows. Because most surfaces in the real world are not smooth, most often incident light undergoes diffuse reflection. Note that Snell's law still holds at each incremental point of incidence of a light ray on an uneven surface.

Diffuse reflection also is the main mechanism that results in scattering of light within biological tissue, which is a *turbid medium* (or random medium) with many different types of intermingled materials that reflect light in all directions. Such diffusely scattered light can be used to probe and image spatial variations in

Fig. 2.11 Illustration of diffuse reflection from an uneven surface

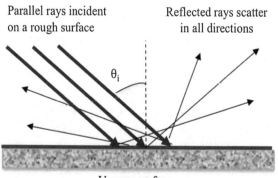

Parallel rays incident on a rough surface

Reflected rays scatter in all directions

θ_i

Uneven surface

macroscopic optical properties of biological tissues. This is the basis of elastic scattering spectroscopy, also known as diffuse reflectance spectroscopy, which is a non-invasive imaging technique for detecting changes in the physical properties of cells in biological tissues. Chapter 9 covers this topic in more detail.

2.5 Interference

All types of waves including lightwaves can interfere with each other if they have the same or nearly the same frequency. When two or more such lightwaves are present at the same time in some region, then the total wavefunction is the sum of the wavefunctions of each lightwave. Thus consider two monochromatic lightwaves of the same frequency with complex amplitudes $U_1(\mathbf{r}) = \sqrt{I_1}\exp(i\varphi_1)$ and $U_2(\mathbf{r}) = \sqrt{I_2}\exp(i\varphi_2)$, as defined in Eq. (2.2), where φ_1 and φ_2 are the phases of the two waves. Superimposing these two lightwaves yields another monochromatic lightwave of the same frequency

$$U(\mathbf{r}) = U_1(\mathbf{r}) + U_2(\mathbf{r}) = \sqrt{I_1}\exp(i\varphi_1) + \sqrt{I_2}\exp(i\varphi_2) \qquad (2.37)$$

where for simplicity the explicit dependence on the position vector $\mathbf{r}$ was omitted on the right-hand side. Then from Eq. (2.5) the intensities of the individual light-waves are $I_1 = |U_1|^2$ and $I_2 = |U_2|^2$ and the intensity I of the composite lightwave is

$$\begin{aligned} I = |U|^2 &= |U_1 + U_2|^2 = |U_1|^2 + |U_2|^2 + U_1{}^*U_2 + U_1U_2{}^* \\ &= I_1 + I_2 + 2\sqrt{I_1 I_2}\cos\varphi \end{aligned} \qquad (2.38)$$

where the phase difference $\varphi = \varphi_1 - \varphi_2$.

The relationship in Eq. (2.38) is known as the *interference equation*. It shows that the intensity of the composite lightwave depends not only on the individual intensities of the constituent lightwaves, but also on the phase difference between the waves. If the constituent lightwaves have the same intensities, $I_1 = I_2 = I_0$, then

$$I = 2I_0(1 + \cos\varphi) \qquad (2.39)$$

If the two lightwaves are in phase so that $\varphi = 0$, then $\cos\varphi = 1$ and $I = 4I_0$, which corresponds to *constructive interference*. If the two lightwaves are 180° out of phase, then $\cos\pi = -1$ and $I = 0$, which corresponds to *destructive interference*.

Example 2.11 Consider the case of two monochromatic interfering light-waves. Suppose the intensities are such that $I_2 = I_1/4$. (a) If the phase difference $\varphi = 2\pi$, what is the intensity I of the composite lightwave? (b) What is the intensity of the composite lightwave if $\varphi = \pi$?

Solution: (a) The condition $\varphi = 2\pi$ means that $\cos \varphi = 1$, so that the two waves are in phase and interfere constructively. From Eq. (2.32) it follows that

$$I = I_1 + I_2 + 2\sqrt{I_1 I_2} = I_1 + I_1/4 + 2(I_1 I_1/4)^{1/2} = (9/4)I_1$$

(b) The condition $\varphi = \pi$ means that $\cos \varphi = -1$, so that the two waves are out of phase and interfere destructively. From Eq. (2.38) it follows that

$$I = I_1 + I_2 - 2\sqrt{I_1 I_2} = I_1 + I_1/4 - 2(I_1 I_1/4)^{1/2} = I_1/4$$

2.6 Optical Coherence

As noted in Sect. 2.2, the complex envelope of a polychromatic waveform typically exists for a finite duration and thus has an associated finite optical frequency bandwidth. Thus, all light sources emit over a finite range of frequencies Δv or wavelengths $\Delta\lambda$, which is referred to as a *spectral width* or *linewidth*. The spectral width most commonly is defined as the full width at half maximum (FWHM) of the spectral distribution from a light source about a central frequency v_0. Equivalently, the emission can be viewed as consisting of a set of finite wave trains. This leads to the concept of *optical coherence*.

The time interval over which the phase of a particular wave train is constant is known as the *coherence time* t_c. Thus the coherence time is the temporal interval over which the phase of a lightwave can be predicted accurately at a given point in space. If t_c is large, the wave has a high degree of temporal coherence. The corresponding spatial interval $l_c = ct_c$ is referred to as the *coherence length*. The importance of the coherence length is that it is the extent in space over which the wave is reasonably sinusoidal so that its phase can be determined precisely. An illustration of the coherence time is shown in Fig. 2.12 for a waveform consisting of random finite sinusoids.

The coherence length of a wave also can be expressed in terms of the linewidth $\Delta\lambda$ through the expression

$$l_c = \frac{4 \ln 2}{\pi} \frac{\lambda_0^2}{\Delta\lambda} \tag{2.40}$$

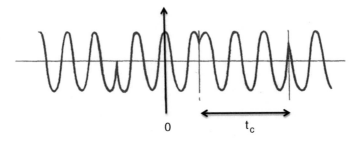

Fig. 2.12 Illustration of the coherence time in random finite sinusoids

Example 2.12 What are the coherence time and coherence length of a white light laser, which emits in the range 380 nm to 700 nm?

Solution: The wavelength bandwidth is $\Delta\lambda = 320$ nm with a center wavelength of 540 nm. Using the relationship in Eq. (2.40) then gives a coherence length $l_c = 804$ nm. The coherence time thus is $t_c = l_c/c = 2.68 \times 10^{-15}$ s $= 2.68$ fs.

2.7 Lightwave-Molecular Dipole Interaction

The interaction effects between light and biological material can be understood by considering the electromagnetic properties of light and of molecules. First consider the structure and electronic properties of molecules. The most common atoms found in biological molecules are carbon, hydrogen, nitrogen, and oxygen. The most abundant atom is oxygen because it is contained in proteins, nucleic acids, carbohydrates, fats, and water. A molecule can be viewed as consisting of positively charged atomic nuclei that are surrounded by negatively charged electrons. Chemical bonding in a molecule occurs because the different constituent atoms share common electron pairs. If the atoms are identical, for example, two oxygen atoms in an O_2 molecule, the common electron pair is usually located between the two atoms. In such a case the molecule is symmetrically charged and is called a *nonpolar molecule*.

If the atoms in a molecule are not identical, then the shared electron pair is displaced toward the atom that has a greater attraction for common electrons. This condition then creates what is known as a *dipole*, which is a pair of equal and opposite charges $+Q$ and $-Q$ separated by a distance d, as shown in Fig. 2.13. The dipole is described in terms of a vector parameter called the *dipole moment*, which has a magnitude μ given by

Fig. 2.13 A dipole is defined as two charges +Q and −Q separated by a distance d

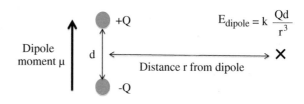

$$\mu = Qd \tag{2.41}$$

A molecule that has a permanent dipole moment is called a *polar molecule*. Examples of polar molecules are water (H_2O), ammonia (NH_3), hydrogen chloride (HCl), and the amino acids arginine, lysine, and tyrosine. Symmetric nonpolar molecules such as oxygen, nitrogen, carbon dioxide (CO_2), methane (CH_4), ammonium (NH_4), carbon tetrachloride (CCl_4), and the amino acids glycine and tryptophan have no permanent dipole moments.

The interaction of the oscillating electric field of a lightwave and the dipole moment of a molecule is a key effect that can help describe light-tissue interactions. In addition, an energy exchange between two oscillating dipoles is used in fluorescence microscopy. When an external electric field interacts with either polar or nonpolar molecules, the field can distort the electron distribution around the molecule. In both types of molecules this action generates a temporary induced dipole moment μ_{ind} that is proportional to the electric field E. This induced dipole moment is given by

$$\mu_{ind} = \alpha E \tag{2.42}$$

where the parameter α is called the *polarizability* of the molecule. Thus when a molecule is subjected to the oscillating electric field of a lightwave, the total dipole moment μ_T is given by

$$\mu_T = \mu + \mu_{ind} = Qd + \alpha E \tag{2.43}$$

As is shown in Fig. 2.13, the resultant electric field E_{dipole} at a distance r from the dipole in any direction has a magnitude

$$E_{dipole} = k\frac{Qd}{r^3} \tag{2.44}$$

where $k = 1/4\pi\varepsilon_0 = 9.0 \times 10^9$ Nm^2/C^2.

Fig. 2.14 Representation of
the dipole moment for a water
molecule

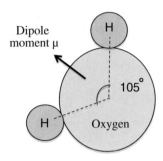

Dipole
moment μ

H

105°

H

Oxygen

Example 2.13 What is the dipole moment of a water molecule?

Solution: As Fig. 2.14 shows, water (H_2O) is an asymmetric molecule in which the hydrogen atoms are situated at a 105° angle relative to the center of the oxygen atom. This structure leads to a dipole moment in the symmetry plane with the dipole pointing toward the more positive hydrogen atoms. The magnitude of this dipole moment has been measured to be

$$\mu = 6.2 \times 10^{-30}\,C \cdot m$$

2.8 Summary

Some optical phenomena can be explained using a wave theory whereas in other cases light behaves as though it is composed of miniature particles called photons. The wave nature explains how light travels through an optical fiber and how it can be coupled between two adjacent fibers, but the particle theory is needed to explain how optical sources generate light and how photodetectors change an optical signal into an electrical signal.

In free space a lightwave travels at a speed c = 3 × 10⁸ m/s, but it slows down by a factor n > 1 when entering a material, where the parameter $\underline{n}$ is the index of refraction (or refractive index) of the material. Example values of the refractive index for materials related to biophotonics are 1.00 for air, about 1.36 for many tissue materials, and between 1.45 and 1.50 for various glass compounds. Thus light travels at about 2.2 × 10⁸ m/s in a biological tissue.

When a light ray encounters a boundary separating two media that have different refractive indices, part of the ray is reflected back into the first medium and the remainder is bent (or refracted) as it enters the second material. As will be discussed in later chapters, these concepts play a major role in describing the amount of optical power that can be injected into a fiber and how lightwaves travel along a fiber.

Other important characteristics of lightwaves for a wide variety of biophotonics microscopy methods, spectroscopy techniques, and imaging modalities are the polarization of light, interference effects, and the properties of coherence.

2.9 Problems

2.1 Consider an electric field represented by the expression

$$E = \left[100e^{i30^\circ}\, e_x + 20e^{-i50^\circ}\, e_y + 140e^{i210^\circ}\, e_z \right]$$

Express this as a measurable electric field as described by Eq. (2.8) at a frequency of 100 MHz.

2.2 A particular plane wave is specified by $y = 8 \cos 2\pi(2t - 0.8z)$, where y is expressed in micrometers and the propagation constant is given in μm^{-1}. Find (a) the amplitude, (b) the wavelength, (c) the angular frequency, and (d) the displacement at time $t = 0$ and $z = 4\ \mu m$. Answers: (a) 8 μm; (b) 1.25 μm; (c) 4π; (d) 2.472 μm.

2.3 Light traveling in air strikes a glass plate at an angle $\theta_1 = 57°$, where θ_1 is measured between the incoming ray and the normal to the glass surface. Upon striking the glass, part of the beam is reflected and part is refracted. (a) If the refracted and reflected beams make an angle of $90°$ with each other, show that the refractive index of the glass is 1.540. (b) Show that the critical angle for this glass is $40.5°$.

2.4 A point source of light is 12 cm below the surface of a large body of water ($n_{water} = 1.33$). What is the radius of the largest circle on the water surface through which the light can emerge from the water into air ($n_{air} = 1.00$)? Answer: 13.7 cm.

2.5 A right-angle prism (internal angles are 45, 45, and $90°$) is immersed in alcohol ($n = 1.45$). What is the refractive index the prism must have if a ray that is incident normally on one of the short faces is to be totally reflected at the long face of the prism? Answer: 2.05.

2.6 Show that the critical angle at an interface between doped silica with $n_1 = 1.460$ and pure silica with $n_2 = 1.450$ is $83.3°$.

2.7 As noted in Sect. 2.4.2, at a certain angle of incidence there is no reflected parallel beam, which is known as Brewster's law. This condition holds when the reflection coefficient r_y given by Eq. (2.26) is zero. (a) Using Snell's law from Eq. (2.22), the condition $n_1 \cos \theta_2 = n_2 \cos \theta_1$ when $r_y = 0$, and the relationship $\sin^2 \alpha + \cos^2 \alpha = 1$ for any angle α, show there is no parallel reflection when $\tan \theta_1 = n_2/n_1$. (b) Show that $\theta_1 + \theta_2 = 90°$ at the Brewster angle.

2.8   Consider the perpendicular and parallel *reflection coefficients* r_x and r_y, given by Eqs. (2.25) and (2.26), respectively. By using Snell's law from Eq. (2.22) and the identity $\sin(\alpha \pm \beta) = (\sin \alpha)(\cos \beta) \pm (\cos \alpha)(\sin \beta)$, eliminate the dependence on n_1 and n_2 to write r_x and r_y as functions of θ_1 and θ_2 only. That is, show that this yields

$$r_x = r_\perp = -\frac{\sin(\theta_1 - \theta_2)}{\sin(\theta_1 + \theta_2)}$$

$$r_y = r_\parallel = \frac{\tan(\theta_1 - \theta_2)}{\tan(\theta_1 + \theta_2)}$$

2.9   Consider the case when light traveling in air ($n_{air} = 1.00$) is incident perpendicularly on a smooth tissue sample that has a refractive index $n_{tissue} = 1.50$. (a) Show that the reflection and transmission coefficients are 0.20 and 0.80, respectively. (b) Show that the reflectance and transmittance values are $R = 0.04$ and $T = 0.96$.

2.10   Show that the reflection coefficients r_x and r_y, given by Eqs. (2.25) and (2.26) can be written in terms of only the incident angle θ_1 and the refractive index ratio $n_{21} = n_2/n_1$ as

$$r_\perp = r_x = \frac{\cos\theta_1 - \left(n_{21}^2 - \sin^2\theta_1\right)^{1/2}}{\cos\theta_1 + \left(n_{21}^2 - \sin^2\theta_1\right)^{1/2}}$$

$$r_\parallel = r_y = \frac{n_{21}^2 \cos\theta_1 - \left(n_{21}^2 - \sin^2\theta_1\right)^{1/2}}{n_{21}^2 \cos\theta_1 + \left(n_{21}^2 - \sin^2\theta_1\right)^{1/2}}$$

2.11   Consider a plane wave that lies in the plane of incidence of an air-glass interface. (a) Show that the values of the reflection coefficients are $r_x = -0.303$ and $r_y = 0.092$ if this lightwave is incident at 45° on the interface. (b) Show that the values of the transmission coefficients are $t_x = 0.697$ and $t_y = 0.728$. Let $n_{air} = 1.00$ and $n_{glass} = 1.50$.

2.12   Consider the case of two monochromatic interfering lightwaves. Suppose the intensities are such that $I_2 = I_1$. (a) If the phase difference $\varphi = 2\pi$, show that the intensity I of the composite lightwave is $4I_1$. (b) Show that the intensity of the composite lightwave is zero when $\varphi = \pi$.

2.13   The frequency stability given by $\Delta v/v$ can be used to indicate the spectral purity of a light source. Consider a Hg^{198} low-pressure lamp that emits at a wavelength 546.078 nm and has a spectral bandwidth $\Delta v = 1000$ MHz. (a) Show that the coherence time is 1 ns. (b) Show that the coherence length is 29.9 cm. (c) Show that the frequency stability is 1.82×10^{-6}. Thus the frequency stability is about two parts per million.

References

1. F. Jenkins, H. White, *Fundamentals of Optics*, 4th edn. (McGraw-Hill, New York, 2002)
2. B.E.A. Saleh, M.C. Teich, *Fundamentals of Photonics*, 2nd edn. (Wiley, Hoboken, NJ, 2007)
3. R. Menzel, *Photonics: Linear and Nonlinear Interactions of Laser Light and Matter*, 2nd edn. (Springer, 2007)
4. A. Ghatak, *Optics* (McGraw-Hill, New York, 2010)
5. C.A. Diarzio, *Optics for Engineers* (CRC Press, Boca Raton, FL, 2012)
6. A. Giambattista, B.M. Richardson, R.C. Richardson, *College Physics*, 4th edn. (McGraw-Hill, Boston, 2012)
7. D. Halliday, R. Resnick, J. Walker, *Fundamentals of Physics*, 10th edn. (Wiley, Hoboken, NJ, 2014)
8. G. Keiser, F. Xiong, Y. Cui, and P. P. Shum, Review of diverse optical fibers used in biomedical research and clinical practice. J. Biomed. Optics. **19**, art. 080902 (Aug. 2014)
9. G. Keiser, *Optical Fiber Communications* (McGraw-Hill, 4th US edn. 2011; 5th international edn. 2015)
10. E. Hecht, *Optics*, 5th edn. (Addison-Wesley, 2016)
11. G.A. Reider, *Photonics* (Springer, 2016)

Chapter 3
Optical Fibers for Biophotonics Applications

Abstract Major challenges in biophotonics applications to the life sciences include how to collect emitted low-power light (down to the nW range) from a tissue specimen and transmit it to a photon detector, how to deliver a wide range of optical power levels to a tissue area or section during different categories of therapeutic healthcare sessions, and how to access a diagnostic or treatment area within a living being with an optical detection probe or a radiant energy source in the least invasive manner. The unique physical and light-transmission properties of optical fibers enable them to help resolve such implementation issues. This chapter provides the background that is necessary to understand how optical fibers function and describes various categories of fibers that are commercially available for use in biophotonics.

The optical power levels that have to be detected or transmitted in a biophotonics process can vary by ten orders of magnitude depending on the particular application. The detected light levels of interest can be in the nanowatt range for spectroscopic applications, whereas optical power being delivered to a biological specimen can be as high as several watts during light therapy sessions or during laser surgery.

Major challenges in biophotonics applications to the life sciences include how to collect emitted low-power light (down to the nW range) from a tissue specimen and transmit it to a photon detector, how to deliver a wide range of optical power levels to a tissue area or section during different categories of therapeutic healthcare sessions, and how to access a diagnostic or treatment area within a living being with an optical detection probe or a radiant energy source in the least invasive manner. Depending on the application, all three of these factors may need to be addressed at the same time.

The unique physical and light-transmission properties of optical fibers enable them to help resolve such implementation issues [1-6]. Consequently, various types of optical fibers are finding widespread use in biophotonics instrumentation for life sciences related clinical and research applications. Each optical fiber structure has certain advantages and limitations for specific uses in different spectral bands that

© Springer Science+Business Media Singapore 2016
G. Keiser, *Biophotonics*, Graduate Texts in Physics,
DOI 10.1007/978-981-10-0945-7_3

are of interest to biophotonics, as Fig. 1.6 illustrates. Therefore, it is essential that biophotonics researchers and implementers know which fiber is best suited for a certain application. This chapter provides the background that is necessary to understand how optical fibers function and describes various categories of fibers that are commercially available for use in biophotonics. Subsequent chapters will illustrate how the diverse types of optical fibers are used for specific biophotonics analysis, imaging, therapy, and measurement applications.

First, Sects. 3.1 and 3.2 discuss the principles of how light propagates in two categories of conventional solid-core fibers. Here the term "conventional" refers to the configuration of optical fibers that are used worldwide in telecom networks. These discussions will be the basis for explaining how light is guided in other types of optical fibers.

Next, Sect. 3.3 describes optical fiber performance characteristics. These include optical signal attenuation as a function of wavelength, sensitivity of the fiber to an increase in power loss as the fiber is progressively bent into small loops, mechanical properties, and optical power-handling capabilities.

Using this background information, Sect. 3.4 through 3.13 then present a variety of other optical fiber structures and materials that are appropriate for use in the wavelength spectrum that is relevant to biomedical research and clinical practice. Table 3.1 summarizes the characteristics of these optical fibers, which can be constructed from materials such as standard silica, UV-resistant silica, halide glasses, chalcogenides, and polymers [5]. The biophotonics applications in this table have been designated by the following three general categories with some basic examples:

1. Light care: healthcare monitoring; laser surfacing or photorejuvenation
2. Light diagnosis: biosensing, endoscopy, imaging, microscopy, spectroscopy
3. Light therapy: ablation, photobiomodulation, dentistry, laser surgery, oncology

3.1 Light Guiding Principles in Conventional Optical Fibers

This section describes the principles of how light propagates in a conventional solid-core fiber [1–6]. Such a fiber normally is a cylindrical dielectric waveguide that operates at optical frequencies. Electromagnetic energy in the form of light is confined within the fiber and is guided in a direction parallel to the fiber axis. The light propagation along such a waveguide can be expressed in terms of a set of guided electromagnetic waves called the *modes* of the waveguide. Each guided mode consists of a pattern of electric and magnetic field distributions that is repeated periodically along the fiber. Only a specific number of modes that satisfy

Table 3.1 Major categories of optical fibers and their applications to biomedical research and clinical practice (*J. Biomed. Opt.* 19(8), 080902 (Aug 28, 2014). doi:10.1117/1.JBO.19.8.080902)

Optical fiber types		Characteristics	Biophotonics applications
Conventional solid-core silica fibers	Multimode	Multimode propagation	Light diagnosis; light therapy
	Single-mode	Single-mode propagation	Light diagnosis
Specialty solid-core fibers	Photosensitive	High photosensitivity to UV radiation; FBG fabrication	Light care; light therapy
	UV-resistant	Low UV sensitivity and reduced attenuation below 300 nm	Light diagnosis
	Bend-loss insensitive	High NA and low bend-loss sensitivity	Light therapy
	Polarization-maintaining	High birefringence; preserve the state of polarization	Light diagnosis
Double-clad fibers		Single-mode core and multimode inner cladding	Light diagnosis
Hard-clad silica fibers		Silica glass core with thin plastic cladding; increased fiber strength; high power transmission	Light diagnosis; light therapy
Coated hollow-core fibers		Low absorption for mid-IR and high optical damage threshold	Light therapy
Photonic crystal fibers		Low loss; transmit high optical power without nonlinear effects	Light diagnosis; light therapy
Plastic optical fibers or Polymer optical fibers		Low cost; fracture resistance; biocompatibility	Light diagnosis
Side-emitting fibers and side-firing fibers		Emit light along the fiber or perpendicular to the fiber axis	Light therapy
Mid-infrared fibers		Efficient IR delivery; large refractive index and thermal expansion	Light diagnosis; light therapy
Optical fiber bundles		Consist of multiple individual fibers	Light diagnosis

Fig. 3.1 Schematic of a
conventional silica fiber
structure

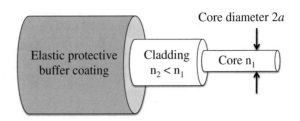

the homogeneous wave equation in the fiber and the boundary condition at the waveguide surfaces can propagate along the fiber.

Figure 3.1 illustrates the structure of a conventional optical fiber. This structure consists of a cylindrical silica-based glass *core* of refractive index n_1, which is surrounded by a glass *cladding* that has a slightly lower refractive index n_2. An external elastic polymer buffer coating protects the fiber from mechanical abrasions and environmental effects such as water, oil, and chemicals that can erode the glass. The refractive index of pure silica decreases with increasing wavelength, for example, it ranges from 1.462 at 500 nm to 1.443 at 1600 nm. The index can be changed slightly by adding impurities such as germanium oxide to the silica during the fiber manufacturing process.

Variations in the material composition and the structure of the conventional solid-core fiber determine how a light signal propagates along a fiber and also influence how the fiber performs in response to environmental perturbations, such as stress, bending, and temperature fluctuations. The two basic fiber types shown in Fig. 3.2 are produced from variations in the material composition of the core. In the *step-index fiber* the refractive index is uniform throughout the core and undergoes an abrupt change (or step) at the cladding boundary. In a *graded-index fiber* the core refractive index decreases as a function of the radial distance from the center of the fiber.

The step-index and graded-index fibers can be further categorized into single-mode and multimode classes. A *single-mode fiber* (SMF) can support only one mode of propagation, whereas many hundreds of modes can propagate in a *multimode fiber* (MMF). Typical sizes of single-and multimode fibers are shown in Fig. 3.2 to illustrate the dimensional scale. A MMF has several advantages compared to a SMF. The larger core radius of a MMF makes it easier to launch optical power into the fiber and to collect light from a biological sample. An advantage of multimode graded-index fibers is that they have larger data rate transmission capabilities than a comparably sized multimode step-index fiber. Single-mode fibers are better suited for delivering a narrow light beam to a specific tissue area and also are required for applications that involve coherence effects between propagating light beams.

In practical conventional fibers the core of radius a has a refractive index n_1, which is nominally equal to 1.48. The core is surrounded by a cladding of slightly lower index n_2, where $n_2 = n_1(1 - \Delta)$. The parameter Δ is called the *core-cladding*

Fig. 3.2 Comparison of conventional single-mode and multimode step-index and graded-index optical fibers

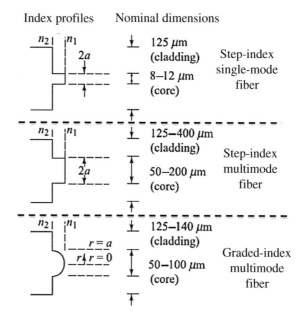

index difference or simply the *index difference*. Typical values of Δ range from 1 to 3 % for MMFs and from 0.2 to 1.0 % for SMFs. Because the core refractive index is larger than the cladding index, electromagnetic energy at optical frequencies is made to propagate along the fiber waveguide through total internal reflections at the core-cladding interface.

The remainder of Sect. 3.1 describes operational characteristics of step-index fibers and Sect. 3.2 describes graded-index fiber structures.

3.1.1 Ray Optics Concepts

When the optical fiber core diameter is much larger than the wavelength of the light, a simple geometrical optics approach based on using light rays (see Sect. 2.1) can explain how light travels along a fiber. For simplicity, the analysis considers only those rays (called *meridional rays*) that are confined to the meridian planes of the fiber, which are the planes that contain the axis of the fiber (the core axis). Figure 3.3 shows a light ray entering the fiber core from a medium of refractive index n at an angle θ_0 with respect to the fiber axis and striking the core-cladding interface inside the fiber at an angle ϕ relative to the normal of the interface. If it strikes this interface at such an angle that it is totally internally reflected, then the ray follows a zigzag path along the fiber core, passing through the axis of the guide after each reflection.

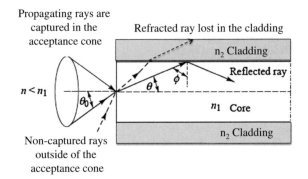

Fig. 3.3 Ray optics picture of the propagation mechanism in an optical fiber

As Sect. 2.4 describes, from Snell's law the minimum or critical angle ϕ_c that supports total internal reflection at the core-cladding interface is given by

$$\sin \phi_c = \frac{n_2}{n_1} \tag{3.1}$$

Rays striking the interface at angles less than ϕ_c will refract out of the core and be lost in the cladding, as the dashed line shows. By applying Snell's law to the air–fiber face boundary, the condition of Eq. (3.1) can be used to determine the maximum entrance angle $\theta_{0,\,max}$, which is called the *acceptance angle* θ_A. This calculation yields the relationship

$$n \sin \theta_{0,max} = n \sin \theta_A = n_1 \sin \theta_c = \left(n_1^2 - n_2^2 \right)^{1/2} \tag{3.2}$$

where $\theta_c = \pi/2 - \phi_c$. If a ray enters the fiber at an angle θ_0 less than θ_A, it will be totally internally reflected inside the fiber at the core-cladding interface. Thus the angle θ_A defines an *acceptance cone* for incoming light. Rays that fall outside of the acceptance cone (for example the ray indicated by the dashed line in Fig. 3.3) will refract out of the core.

Equation (3.2) also defines the *numerical aperture* (NA) of a step-index fiber for meridional rays:

$$NA = n \sin \theta_A = \left(n_1^2 - n_2^2 \right)^{1/2} \approx n_1 \sqrt{2\Delta} \tag{3.3}$$

The approximation on the right-hand side of Eq. (3.3) is valid because the parameter Δ is much less than 1. Because it is related to the acceptance angle θ_A, the NA is used to describe the light acceptance or gathering capability of a multimode fiber and to calculate source-to-fiber optical power coupling efficiencies. The NA value generally is listed on optical fiber data sheets.

Example 3.1 Consider a multimode step-index silica fiber that has a core refractive index $n_1 = 1.480$ and a cladding index $n_2 = 1.460$. Find (a) the critical angle, (b) the numerical aperture, and (c) the acceptance angle.

Solution:

(a) From Eq. (3.1), the critical angle is given by

$$\varphi_c = \sin^{-1}\frac{n_2}{n_1} = \sin^{-1}\frac{1.460}{1.480} = 80.5°$$

(b) From Eq. (3.3) the numerical aperture is

$$NA = \left(n_1^2 - n_2^2\right)^{1/2} = 0.242$$

c) From Eq. (3.3) the acceptance angle in air (n = 1.00) is

$$\theta_A = \sin^{-1}NA = \sin^{-1}0.242 = 14.0°$$

Example 3.2 Consider a multimode step-index fiber that has a core refractive index of 1.480 and a core-cladding index difference of 2.0 % ($\Delta = 0.020$). Find (a) the numerical aperture, (b) the acceptance angle, and (c) the critical angle.

Solution: The cladding index is $n_2 = n_1(1 - \Delta) = 1.480(0.980) = 1.450$.

(a) From Eq. (3.3) the numerical aperture is

$$NA = n_1\sqrt{2\Delta} = 1.480(0.04)^{1/2} = 0.296$$

(b) Using Eq. (3.3) the acceptance angle in air (n = 1.00) is

$$\theta_A = \sin^{-1}NA = \sin^{-1}0.296 = 17.2°$$

(c) From Eq. (3.1), the critical angle is given by

$$\varphi_c = \sin^{-1}\frac{n_2}{n_1} = \sin^{-1}0.980 = 78.5°$$

3.1.2   Modal Concepts

Whereas the geometrical optics ray approach gives a simple picture of how light is guided along a fiber, mode theory is needed to explain concepts such as mode coupling, signal dispersion, and coherence or interference phenomena. Figure 3.4 shows the oscillating field patterns of three of the lower-order transverse electric (TE) modes. The mode order is equal to the number of times the field value is zero within the guiding core, as is illustrated by the solid dots in Fig. 3.4. As is shown by the plots, the electric fields of the guided modes are not completely confined to the core but extend partially in the cladding. The fields oscillate harmonically in the fiber core and decay exponentially in the cladding region. The exponentially decaying field is referred to as an *evanescent field*. The fields of the low-order modes are concentrated near the fiber axis and do not penetrate far into the cladding region. The fields of the higher-order modes are located more strongly near the core edges and penetrate farther into the cladding.

As the core radius a shown in Fig. 3.1 is made progressively smaller, all the higher-order modes except the *fundamental mode* (the zeroth-order linearly polarized mode designated by LP_{01}) shown in Fig. 3.4 will gradually get cut off and consequently will not propagate in the fiber. A fiber in which only the fundamental mode can propagate is called a *single-mode fiber*. An important parameter related to the cutoff condition is the *V number* (also called the *normalized frequency*) defined by

$$V = \frac{2\pi a}{\lambda} \left(n_1^2 - n_2^2 \right)^{1/2} = \frac{2\pi a}{\lambda} NA \approx \frac{2\pi a}{\lambda} n_1 \sqrt{2\Delta} \qquad (3.4)$$

where the approximation on the right-hand side comes from Eq. (3.3). The V number is a dimensionless parameter that designates how many modes can propagate in a specific optical fiber. Except for the fundamental LP_{01} mode, a higher-order mode can exist only for values of V greater than 2.405 (with each

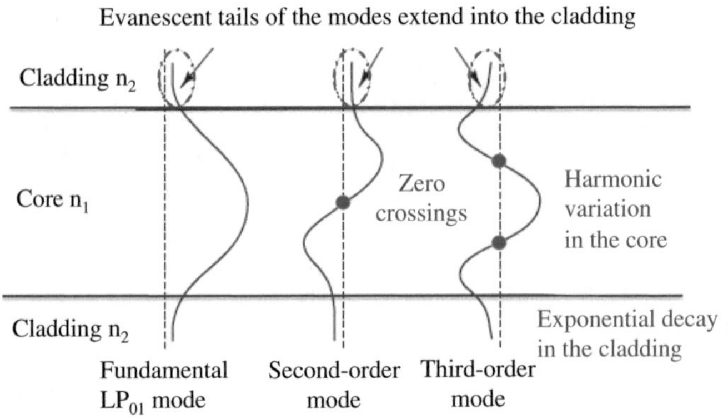

Fig. 3.4 Electric field distributions of lower-order guided modes in an optical fiber (longitudinal cross-sectional view)

mode having a different V limit). The wavelength at which no higher-order modes exist in the fiber is called the *cutoff wavelength* λ_c. Only the fundamental LP_{01} mode exists in single-mode fibers.

Example 3.3 A step-index fiber has a normalized frequency V = 26.6 at a 1300-nm wavelength. If the core radius is 25 μm, what is the numerical aperture?

Solution: From Eq. (3.4) the numerical aperture is

$$NA = V\frac{\lambda}{2\pi a} = 26.6\frac{1.30\,\mu m}{2\pi \times 25\,\mu m} = 0.220$$

The V number also designates the number of modes M in a step-index MMF when V is large. An estimate of the total number of modes supported in a MMF is

$$M = \frac{1}{2}\left(\frac{2\pi a}{\lambda}\right)^2 (n_1^2 - n_2^2) = \frac{V^2}{2} \qquad (3.5)$$

Example 3.4 Consider a step-index MMF with a 62.5-μm core diameter and a core-cladding index difference Δ of 1.5 %. If the core refractive index is 1.480, estimate the V number of the fiber and the total number of modes supported in the fiber at a wavelength of 850 nm.

Solution: From Eq. (3.4) the V number is

$$V \approx \frac{2\pi a}{\lambda}n_1\sqrt{2\Delta} = \frac{2\pi \times 31.25\,\mu m \times 1.480}{0.850\,\mu m}\sqrt{2 \times 0.015} = 59.2$$

Using Eq. (3.5), the total number of modes is

$$M = \frac{V^2}{2} = 1752$$

Example 3.5 Consider a multimode step-index optical fiber that has a core radius of 25 μm, a core index of 1.48, and an index difference Δ = 0.01. How many modes does the fiber sustain at the wavelengths 860, 1310, and 1550 nm?

Solution:

(a) First, from Eq. (3.4), at an operating wavelength of 860 nm the value of V is

$$V \approx \frac{2\pi a}{\lambda} n_1 \sqrt{2\Delta} = \frac{2\pi \times 25\,\mu m \times 1.480}{0.860\,\mu m} \sqrt{2 \times 0.010} = 38.2$$

Using Eq. (3.5), the total number of modes at 860 nm is

$$M = \frac{V^2}{2} = 729$$

(b) Similarly, V = 25.1 and M = 315 at 1310 nm.
(c) Finally, V = 21.2 and M = 224 at 1550 nm.

As is shown in Fig. 3.4, the field of a guided mode extends partly into the cladding. This means that a fraction of the power in any given mode will flow in the cladding. As the V number approaches the cutoff condition for any particular mode, more of the power of that mode is in the cladding. At the cutoff point, all the optical power of the mode resides in the cladding. Far from cutoff—that is, for large values of V—the fraction of the average optical power propagating in the cladding can be estimated by

$$\frac{P_{clad}}{P} \approx \frac{4}{3\sqrt{M}} \qquad\qquad (3.6)$$

where P is the sum of the optical powers in the core and in the cladding.

Example 3.6 Consider a multimode step-index optical fiber that has a core radius of 25 μm, a core index of 1.48, and an index difference Δ = 0.01. Find the percentage of optical power that propagates in the cladding at 840 nm.

Solution: From Eq. (3.4) at an operating wavelength of 840 nm the value of V is

$$V \approx \frac{2\pi a}{\lambda} n_1 \sqrt{2\Delta} = \frac{2\pi \times 25\,\mu m \times 1.480}{0.840\,\mu m} \sqrt{2 \times 0.010} = 39$$

Using Eq. (3.5), the total number of modes is

$$M = \frac{V^2}{2} = 760$$

From Eq. (3.6) it follows that

$$\frac{P_{clad}}{P} \approx \frac{4}{3\sqrt{M}} = 0.05$$

In this case, approximately 5 % of the optical power propagates in the cladding. If Δ is decreased to 0.03 in order to lower the signal dispersion effects, then there are 242 modes in the fiber and about 9 % of the power propagates in the cladding.

3.1.3   Mode Field Diameter

The geometric distribution of optical power in the fundamental mode in a SMF is needed when predicting performance characteristics such as splice loss between two fibers, bending loss, and cutoff wavelength. The parameter describing this optical power distribution is the *mode field diameter* (MFD), which is analogous to the core diameter in a MMF. The MFD is a function of the optical wavelength, the fiber core radius, and the refractive index profile of the fiber.

A standard method to find the MFD is to measure the far-field intensity distribution $E^2(r)$ at the fiber output and then calculate the MFD using the equation [6]

$$\text{MFD} = 2w_0 = 2\left[\frac{2\int_0^\infty E^2(r)r^3 dr}{\int_0^\infty E^2(r)r dr}\right]^{1/2} \tag{3.7}$$

where the parameter $2w_0$ (with w_0 being called the *spot size* or the *mode field radius*) is the full width of the far-field distribution. For calculation simplicity for values of V lying between 1.8 and 2.4 the field distribution can be estimated by a Gaussian function [6]

$$E(r) = E_0 \exp\left(-r^2/w_0^2\right) \tag{3.8}$$

where r is the radius and E_0 is the electric field at zero radius, as shown in Fig. 3.5. Then the MFD is given by the $1/e^2$ width of the optical power. An approximation to

Fig. 3.5 Distribution of light in a single-mode fiber above its cutoff wavelength; for a Gaussian distribution the MFD is given by the $1/e^2$ width of the optical power

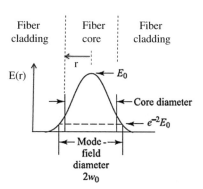

the relative spot size w_0/a, with an accuracy better than 1 % for a SMF, is given by
[6]

$$\frac{w_0}{a} = 0.65 + 1.619V^{-3/2} + 2.879V^{-6} \tag{3.9}$$

SMFs typically are designed with V > 2.0 to prevent high cladding losses but somewhat less than 2.4 to ensure that only one mode propagates in the fiber.

Example 3.7 Suppose a certain single-mode step-index fiber has an MFD = 11.2 μm and V = 2.25. What is the core diameter of this fiber?

Solution: From Eq. (3.7) w_0 = MFD/2 = 5.6 μm. Using Eq. (3.9) then yields

$$a = w_0/\left(0.65 + 1.619V^{-3/2} + 2.879V^{-6}\right)$$
$$= \frac{5.6\,\mu m}{0.65 + 1.619(2.25)^{-3/2} + 2.879(2.25)^{-6}}$$
$$= \frac{5.6\,\mu m}{1.152} = 4.86\,\mu m$$

Thus the core diameter is $2a$ = 9.72 μm.

3.2   Graded-Index Optical Fibers

3.2.1   Core Index Structure

As Fig. 3.2 illustrates, in a graded-index fiber the core refractive index decreases continuously with increasing radial distance r from the center of the fiber but is generally constant in the cladding [5]. The most commonly used construction for the refractive-index variation in the core is the power law relationship

$$n(r) = n_1\left[1 - 2\Delta\left(\frac{r}{a}\right)^\alpha\right]^{1/2} \text{ for } 0 \leq r \leq a \tag{3.10}$$
$$= n_1(1 - 2\Delta)^{1/2} \approx n_1(1 - \Delta) = n_2 \text{ for } r \geq a$$

Here, r is the radial distance from the fiber axis, a is the core radius, n_1 is the refractive index at the core axis, n_2 is the refractive index of the cladding, and the dimensionless parameter α defines the shape of the index profile. The index difference Δ for the graded-index fiber is given by [5]

$$\Delta = \frac{n_1^2 - n_2^2}{2n_1^2} \approx \frac{n_1 - n_2}{n_1} \tag{3.11}$$

The approximation on the right-hand side reduces this expression for Δ to that of the step-index fiber. Thus, the same symbol is used in both cases. For $\alpha = \infty$, inside the core Eq. (3.10) reduces to the step-index profile $n(r) = n_1$.

3.2.2   Graded-Index Numerical Aperture

Whereas for a step-index fiber the NA is constant across the core end face, in graded-index fibers the NA is a function of position across the core end face. Thus determining the NA is more complex for graded-index fibers. Light incident on the fiber core end face at a position r will be a guided mode only if it is within the *local numerical aperture* NA(r) at that point. This parameter is defined as [6]

$$NA(r) = \left[n^2(r) - n_2^2\right]^{1/2} \approx NA(0)\sqrt{1 - (r/a)^\alpha} \ \text{for} \ r \le a$$
$$= 0 \ \text{for} \ r > a \tag{3.12}$$

where the *axial numerical aperture* is defined as

$$NA(0) = \left[n^2(0) - n_2^2\right]^{1/2} = \left(n_1^2 - n_2^2\right)^{1/2} \approx n_1\sqrt{2\Delta} \tag{3.13}$$

Thus, the numerical aperture of a graded-index fiber decreases from NA(0) to zero for values of r ranging from the fiber axis to the core edge. In a graded-index fiber the number of bound modes M_g is [1]

$$M_g = \frac{\alpha}{\alpha + 2}\left(\frac{2\pi a}{\lambda}\right)^2 n_1^2\Delta \approx \frac{\alpha}{\alpha + 2}\frac{V^2}{2} \tag{3.14}$$

Typically a parabolic refractive index profile given by $\alpha = 2.0$ is selected by fiber manufacturers. In this case, $M_g = V^2/4$, which is half the number of modes supported by a step-index fiber (for which $\alpha = \infty$) that has the same V value.

Example 3.8 Consider a 50-μm diameter graded-index fiber that has a parabolic refractive index profile ($\alpha = 2$). If the fiber has a numerical aperture NA = 0.22, how many guided modes are there in the fiber at a wavelength of 1310 nm?

Solution: First, from Eq. (3.4)

$$V = \frac{2\pi a}{\lambda} NA = \frac{2\pi \times 25\,\mu m}{1.31\,\mu m} \times 0.22 = 26.4$$

Then from Eq. (3.14) the total number of modes for $\alpha = 2$ is

$$M_g \approx \frac{\alpha}{\alpha + 2} \frac{V^2}{2} = \frac{V^2}{4} = 174$$

3.2.3   Cutoff Condition in Graded-Index Fibers

Similar to step-index fibers, graded-index fibers can be designed as single-mode fibers at a desired operational wavelength. An empirical expression (an expression based on observation) of the V parameter at which the second lowest order mode is cut off for graded-index fibers has been shown to be [1]

$$V_{cutoff} = 2.405\sqrt{1 + \frac{2}{\alpha}} \tag{3.15}$$

Equation (3.15) shows that for a graded-index fiber the value of V_{cutoff} decreases as the profile parameter α increases [5]. This equation also shows that the critical value of V for the cutoff condition in parabolic graded-index fibers ($\alpha = 2$) is a factor of $\sqrt{2}$ larger than for a similar-sized step-index fiber. Furthermore, from the definition of V given by Eq. (3.4), the numerical aperture of a graded-index fiber is larger than that of a step-index fiber of comparable size.

3.3   Performance Characteristics of Generic Optical Fibers

When considering what fiber to use in a particular biophotonics system application, some performance characteristics that need to be taken into account are optical signal attenuation as a function of wavelength, optical power-handling capability, the degree of signal loss as the fiber is bent, and mechanical properties of the optical fiber [5–11].

3.3.1   Attenuation Versus Wavelength

Attenuation of an optical signal is due to absorption, scattering, and radiative power losses as light travels along a fiber. For convenience of power-budget calculations, attenuation is measured in units of decibels per kilometer (dB/km) or decibels per meter (dB/m). Recall that Sect. 1.4 gives a discussion of decibels. A variety of materials that exhibit different light-attenuation characteristics in various spectral bands are used to make the diverse types of optical fibers employed in biophotonics applications. The basic reason for such a selection is that each material type exhibits different attenuation characteristics as a function of wavelength. For example, the silica (SiO_2) glass material used in conventional optical fibers for telecom applications has low losses in the 800-to1600-nm telecom spectral range, but the loss is much greater at shorter and longer wavelengths. Thus, fibers materials with other attenuation characteristics are needed for biophotonics applications at wavelengths outside of the telecom spectral band.

3.3.2 Bend-Loss Insensitivity

Specially designed fibers with a moderately higher numerical aperture (NA) than in a conventional single-mode fiber are less sensitive to bending loss [5]. Bend-insensitive or bend-tolerant fibers are available commercially to provide optimum low bending loss performance at specific operating wavelengths, such as 820 or 1550 nm. These fibers typically have an 80-μm cladding diameter. In addition to low bending losses, this smaller outer diameter yields a much smaller coil volume compared with a standard 125-μm cladding diameter when a length of this low-bend-loss fiber is coiled up within a miniature optoelectronic device package or in a compact biophotonics instrument.

3.3.3 Mechanical Properties

A number of unique mechanical properties make optical fibers attractive for biomedical applications [5]. One important mechanical factor is that optical fibers consist of a thin highly flexible medium, which allows various minimally invasive medical treatment or diagnostic procedures to take place in a living body. As described in subsequent chapters, these applications can include various endoscopic procedures, cardiovascular surgery, and microsurgery.

A second important mechanical characteristic is that by monitoring or sensing some intrinsic physical variation of an optical fiber, such as elongation or refractive index changes, one can create fiber sensors to measure many types of external physical parameter changes [5]. For example, if a varying external parameter elongates the fiber or induces refractive index changes at the cladding boundary, this effect can modulate the intensity, phase, polarization, wavelength, or transit time of light in the fiber. The degree of light modulation then is a direct measure of changes in the external physical parameter. For biophotonics applications the external physical parameters of interest include pressure, temperature, stress, strain, and the molecular composition of a liquid or gas surrounding the fiber. Operational details and examples of these biosensor applications are presented in Chap. 7.

3.3.4 Optical Power-Handling Capability

In some biomedical photonics applications, such as imaging and fluorescence spectroscopy, the optical fibers carry power levels of less than 1 μW [5]. In other situations the fibers must be able to transmit optical power levels of 10 W and higher. A principal high-power application is laser surgery, which includes bone ablation, cardiovascular surgery, cosmetic surgery, dentistry, dermatology, eye surgery, and oncology surgery.

Hard-clad silica optical fibers with fused silica cores that have very low contaminants are described in Sect. 3.4. These fibers are capable of conducting high optical power from either a continuous wave (CW) or pulsed laser [5]. Other fibers that are capable of transmitting high optical power levels include conventional hollow-core fibers (see Sect. 3.5), photonic crystal fibers (see Sect. 3.6), and germanate glass fibers (see Sect. 3.9). The challenge is the launching of high optical power levels into a fiber. Artifacts such as dust or scratches on the end face of the fiber can form absorption sites that generate elevated temperature levels at the fiber tip. In standard connectors where the fibers are glued into the connector housing, these temperature levels can cause the surrounding epoxy to break down and give off gases. The gases ignite and burn onto the tip of the fiber, thereby causing catastrophic damage to the fiber and the connector. To handle high power levels, various manufacturers have developed special fiber optic patch cords that have carefully prepared fiber end faces and specially designed fiber optic connectors that greatly reduce the susceptibility to thermal damage.

3.4 Conventional Solid-Core Fibers

As a result of extensive development work for telecom networks, conventional solid-core silica-based optical fibers are highly reliable and are widely available in a variety of core sizes [5]. These fibers are used throughout the world in telecom networks and in many biophotonics applications. Figure 3.6 shows the optical signal attenuation per kilometer as a function of wavelength. The shape of the attenuation curve is due to three factors. First, intrinsic absorption due to electronic absorption bands causes high attenuations in the ultraviolet region for wavelengths less than about 500 nm. Then the Rayleigh scattering effect starts to dominate the attenuation for wavelengths above 500 nm, but diminishes rapidly with increasing wavelength because of its $1/\lambda^4$ behavior. Thirdly, intrinsic absorption associated with atomic vibration bands in the basic fiber material increases with wavelength and is the dominant attenuation mechanism in the infrared region above about

Fig. 3.6 Typical attenuation curve of a silica fiber as a function of wavelength (*J. Biomed. Opt.* 19(8), 080902 (Aug 28, 2014). doi:10.1117/1.JBO.19.8.080902)

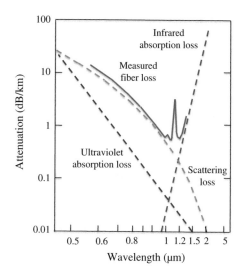

1500 nm. As shown in Fig. 3.6, these attenuation mechanisms produce a low-loss region in silica fibers in the spectral range of 700–1600 nm, which matches the low-absorption biophotonics window illustrated in Fig. 1.6. The attenuation spike around 1400 nm is due to absorption by residual water ions in the silica material. Greatly reducing these ions during the fiber manufacturing process results in a *low-water-content fiber* or *low-water-peak fiber* in which the attenuation spike has been greatly reduced.

Standard commercially available multimode fibers have core diameters of 50, 62.5, 100, 200 μm, or larger. These multimode fibers are used in applications such as light delivery to specific tissue areas, photobiomodulation, optical fiber probes, and photodynamic therapy (PDT). Single-mode fibers have core diameters around 10 μm, the exact value depending on the wavelength of interest. Applications of single-mode optical fibers include uses in clinical analytical instruments, in endoscopes or catheters, in various imaging systems, and in healthcare sensors.

3.5   Specialty Solid-Core Fibers

Specialty solid-core fibers are custom-designed for functions such as the following [5]:

- Manipulating lightwave signals to achieve some type of optical signal-processing function
- Extending the spectral operating range of the fiber
- Sensing variations in a physical parameter such temperature or pressure
- Analyzing biomedical fluids

Incorporation of such features into an optical fiber is achieved through either material or structural variations [5]. For biophotonics applications the main specialty solid-core fiber types are photosensitive fibers for creating internal gratings, fibers resistant to darkening from ultraviolet light, bend-loss insensitive fibers for circuitous routes inside bodies, and polarization-preserving optical fibers for imaging and for fluorescence analyses in spectroscopic systems. The following subsections describe the characteristics of these optical fibers.

3.5.1   Photosensitive Optical Fiber

A photosensitive fiber is designed so that its refractive index changes when it is exposed to ultraviolet light [5]. For example, doping the fiber core material with germanium and boron ions can provide this sensitivity. The main application for such a fiber is to create a short *fiber Bragg grating* (FBG) in the fiber core, which is a periodic variation of the refractive index along the fiber axis [12–14].

This index variation is illustrated in Fig. 3.7, where n_1 is the refractive index of the core of the fiber, n_2 is the index of the cladding, and Λ is the period of the grating, that is, the spacing between the maxima of the index variations [5]. If an incident optical wave at a wavelength λ_B (which is known as the *Bragg wavelength*) encounters a periodic variation in refractive index along the direction of propagation, λ_B will be reflected back if the following condition is met:

$$\lambda_B = 2n_{eff}\Lambda \qquad (3.16)$$

Here n_{eff} is the effective refractive index, which has a value falling between the refractive indices n_1 of the core and n_2 of the cladding. When a specific wavelength λ_B meets this condition, the grating reflects this wavelength and all others will pass through. Fiber Bragg gratings are available in a selection of Bragg wavelengths with spectral reflection bandwidths at a specific wavelength varying from a few picometers to tens of nanometers.

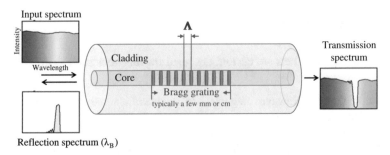

Fig. 3.7 A periodic index variation in the core of a single-mode fiber creates a fiber Bragg grating (*J. Biomed. Opt.* 19(8), 080902 (Aug 28, 2014). doi:10.1117/1.JBO.19.8.080902)

As described in Chap. 7, an important biophotonics application of a FBG is to sense a variation in a physical parameter such as temperature or pressure [5]. For example, an external factor such as strain will slightly stretch the fiber. This stretching will lengthen the period Λ of the FBG and thus will change the value of the specific reflected wavelength. Similarly, rises or drops in temperature will lengthen or shorten the value of Λ, respectively, thereby changing the value of λ_B, which then can be related to the change in temperature.

Example 3.9 Consider a FBG that was made to reflect a wavelength $\lambda_B = 1544.2$ nm at 20 °C. Suppose that the Bragg wavelength changes linearly with temperature with an incremental wavelength change $\Delta\lambda_B = (0.02$ nm/°C$)\Delta T$ per incremental temperature change ΔT. (a) Find the Bragg wavelength when $T = 35$ °C. (b) What is the Bragg wavelength when $T = 10$ °C?

Solution: The linear dependence of λ_B on temperature can be expressed as

$$\lambda_B(T) = 1544.2\,\text{nm} + (0.02\,\text{nm/}^\circ\text{C})(T - 20\,^\circ\text{C})$$

(a) At $T = 35$ °C the Bragg wavelength becomes

$$\lambda_B(35\,^\circ\text{C}) = 1544.2\,\text{nm} + (0.02\,\text{nm/}^\circ\text{C})(35 - 20\,^\circ\text{C}) = 1544.5\,\text{nm}$$

(b) At $T = 10$ °C the Bragg wavelength becomes

$$\lambda_B(10\,^\circ\text{C}) = 1544.2\,\text{nm} + (0.02\,\text{nm/}^\circ\text{C})(10 - 20\,^\circ\text{C}) = 1544.0\,\text{nm}$$

3.5.2 Fibers Resistant to UV-Induced Darkening

Conventional solid-core germanium-doped silica fibers are highly sensitive to ultraviolet (UV) light [5]. The intrinsic attenuation is higher in the UV region compared to the visible and near-infrared spectra, and there are additional losses due to UV-absorbing material defects, which are created by high-energy UV photons. These additional UV-induced darkening losses are known as *solarization* and occur strongly at wavelengths less than about 260 nm. Thus although newly manufactured conventional silica fibers offer low attenuation in the 214–254 nm range, upon exposure to an unfiltered deuterium lamp the transmission of these fibers drops to about 50 % of the original value within a few hours of continuous UV irradiation. Consequently, conventional silica optical fibers can be used only for applications above approximately 300 nm.

However, currently special material processing methods, such as fluorine doping of silica, have resulted in fibers with moderate (around 50 % attenuation) to minimal (a few percent additional loss) UV sensitivity below 260 nm [15–17]. When such fibers are exposed to UV light the transmittance first decreases rapidly but then stabilizes at an asymptotic attenuation value. The specific asymptotic attenuation level and the time to reach this level depend on the fiber manufacturing process and on the UV wavelength being used. Shorter wavelengths result in higher attenuation changes and it takes longer to reach the asymptotic attenuation value. Solarization-resistant fibers for operation in the 180 to 850-nm range are available with core diameters ranging from 50 to 1000 μm and a numerical aperture of 0.22.

3.5.3   Bend Insensitive Fiber

In many medical applications of optical fibers within a living body, the fibers need to follow a sinuous path with sharp bends through arteries that snake around bones and organs [5]. Special attention must be paid to this situation, because optical fibers exhibit radiative losses whenever the fiber undergoes a bend with a finite radius of curvature. For slight bends this factor is negligible. However, as the radius of curvature decreases, the losses increase exponentially until at a certain critical radius the losses become extremely large. As shown in Fig. 3.8, the bending loss becomes more sensitive at longer wavelengths. A fiber with a small bend radius might be transmitting well at 1310 nm, for example, giving an additional loss of 1 dB at a 1-cm bending radius. However, for this bend radius it could have a significant loss at 1550 nm resulting in an additional loss of about 100 dB for the conventional fiber.

In recent years the telecom industry started installing optical fiber links within homes and businesses. Thus bend-loss insensitive fibers were developed because

Fig. 3.8 Generic bend loss sensitivity for conventional fibers at 1310 and 1550 nm

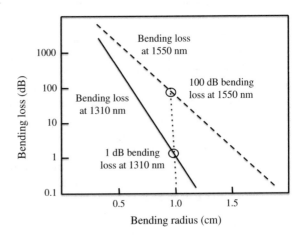

optical fibers in such indoor installations must tolerate numerous sharp bends, These same fibers also can be used for medical applications [18–20]. These bend-loss insensitive fibers have a moderately higher numerical aperture (NA) than conventional single-mode fibers. Increasing the NA reduces the sensitivity of the fiber to bending loss by confining optical power more tightly within the core than in conventional single-mode fibers.

Various manufacturers offer a bend-loss insensitive fiber that has a lower single-mode cutoff wavelength, a nominally 50 % higher index difference value Δ, and a 25 % higher NA than conventional telecom fibers [5]. The higher NA of low-bend-loss fibers allows an improved coupling efficiency from laser diode sources to planar waveguides. Generally for bend radii of greater than 20 mm, the bending-induced loss is negligibly small. Fibers are available in which the maximum bending induced loss is less than 0.2 dB due to 100 turns on a 10-mm mandrel. A factor to keep in mind is that at operating wavelengths in the near infrared, the smaller mode field diameter of low-bend-loss fibers can induce a mode-mismatch loss when interconnecting these fibers with standard single-mode fibers. However, carefully made splices between these different fibers typically results in losses less than 0.1 dB.

3.5.4 Polarization-Maintaining Fiber

In a conventional single-mode fiber the fundamental mode consists of two orthogonal polarization modes (see Chap. 2) [5]. These modes may be chosen arbitrarily as the horizontal and vertical polarizations in the x direction and y direction, respectively, as shown in Fig. 3.9. In general, the electric field of the light

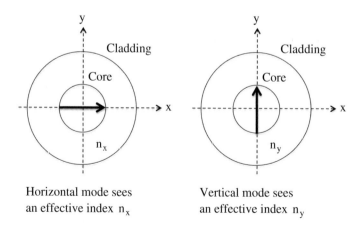

Horizontal mode sees an effective index n_x

Vertical mode sees an effective index n_y

Fig. 3.9 Two polarization states of the fundamental mode in a single-mode fiber

propagating along the fiber is a linear superposition of these two polarization modes and depends on the polarization of the light at the launching point into the fiber.

In ideal fibers with perfect rotational symmetry, the two modes are degenerate (that is, they have the same resonant frequency) with equal propagation constants ($\beta_x = n_x 2\pi/\lambda = \beta_y = n_y 2\pi/\lambda$, where n_x and n_y are the effective refractive indices along the x and y axes, respectively) [5]. Thus, any polarization state injected into the fiber will propagate unchanged. In actual fibers there are imperfections, such as asymmetrical lateral stresses, noncircular cores, and variations in refractive-index profiles. These imperfections break the circular symmetry of the ideal fiber and lift the degeneracy of the two modes, so that now $\beta_x \neq \beta_y$. The modes then propagate with different phase velocities, and the difference between their effective refractive indices is called the *fiber birefringence*,

$$B_f = \frac{\lambda}{2\pi}\left(\beta_x - \beta_y\right) \tag{3.17}$$

If light is injected into the fiber so that both modes are excited at the same time, then one mode will be delayed in phase relative to the other as they propagate [5]. When this phase difference is an integral multiple of 2π, the two modes will beat at this point and the input polarization state will be reproduced. The length over which this beating occurs is the *fiber beat length* $L_B = \lambda/B_f$.

In conventional fibers, the small degrees of imperfections in the core will cause the state of polarization to fluctuate as a light signal propagates through the fiber [5]. In contrast, *polarization-maintaining fibers* have a special design that preserves the state of polarization along the fiber with little or no coupling between the two modes. Figure 3.10 illustrates the cross-sectional geometry of two different popular polarization-maintaining fibers. The light circles represent the core and the cladding and the dark areas are built-in stress elements that are made from a different type of glass. The goal in each design is to use stress-applying parts to create slow and fast

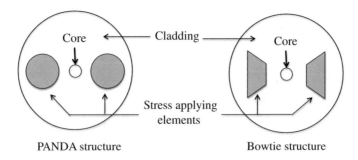

Core ← Cladding → Core

Stress applying
elements

PANDA structure   Bowtie structure

Fig. 3.10 Cross-sectional geometry of two different polarization-maintaining fibers

axes in the core. Each of these axes will guide light at a different velocity. Crosstalk between the two axes is suppressed so that polarized light launched into either of the axial modes will maintain its state of polarization as it travels along the fiber. These fibers are used in special biophotonics applications such as fiber optic sensing and interferometry where polarization preservation is essential [21].

The structural arrangement of the bowtie geometry shown in Fig. 3.10 creates an extreme birefringence in a fiber [5]. Such a birefringence allows one and only one polarization state of the fundamental mode to propagate with all other polarization modes being greatly suppressed. In these fibers, single-polarization guidance occurs in only a limited wavelength range of about 100 nm. Outside of that spectral range, either both of the polarization states or no light at all may be guided. These fibers are used in special fiber optic biosensing applications where it is desirable to monitor a single state of polarization.

Example 3.10 A single-mode optical fiber has a beat length of 8 cm at 1310 nm. What is the birefringence?

Solution: From Eq. (3.17) and the condition that $L_B = \lambda/B_f$ the modal birefringence is

$$B_f = \frac{\lambda}{L_B} = \frac{1.31 \times 10^{-6}\,\text{m}}{8 \times 10^{-2}\,\text{m}} = 1.64 \times 10^{-5}$$

This value is characteristic of an intermediate type fiber, because birefringence can vary from $B_f = 1 \times 10^{-3}$ (for a typical high birefringence fiber) to $B_f = 1 \times 10^{-8}$ (for a typical low birefringence fiber).

3.6 Double-Clad Fibers

A double-clad fiber (DCF) originally was created to help construct optical fiber lasers and now is being applied in the medical imaging field [22–24]. More details on these imaging applications are given in Chaps. 9 and 10. As indicated in Fig. 3.11, a DCF consists of a core region, an inner cladding, and an outer cladding arranged concentrically [5]. Typical dimensions of a commercially available DCF are a 9-μm core diameter, an inner cladding diameter of 105 μm, and an outer cladding with a 125-μm diameter. Light transmission in the core region is single-mode, whereas in the inner cladding it is multimode. The index of a DCF is decreasingly cascaded from the core center to the outer cladding boundary. The integration of both single-mode and multimode transmissions allows using a single optical fiber for the delivery of the illumination light (using the single-mode core) and collection of the tissue-reflected light (using the multimode inner cladding).

Fig. 3.11 Cross-sectional representation of a typical DCF and its index profile

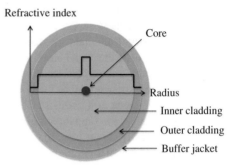

3.7 Hard-Clad Silica Fibers

A hard-clad silica (HCS) optical fiber is a MMF structure consisting of a silica glass core that is covered by a thin hard plastic cladding, which increases fiber strength [25]. These fibers also feature bend insensitivity, long-term reliability, ease of handling, and resistance to harsh chemicals and humid environments [5]. Core diameters of commercially available HCS fibers range from 200 to 1500 μm. Table 3.2 lists some performance parameters of three selected HCS fibers. A common HCS fiber for industrial and medical applications has a core diameter of 200 μm and a cladding diameter of 230 μm, which results in a very strong optical fiber with low attenuation (less than 10 dB/km or 0.01 dB/m at 820 nm), a numerical aperture of 0.39, negligible bending-induced loss for bend diameters less than 40 mm, and an extremely high core-to-clad ratio to enable efficient light coupling into and out of the fiber. The fibers are available in both high OH and low

Table 3.2 General specifications of selected HCS fibers (*J. Biomed. Opt.* 19(8), 080902 (Aug 28, 2014). doi:10.1117/1.JBO.19.8.080902)

	HCS fiber 1	HCS fiber 2	HCS fiber 3
Core diameter (μm)	200 ± 5	600 ± 10	1500 ± 30
Cladding diameter (μm)	225 ± 5	630 ± 10	1550 ± 30
Wavelength range (high OH content) (nm)	300–1200	300–1200	300–1200
Wavelength range (low OH content) (nm)	400–2200	400–2200	400–2200
Maximum power capability (CW) (kW)	0.2	1.8	11.3
Maximum power capability (pulsed) (MW)	1.0	9.0	56.6
Maximum long term bend radius (mm)	40	60	150
Max attenuation at 850 nm	10 dB/km (0.010 dB/m)	12 dB/km (0.012 dB/m)	18 dB/km (0.018 dB/m)

OH content for operation in the UV, visible, and near-IR regions. The mechanical, optical, and structural properties of HCS fibers are especially useful in applications such as laser delivery, endoscopy, photodynamic therapy, and biosensing systems.

3.8   Coated Hollow-Core Fibers

Because a conventional solid-core silica fiber exhibits tremendously high material absorption above about 2 μm, internally coated hollow-core fibers provide one alternative solution for the delivery of mid-infrared (2–10 μm) light to a localized site [5]. Section 3.12 describes middle-infrared fibers as another alternative for operation above 2 μm. Examples of light sources operating above 2 μm include CO_2 (10.6 μm) and Er: YAG lasers (2.94 μm), which have wide applications in urology, dentistry, otorhino-laryngology, and cosmetic surgery. In addition, hollow-core fibers are useful in the ultraviolet region, where silica-based fibers also exhibit high transmission losses.

As shown in Fig. 3.12, a hollow-core fiber is composed of a glass tube with metal and dielectric layers deposited at the inner surface plus a protection jacket on the outside [26–29]. In the fabrication process of these fibers, a layer of silver (Ag) is deposited on the inside of a glass tube, which then is covered with a thin dielectric film such as silver iodide (AgI) [5]. Light is transmitted along the fiber through mirror-type reflections from this inner metallic layer. The thickness of the dielectric layer (normally less than 1 μm) is selected to give a high reflectivity at a particular infrared wavelength or a band of wavelengths. Tubes of other materials, such as plastic and metal also are employed as hollow-core waveguides, but glass hollow-core fiber provides more flexibility and better performance. The bore sizes (diameters of the fiber hole) can range from 50 to 1200 μm. However, since the loss of all hollow-core fibers varies as $1/r^3$, where r is the bore radius, the more flexible smaller bore hollow-core fibers have higher losses.

Compared to solid-core fibers, the optical damage threshold is higher in hollow-core fibers, there are no cladding modes, and there is no requirement for angle cleaving or anti-reflection coating at the fiber end to minimize laser feedback effects [5]. To direct the light projecting from the hollow-core fiber into a certain direction, sealing caps with different shapes (e.g., cone and slanted-end) at the distal

Fig. 3.12 The cross-section of a typical hollow-core fiber

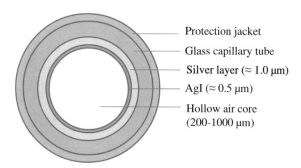

Protection jacket

Glass capillary tube

Silver layer (≈ 1.0 μm)

AgI (≈ 0.5 μm)

Hollow air core (200-1000 μm)

end (the light exit end) of the fiber have been proposed and demonstrated by using a
fusing and polishing technique [27].

Example 3.11 The bore diameters in coated hollow-core fibers can range
from 50 to 1200 μm. In these fibers the loss varies as $1/r^3$, where r is the bore
radius. Compare the losses of 300-μm and 800-μm hollow-core fibers relative
to a 1200-μm bore sized fiber.

Solution: First, assume that the loss is 0.2 dB/m for a fiber with a bore size
r_{ref} = 1200-μm.

(a) Then for a 300-μm hollow-core fiber the loss is

$$\text{Loss at } 300 \, \mu m = (0.2 \, dB/m)(r_{ref}/r)^3 = (0.2 \, dB/m)(1200/300)^3$$
$$= 12.8 \, dB/m$$

(b) For an 800-μm hollow-core fiber the loss is

$$\text{Loss at } 800 \, \mu m = (0.2 \, dB/m)(r_{ref}/r)^3 = (0.2 \, dB/m)(1200/800)^3$$
$$= 0.68 \, dB/m$$

3.9   Photonic Crystal Fibers

A *photonic crystal fiber* (PCF) has a geometric arrangement of internal air holes
that run along the entire length of the fiber [30–35]. A PCF also is referred to as a
microstructured fiber or a *holey fiber*. The core of the PCF can be solid or hollow as
shown in Fig. 3.13 by the cross-sectional images of two typical PCF structures [5].

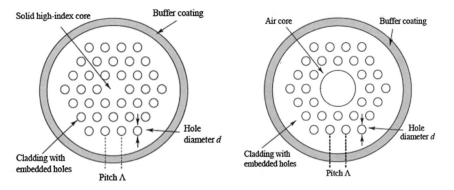

Fig. 3.13 Sample structural arrangements of air holes in solid-core (*left*) and hollow-core (*right*)
PCF (*J. Biomed. Opt.* 19(8), 080902 (Aug 28, 2014). doi:10.1117/1.JBO.19.8.080902)

This structural arrangement creates an internal microstructure, which offers another dimension of light control in the fiber compared to a conventional solid-core fiber. The arrangement, size, and spacing (known as the *pitch*) of the holes in the microstructure and the refractive index of its constituent material determine the light-guiding characteristics of photonic crystal fibers. Depending on the PCF structure, light is guided along the fiber by either total internal reflection or by a photonic bandgap effect.

The fact that the core can be made of pure silica, gives the PCF a number of operational advantages over conventional fibers, which typically have a germanium-doped silica core [5]. These include very low losses, the ability to transmit high optical power levels, and a strong resistance to darkening effects from nuclear radiation. The fibers can support single-mode operation over wavelengths ranging from 300 nm to more than 2000 nm. The mode field area of a PCF can be greater than 300 μm^2 compared to the 80-μm^2 area of conventional single-mode fibers. This allows the PCF to transmit high optical power levels without encountering the nonlinear effects exhibited by conventional fibers.

For biophotonics applications, some hollow-core PCFs are being used in place of the coated hollow-core fiber mentioned in Sect. 3.8 for delivering mid-infrared light with a broadened transmission window and reduced transmission or bending loss. Photonic crystal fibers also are being used in a wide variety of optical fiber-based biosensors. Chap. 7 discusses several of these biosensor applications.

3.10   Plastic Fibers

A *plastic optical fiber* or *polymer optical fiber* (POF) is an alternative to glass optical fibers in areas such as biomedical sensors [36–40]. Most POFs are made of polymethylmethacrylate (PMMA) with a refractive index of around 1.492, which is slightly higher than the index of silica [5]. The size of a POF is normally larger than typical silica fibers, with diameters ranging up to 0.5 mm. In addition to POF with large core diameters, currently both multimode and single-mode (SM) plastic optical fibers are commercially available. The standard core sizes of multimode POF include 50-μm and 62.5-μm diameters, which are compatible with the core diameters of standard multimode glass telecom fibers.

The development of SM POF structures enables the creation of FBGs inside of plastic fibers, which therefore provides more possibilities for POF-based biosensing [5]. Together with the advantages of low cost, inherent fracture resistance, low Young's modulus, and biocompatibility, POF has become a viable alternative for silica fibers in the areas of biomedical applications. For example, some special fiber sensor designs, such as the exposed-core technique now being used in PCF, were first realized using polymer optical fibers. Although most POFs have a higher refractive index than silica, single-mode perfluorinated POF with a refractive index of 1.34 has been fabricated. This low-index fiber allows the potential for improvement in the performance of biosensors, because it results in a stronger optical coupling between a light signal and the surrounding biosensor analyte. More details on biosensors are given in Chap. 7.

3.11 Side-Emitting or Glowing Fibers

In applications such as telecom links, it is necessary that optical fibers transport light to the intended destination with low optical power loss and small signal distortion. However, a different type of optical fiber that emits light along its entire length has been used in applications such as decorative lamps, submersible lighting, event lighting, and lighted signs and also these fibers are being incorporated into biomedical applications [41–46]. Such a *side-emitting fiber* or *glowing fiber* acts as an extended emitting optical source [5].

Another embodiment for lateral emission of light is the *side-firing fiber* [5]. As described further in Sect. 7.1, this structure consists of an angled end face at the distal end of the fiber. This enables the light to be emitted almost perpendicular to the fiber axis. Uses of side-emitting and side-firing fibers for biomedical procedures include photodynamic therapy, therapy for atrial fibrillation, treatment of prostate enlargements, and dental treatments.

A variety of fiber materials and construction designs have been used to fabricate side-emitting fibers [5]. Both plastic and silica fibers are available. One popular method of achieving the fiber glowing process is to add scattering materials into either the core or cladding of the fiber. Thereby a side-emitting effect is created through the scattering of light from the fiber core into the cladding and then into the medium outside of the fiber. As Fig. 3.14 shows, one implementation of a side-emitting fiber is to attach a length of this type of fiber (e.g., 10–70 mm for short diffusers or 10–20 cm for long diffusers) to a longer delivery fiber (nominally 2–2.5 m), which runs from the laser to the side-emitting unit.

As a result of scattering effects, the light intensity in a side-emitting fiber decreases exponentially with distance along the fiber [5]. Therefore, the intensity of the side-emitted light will decrease as a function of distance along the fiber. If it is

Fig. 3.14 The *red* input light is scattered radially along the length of the side-emitting fiber (*J. Biomed. Opt.* 19(8), 080902 (Aug 28, 2014). doi:10.1117/1.JBO.19.8.080902)

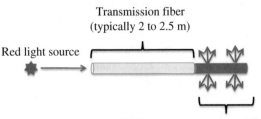

Transmission fiber (typically 2 to 2.5 m)

Red light source

Diffuser segment emits red light along its length (10 to 70 mm)

assumed that the side-emitting effect is much stronger than light absorption and other losses in the fiber, then the side-glowing radiation intensity I_S that is emitted in any direction per steradian at a distance x from the light input end is

$$I_S = \frac{I_0}{4\pi} \exp(-kx) \tag{3.18}$$

where I_0 is the input radiation intensity and k is the side-scattering efficiency coefficient. Typical values of k range from 0.010 to 0.025 m^{-1}.

In biomedical applications where the glowing fiber lengths are on the order of 10 or 20 cm, the effect of an exponential decrease in side-emitted light intensity may not be a major problem [5]. However, there are two methods for creating a more uniform longitudinal distribution of the side-emitted light. One method is to inject light into both ends of the fiber simultaneously, if it is possible to do so. In other applications, a reflector element attached to the distal end of the fiber can produce a relatively uniform emitted light distribution of the combined transmitted and reflected light if the fiber is not too long.

Example 3.12 Consider a 10 cm long side-emitting fiber diffuser that has a side-scattering efficiency coefficient of $k = 0.02$ m^{-1}. What is the emitted intensity at the 10-cm point relative to the intensity at the diffuser input point?

Solution: From Eq. (3.18)

$$I_s(10\,cm) = I_s(0\,cm)\exp(-kx) = I_s(0\,cm)\exp[-(0.02\,m^{-1})(0.1\,m)]$$
$$= 0.998\ I_s(0\,cm)$$

Thus at 10 cm the emitted intensity is 99.8 % of that at the diffuser input point.

3.12   Middle-Infrared Fibers

A *middle-infrared fiber*, or simply an *infrared fiber* (IR fiber), transmits light efficiently at wavelengths greater than 2 μm [47–53]. Typically in biomedical photonics, IR fibers are used in lengths less than 2 to 3 m for a variety of fiber optic sensors and optical power delivery applications [5]. Based on the fiber material and structure used for their fabrication, IR fibers can be classified into glass, crystalline, photonic crystal fibers, and hollow-core waveguide categories. This section describes glass and crystalline fiber classes. Hollow-core waveguides and photonic crystal fibers are described in Sects. 3.8 and 3.9, respectively.

Glass materials for IR fibers include heavy metal fluorides, chalcogenides, and heavy metal germinates [5]. *Heavy metal fluoride glasses* (HMFG) are the only materials that transmit light from ultra-violet to mid-IR without any absorption peak. Commercial HMFG fibers include InF_3 (with attenuations of <0.25 dB/m between 1.8 and 4.7 μm) and ZrF_4 (with attenuations of <0.25 dB/m between 1.8 and 4.7 μm). Such optical losses allow for practical short and medium length applications of a few meters. The reliability of HMFG fibers depends on protecting the fiber from moisture and on manufacturing schemes to reduce surface crystallization. Thus HMFG fibers are a good choice for the 1.5-to-4.0-μm spectral region, which is increasingly used in medical applications.

Chalcogenide glasses are based on the chalcogen elements sulfur (S), selenium (Se), and tellurium (Te) together with the addition of other elements such as germanium (Ge), arsenic (As), and antimony (Sb) [5]. These glasses are very stable, durable, and insensitive to moisture. Chalcogenide fibers are able to transmit longer wavelengths in the IR than fluoride glass fibers. Longer wavelengths are transmitted through the addition of heavier elements. All chalcogenide fibers have strong extrinsic absorption resulting from contaminants such as hydrogen. Although the losses of chalcogenides are generally higher than the fluoride glasses, these losses are still adequate for about 2-m transmissions in the spectral band ranging from 1.5 to 10.0-μm. Some typical loss values are less than 0.1 dB/m at the widely used 2.7-μm and 4.8-μm wavelengths. The maximum losses for the three common chalcogenides are as follows:

(a) Sulfides: <1 dB/m over 2 to 6 μm
(b) Selenides: <2 dB/m over 5 to 10 μm
(c) Tellurides: <2 dB/m over 5.5 to 10 μm

Applications of chalcogenide fibers include Er:YAG (2.94 μm) and CO_2 (10.6 μm) laser power delivery, microscopy and spectroscopy analyses, and chemical sensing.

Germanate glass (GeO_2) fibers generally contain heavy metal oxides (for example, PbO, Na_2O, and La_2O_3) to shift the IR absorption edge to longer wavelengths [5]. The advantage of germanate fibers over HMFG fibers is that GeO_2 glass has a higher laser damage threshold and a higher glass transition temperature, thus making germanate glass fibers more mechanically and thermally stable than, for example, tellurite glasses. GeO_2 glass has an attenuation of less than 1 dB/m in the spectrum ranging from 1.0 to 3.5 μm. A key application of germanate fibers is for laser power delivery from the laser to a patient from Ho:YAG (2.12 μm) or Er:YAG (2.94 μm) lasers, where the fiber losses nominally are 0.25 and 0.75 dB/m, respectively. These fibers can handle up to 20 watts of power for medical procedures in dermatology, dentistry, ophthalmology, orthopedics, and general surgery.

Crystalline IR fibers can transmit light at longer wavelengths than IR glass fibers, for example, transmission up to 18 μm is possible [5]. These fibers can be classified into single-crystal and polycrystalline fiber types. Although there are many varieties of halide crystals with excellent IR transmission features, only a few have been fabricated into optical fibers because most of the materials do not meet the required physical property specifications needed to make a durable fiber.

Polycrystalline silver-halide fibers with AgBr cores and AgCl claddings have shown to have excellent crystalline IR fiber properties. There are several extrinsic absorption bands for Ag-halide fibers, for example, 3 and 6.3 μm due to residual water ions. The attenuation of Ag-halide fibers is normally less than 1 dB/m in the 5-to-12 μm spectral band and somewhat higher in the 12-to-18 μm spectral band. At 10.6 μm the loss is 0.3–0.5 dB/m and can be as low as 0.2 dB/m at this wavelength.

3.13 Optical Fiber Bundles

To achieve greater throughput with flexible glass or plastic fibers, multiple fibers are often arranged in a bundle [5, 54–56]. Each fiber acts as an independent waveguide that enables light to be carried over long distances with minimal attenuation. A typical large fiber bundle array can consist of a few thousand to 100 thousand individual fibers, with an overall bundle diameter of <1 mm and an individual fiber diameter between 2 and 20 μm. Such fiber bundles are mainly used for illumination purposes. In addition to flexibility, fiber bundles have other potential advantages in illumination systems:

1. Illuminate multiple locations with one source by splitting the bundle into two or more branches
2. Merge light from several sources into a single output
3. Integrate different specialty fibers into one bundle

For bundles with a large number of fibers, the arrangement of individual fibers inside the bundle normally occurs randomly during the manufacturing process [5]. However, for certain biomedical applications, it is required to arrange a smaller number of fibers in specific patterns. Optical fibers can be bundled together in an aligned fashion such that the orientations of the fibers at both ends are identical. Such a fiber bundle is called a *coherent fiber bundle* or an *ordered fiber bundle*. Here the term "coherent" refers to the correlation between the spatial arrangement of the fibers at both ends and does not refer to the correlation of light signals. As a result of the matched fiber arrangements on both ends of the bundle, any pattern of illumination incident at the input end of the bundle is maintained when it ultimately emerges from the output end.

Figure 3.15 illustrates illumination and coherent fiber bundle configurations [5]. As shown in Fig. 3.15a, the optical fibers are oriented randomly on both ends in a bundle configuration that contains hundreds or thousands of fibers. Figure 3.15b shows one possibility of an ordered fiber arrangement in a bundle. In this hexagonal packaging scheme, the central fiber can be used for illuminating a target tissue area and the surrounding fibers can be used for fluorescent, reflected, or scattered light returning from the tissue. The discussions in Chap. 7 give further examples of fiber arrangements in optical probes.

(a)   **(b)**

Randomly oriented   Collection   Illuminating
optical fibers   fibers   fiber

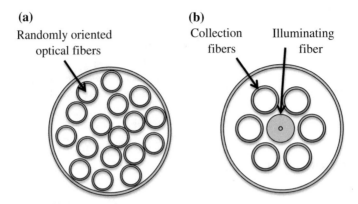

Fig. 3.15 **a** Randomly arranged fibers in a bundle containing many fibers; **b** Coherent bundle cable with identical arrangements of fibers on both ends (*J. Biomed. Opt.* 19(8), 080902 (Aug 28, 2014). doi:10.1117/1.JBO.19.8.080902)

When using fiber bundles, there is a trade-off between light collection efficiency and good image resolution [5]. A large core diameter enables high light transmission but poor image resolution. A thicker cladding avoids crosstalk among individual fibers but limits light collection and image resolution, because more of the light-emitting area is blocked when a thicker cladding is used. In practice, the core diameter of individual fibers is 10–20 μm and the cladding thickness is around 1.5–2.5 μm. Coherent fiber bundles are the key components in a variety of fiber optic endoscopes. Current sophisticated ear, nose, throat, and urological procedures utilize high-resolution flexible image bundles for image transfer.

3.14 Summary

The extensive and rapidly growing use of photonics technology for basic life sciences research and for biomedical diagnosis, therapy, imaging, and surgery has been assisted greatly through the use of a wide variety of optical fibers. Table 3.1 gives a summary of various fibers. The biophotonics applications of these fibers include

- Light care: healthcare monitoring; laser surfacing or photorejuvenation
- Light diagnosis: biosensing, endoscopy, imaging, microscopy, spectroscopy
- Light therapy: ablation, photobiomodulation, dentistry, laser surgery, oncology

Major challenges in biophotonics applications to the life sciences include

- How to collect emitted low-power light (down to the nW range) from a tissue specimen and transmit it to a photon detector

- How to deliver a wide range of optical power levels to a tissue area or section during different categories of therapeutic healthcare sessions
- How to access a diagnostic or treatment area within a living being with an optical detection probe or a radiant energy source in the least invasive manner.

Depending on the application, all three of these factors may need to be addressed at the same time. Because the unique physical and light-transmission properties of optical fibers can help resolve such implementation issues, various types of optical fibers are finding widespread use in biophotonics. Each optical fiber structure has certain advantages and limitations for specific uses in different spectral bands that are of interest to biophotonics (as Fig. 1.6 illustrates). The descriptions in this chapter of these characteristics give the essential information that biophotonics researchers and implementers need to know when selecting a fiber for a certain application.

3.15 Problems

3.1 Consider the interface between the core of a fiber and its cladding that have refractive indices of n_1 and n_2, respectively. (a) If n_2 is smaller than n_1 by 1 % and if $n_1 = 1.450$, show that $n_2 = 1.435$. (b) Show that the critical angle is $\varphi_c = 81.9°$.

3.2 Calculate the numerical aperture of a step-index fiber having a core index $n_1 = 1.48$ and a cladding index $n_2 = 1.46$. What is the acceptance angle θ_A for this fiber if the outer medium is air with $n = 1.00$?

3.3 Consider a multimode step-index optical fiber that has a core diameter of 62.5 μm, a core index of 1.48, and an index difference $\Delta = 0.015$. Show that at 1310 nm (a) the value of V is 38.4 and (b) the total number of modes is 737.

3.4 A step-index multimode fiber with a numerical aperture of 0.20 supports approximately 1000 modes at a wavelength of 850 nm.

(a) Show that the diameter of its core is 60.5 μm.
(b) Show that the fiber supports 414 modes at 1320 nm.
(c) Show that the fiber supports 300 modes at 1550 nm.

3.5 Consider a step-index fiber having a 25-μm core radius, $n_1 = 1.48$, and $n_2 = 1.46$.

(a) Show that the normalized frequency at 820 nm is 46.5.
(b) Verify that 1081 modes propagate in this fiber at 820 nm.
(c) Verify that 417 modes propagate in this fiber at 1320 nm.
(d) Verify that 303 modes propagate in this fiber at 1550 nm.
(e) What percent of the optical power flows in the cladding in each case? (Answer: 4.1 % at 820 nm; 6.6 % at 1320 nm; 7.8 % at 1550 nm).

3.6   Consider a fiber with a 25-μm core radius, a core index $n_1 = 1.48$, and $\Delta = 0.01$.

(a) Show that the value of V is 25 at 1320 nm and that 312 modes propagate in the fiber.
(b) Show that 7.5 % of the optical power flows in the cladding.
(c) If the core-cladding difference is reduced to $\Delta = 0.003$, show that the fiber supports 94 modes and that now 13.7 % of the optical power flows in the cladding.

3.7   Consider a step-index fiber that has a 5-μm core radius, an index difference $\Delta = 0.002$, and a core index $n_1 = 1.480$.

(a) By calculating the V number, verify that at 1310 nm this is a single-mode fiber.
(b) Verify that at 820 nm the fiber is not single-mode because $V = 3.59$.

3.8   Consider a 62.5-μm core diameter graded-index fiber that has a parabolic index profile ($\alpha = 2$). Suppose the fiber has a numerical aperture $NA = 0.275$.

(a) Show that the V number for this fiber at 850 nm is 63.5.
(b) Show that the number of guided modes is 1008 at 850 nm.

3.9   A graded-index fiber with a parabolic index profile ($\alpha = 2$) has a core index $n_1 = 1.480$ and an index difference $\Delta = 0.010$.

(a) Using Eqs. (3.4) and (3.15) show that the maximum value of the core radius for single-mode operation at 1310 nm is 3.39 μm.
(b) Show that the maximum value of the core radius for single-mode operation at 1550 nm is 4.01 μm.

3.10   Calculate the numerical apertures of (a) a plastic step-index fiber having a core refractive index of $n_1 = 1.60$ and a cladding index of $n_2 = 1.49$, (b) a step-index fiber having a silica core ($n_1 = 1.458$) and a silicone resin cladding ($n_2 = 1.405$). (Answer: 0.58; 0.39).

3.11   Consider a FBG that has the following parameter values: $n_{eff} = 1.479$ and $\Lambda = 523$ nm at 20 °C.

(a) Show that this FBG will reflect a wavelength $\lambda_B = 1547$ nm at 20 °C.
(b) Suppose that the Bragg wavelength changes linearly with temperature with an incremental wavelength change $\Delta\lambda_B = 23.4$ pm/°C per incremental temperature change ΔT. Show that the Bragg wavelength is 1547.351 nm when $T = 35$ °C.
(c) Show that the Bragg wavelength is 1546.766 nm when $T = 10$ °C?

3.12   Consider an optoelectronic device page that has a 30-cm length of fiber coiled up inside. Show that the volume saved by using a reduced-cladding fiber with a cladding diameter of 80 μm is 2.2 mm^3 smaller compared to a fiber with a 125-μm outer cladding diameter.

References

1. B.E.A. Saleh, M.C. Teich, *Fundamentals of Photonics*, 2nd edn. (Wiley, Hoboken, NJ, 2007)
2. C.L. Chen, *Foundations of Guided-Wave Optics* (Wiley, Hoboken, NJ, 2007)
3. W.H. Hayt Jr., J.A. Buck, *Engineering Electromagnetics*, 8th edn. (McGraw-Hill, New York, 2012)
4. S.O. Kasap, *Optoelectronics and Photonics: principles and Practices*, 2nd edn. (Prentice-Hall, Englewood Cliffs, New Jersey, 2013)
5. G. Keiser, F. Xiong, Y. Cui, P.P. Shum, Review of diverse optical fibers used in biomedical research and clinical practice. J. Biomed. Opt. **19**, 080902 (2014)
6. G. Keiser, *Optical Fiber Communications*, McGraw-Hill, 4th US edn, 2011; 5th international edn (2015)
7. S.T. Jung, D.H. Shin, Y.H. Lee, Near-field fiber tip to handle high input power more than 150 mW. Appl. Phys. Lett. **77**(17), 2638–2640 (2000)
8. M. De Rosa, J. Carberry, V. Bhagavatula, K. Wagner, C. Saravanos, High-power performance of single-mode fiber-optic connectors. J. Lightw. Technol. **20**(5), 879–885 (2002)
9. K. Hogari, K. Kurokawa, I. Sankawa, Influence of high-optical power light launched into optical fibers in MT connector. J. Lightw. Technol. **21**(12) (2003)
10. Y. Shuto, S. Yanagi, S. Asakawa, M. Kobayashi, R. Nagase, Fiber fuse generation in single-mode fiber-optic connectors. IEEE Photon. Technol. Lett. **16**(1), 174–176 (2004)
11. A.A. Stolov, B.E. Slyman, D.T. Burgess, A.S. Hokansson, J. Li, R.S. Allen, Effects of sterilization methods on key properties of specialty optical fibers used in medical devices. Proceedings of the SPIE, vol. 8576, p. 857606 (2013)
12. R. Kashyap, *Fiber Bragg Gratings*, 2nd edn. (Academic Press, New York, 2010)
13. E. Al-Fakih, N.A. Abu Osman, F.R.M. Adikan, The use of fiber Bragg grating sensors in biomechanics and rehabilitation applications: the state-of-the-art and ongoing research topics. Sensors **12**, 12890–12926 (2012)
14. L. Dziuda, F.W. Skibniewski, M. Krej, P.A. Baran, Fiber Bragg grating-based sensor for monitoring respiration and heart activity during magnetic resonance imaging examinations. J. Biomed. Opt. **18**(5), 057006 (2013)
15. V. Khalilov, J.H. Shannon, R.J. Timmerman, Improved deep UV fiber for medical and spectroscopy applications. Proceedings of the SPIE, vol. 8938, p. 89380A (2014)
16. T. Tobisch, H. Ohlmeyer, H. Zimmermann, S. Prein, J. Krichhof, S. Unger, M. Belz, K.F. Klein, Improvement of optical damage in specialty fiber at 266 nm wavelength. Proceedings of the SPIE, vol. 8938, p. 89380G (2014)
17. F. Gebert, M.H. Frosz, T. Weiss, Y. Wan, A. Ermolov, N.Y. Joly, P.O. Schmidt, P.St. J. Russell, Damage-free single-mode transmission of deep-UV light in hollow-core PCF, Opt. Express **22**, 15388–15396 (2014)
18. M.-J. Li, P. Tandon, D.C. Bookbinder, S.R. Bickham, M.A. McDermott, R.B. Desorcie, D.A. Nolan, J.J. Johnson, K.A. Lewis, J.J. Englebert, Ultra-low bending loss single-mode fiber for FTTH. J. Lightw. Technol. **27**(3), 376–382 (2009)
19. T. Matsui, K. Nakajima, Y. Goto, T. Shimizu, T. Kurashima, Design of single-mode and low-bending-loss hole-assisted fiber and its MPI characteristics. J. Lightw. Technol. **29**(17), 2499–2505 (2011)
20. D. Kusakari, H. Hazama, R. Kawaguchi, K. Ishii, K. Awazu, Evaluation of the bending loss of the hollow optical fiber for application of the carbon dioxide laser to endoscopic therapy. Opt. Photon. J. **3**, 14–19 (2013)
21. V.V. Tuchin, Polarized light interaction with tissues. J. Biomed. Opt. **21**(7), 071114 (2016)
22. S. Lemire-Renaud, M. Strupler, F. Benboujja, N. Godbout, C. Boudoux, Double-clad fiber with a tapered end for confocal endomicroscopy. Biomed. Opt. Exp. **2**, 2961–2972 (2011)
23. S. Liang, A. Saidi, J. Jing, G. Liu, J. Li, J. Zhang, C. Sun, J. Narula, Z. Chen, Intravascular atherosclerotic imaging with combined fluorescence and optical coherence tomography probe based on a double-clad fiber combiner. J. Biomed. Opt. **17**, 070501 (2012)

24. K. Beaudette, H.W. Bac, W.-J. Madore, M. Villiger, N. Gadbout, B.E. Bouma, C. Boudoux, Laser tissue coagulation and concurrent optical coherence tomography through a double-clad fiber coupler. Biomed. Opt. Exp. **6**, 1293–1303 (2015)
25. B.J. Skutnik, B. Foley, K. Moran, Hard plastic clad silica fibers for near UV applications. Proc. SPIE **5691**, 23–29 (2005)
26. T. Watanabe, Y. Matsuura, Side-firing sealing caps for hollow optical fibers. Lasers Surg. Med. **38**(8), 792–797 (2006)
27. F. Yu, W.J. Wadsworth, J.C. Knight, Low loss silica hollow core fibers for 3–4 μm spectral region. Opt. Express **20**(10), 11153–11158 (2012)
28. C.M. Bledt, J.A. Harrington, J.M. Kriesel, Loss and modal properties of Ag/AgI hollow glass waveguides. Appl. Opt. **51**, 3114–3119 (2012)
29. T. Monti, G. Gradoni, Hollow-core coaxial fiber sensor for biophotonic detection. J. Sel. Topics Quant. Electron. **20**(2), 6900409 (2014)
30. P. John Russell, Photonic crystal fibers. J. Lightw. Technol. **24**(12), 4729–4749 (2006)
31. F. Poli, A. Cucinotta, S. Selleri, *Photonic Crystal Fibers* (Springer, New York, 2007)
32. M. Large, L. Poladian, G. Barton, M.A. van Eijkelenborg, *Microstructured Polymer Optical Fibres* (Springer, New York, 2008)
33. D. Threm, Y. Nazirizadeh, M. Gerken, Photonic crystal biosensor towards on-chip integration. J. Biophotonics **5**(8–9), 601–616 (2012)
34. P. Ghenuche, S. Rammler, N.Y. Joly, M. Scharrer, M. Frosz, J. Wenger, P.St. John Russell, H. Rigneault, Kagome hollow-core photonic crystal fiber probe for Raman spectroscopy. Opt. Lett. **37**(21), 4371–4373 (2012)
35. T. Gong, N. Zhang, K.V. Kong, D. Goh, C. Ying, J.-L. Auguste, P.P. Shum, L. Wei, G. Humbert, K.-T. Yong, M. Olivo, Rapid SERS monitoring of lipid-peroxidation-derived protein modifications in cells using photonic crystal fiber sensor, J. Biophotonics (2015)
36. J. Zubia, J. Arrue, Plastic optical fibers: an introduction to their technological processes and applications. Opt. Fiber Technol. **7**(2), 101–140 (2001)
37. O. Ziemann, J. Krauser, P.E. Zamzow, W. Daum, *POF Handbook*, 2nd edn. (Springer, Berlin, 2008)
38. G. Zhou, C.-F.J. Pun, H.-Y. Tam, A.C.L. Wong, C. Lu, P.K.A. Wai, Single-mode perfluorinated polymer optical fibers with refractive index of 1.34 for biomedical applications. IEEE Photon. Technol. Lett. **22**(2), 106–108 (2010)
39. Y. Koike, K. Koike, Progress in low-loss and high-bandwidth plastic optical fibers, J. Polym. Sci. B Polym. Phys. **49**, 2–17 (2011)
40. L. Bilro, N. Alberto, J.L. Pinto, R. Nogueira, Optical sensors based on plastic fibers. Sensors **12**, 12184–12207 (2012)
41. J. Spigulis, Side-emitting fibers brighten our world. Opt. Photonics News **16**, 36–39 (2005)
42. J. Shen, C. Chui, X. Tao, Luminous fabric devices for wearable low-level light therapy. Biomed. Opt. Express **4**(12), 2925–2937 (2013)
43. M. Krehel, M. Wolf, L.F. Boesel, R.M. Rossi, G.-L. Bona, L.J. Scherer, Development of a luminous textile for reflective pulse oximetry measurements. Biomed. Opt. Express **5**(8), 2537–2547 (2014)
44. I. Peshkoa, V. Rubtsovb, L. Vesselovc, G. Sigala, H. Laks, Fiber photo-catheters for laser treatment of atrial fibrillation. Opt. Lasers Eng. **45**, 495–502 (2007)
45. R. George, L.J. Walsh, Performance assessment of novel side firing flexible optical fibers for dental applications. Lasers Surg. Med. **41**, 214–221 (2009)
46. R. Mishra, A. Shukla, D. Kremenakova, J. Militky, Surface modification of polymer optical fibers for enhanced side emission behavior. Fibers Polym. **14**, 1468–1471 (2013)
47. J.A. Harrington, *Infrared fibers and their applications* (SPIE Press, 2004)
48. M. Saad, Heavy metal fluoride glass fibers and their applications. Proc. SPIE **8307**, 83070N (2011)
49. J. Bei, T.M. Monro, A. Hemming, H. Ebendorff-Heidepriem, Fabrication of extruded fluoroindate optical fibers. Opt. Mater. Express **3**(3), 318–328 (2013)

50. B.J. Eggleton, B. Luther-Davies, K. Richardson, Chalcogenide photonics. Nat. Photonics **5**, 141–148 (2011)
51. D. Lezal, J. Pedlikova, P. Kostka, J. Bludska, M. Poulain, J. Zavadil, Heavy metal oxide glasses: preparation and physical properties. J. Non-Cryst. Solids **284**, 288–295 (2001)
52. C.A. Damin, A.J. Sommer, Characterization of silver halide fiber optics and hollow silica waveguides for use in the construction of a mid-infrared attenuated total reflection Fourier transform infrared (ATR FT-IR) spectroscopy probe. Appl. Spectros. **67**(11), 1252–1263 (2013)
53. S. Israeli, A. Katzir, Attenuation, absorption, and scattering in silver halide crystals and fibers in the mid-infrared. J. Appl. Phys. **115**, 023104 (2014)
54. E. Rave, A. Katzir, Ordered bundles of infrared transmitting silver halide fibers: attenuation, resolution and crosstalk in long and flexible bundles. Opt. Eng. **41**, 1467–1468 (2002)
55. H.H. Gorris, T.M. Blicharz, D.R. Walt, Optical-fiber bundles. The FEBS J. **274**(21), 5462–5470 (2007)
56. J.D. Enderle, J.D. Bronzino, Biomedical optics and lasers. Chap. 17 in *Introduction to Biomedical Engineering*, 3rd edn. (Academic Press, New York, 2012)

Chapter 4
Fundamentals of Light Sources

Abstract A broad selection of light sources is available for the biophotonics UV, visible, or infrared regions. These sources include arc lamps, light emitting diodes, laser diodes, superluminescent diodes, and various types of gas, solid-state, and optical fiber lasers. This chapter first defines terminology used in radiometry, which deals with the measurement of optical radiation. Understanding this terminology is important when determining and specifying the degrees of interaction of light with tissue. Next the characteristics of optical sources for biophotonics are described. This includes the spectrum over which the source emits, the emitted power levels as a function of wavelength, the optical power per unit solid angle emitted in a given direction, the light polarization, and the coherence properties of the emission. In addition, depending on the operating principles of the light source, it can emit light in either a continuous mode or a pulsed mode.

Many categories of light sources with diverse sizes, shapes, operational configurations, light output powers, and emitting in either a continuous or a pulsed mode are used in biophotonics. These sources can be selected for emissions with different spectral widths in the UV, visible, or infrared regions. Each light source category has certain advantages and limitations for specific life sciences and medical research, diagnostic, imaging, therapeutic, or health-status monitoring applications. The characteristics of the optical radiation emitted by any particular light source category can vary widely depending on the physics of the photon emission process and on the source construction and its material. The decision of which optical source to use for a given application depends on the characteristics of the optical radiation that is emitted by the source and the way this radiation interacts with the specific biological substance or tissue being irradiated. That is, as Chap. 6 describes, the absorption and scattering characteristics of light in biological tissues and fluids (e.g., skin, brain matter, bone, blood vessels, and eye-related tissue) are dependent on the wavelength region, and the interactions of light with healthy or diseased cells can vary significantly.

© Springer Science+Business Media Singapore 2016
G. Keiser, *Biophotonics*, Graduate Texts in Physics,
DOI 10.1007/978-981-10-0945-7_4

An important factor to keep in mind is that biological tissue has a multilayered characteristic from both compositional and functional viewpoints. Specific biological processes and diseases occur at different levels within this multilayered structure. Thus when selecting an optical source, it is necessary to ensure that the specific wavelength of the light that is aimed at the targeted biological process or disease can penetrate the tissue down to the desired layer.

The medical field also uses many types of surgical lights and a variety of lamps for illumination and visual diagnostic purposes in operating rooms, medical offices, and health clinics. In general these sources do not relate directly to biophotonics, so they will not be covered in this chapter. A key factor in selecting an optical source is the range of wavelengths over which the source emits and the radiant power per unit wavelength over the emission spectrum. The radiant power spectrum can range from being nearly monochromatic (as emitted by certain lasers) to being broadband, that is, the emission covers a wide spectrum, which is the characteristic of a light source such as an incandescent lamp. Other types of sources, for example light-emitting diodes, have spectral ranges between these two extremes. Further details on specific optical sources can be found on manufacturers' data sheets.

For biophotonics the radiometric wavelengths of interest range from about 190 nm to 10 µm, which covers the ultraviolet, visible, and infrared regions. A selection of sample light sources and their emission wavelengths are shown in Fig. 4.1 and summarized in Table 4.1. To cover a wide range of applications, the source types include arc lamps, semiconductor light-emitting diodes (LEDs), superluminescent diodes, and various lasers including excimer, gas, liquid, solid-state crystal, semiconductor, and optical fiber lasers. The following sections describe these sources and their operating characteristics.

In this chapter, Sect. 4.1 first defines terminology used in radiometry, which deals with the measurement of optical radiation. Understanding the significance of this terminology is important when determining and specifying the degrees of interaction of light with tissue. Then the material in Sects. 4.2–4.4 describes a selection of optical sources used in biophotonics and presents the characteristics of

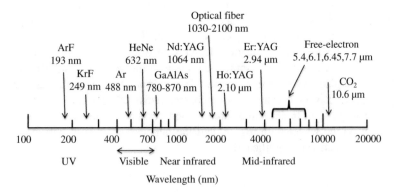

Fig. 4.1 Examples of light sources used across the biophotonics spectrum

Table 4.1 Characteristics of representative light sources used in biophotonics; *YAG* yttrium-aluminum-garnet

Lasing material	Laser type	Wavelength
Argon fluoride (ArF)	Excimer	193 nm
Krypton chloride (KrCl)	Excimer	222 nm
Krypton fluoride (KrF)	Excimer	248 nm
Xenon chloride (XeCl)	Excimer	308 nm
Xenon fluoride (XeFl)	Excimer	351 nm
Argon (Ar)	Gas	488.0 and 514.5 nm
Krypton (Kr)	Gas	530.9 and 568.2 nm
KTP/Nd:YAG	Solid state	532.0 nm
Helium–neon (He–Ne)	Gas	632.8 nm
Pulsed dyes	Liquid	400–500, 550–700 nm
Ruby	Solid state	694.3 nm
Alexandrite	Solid state	700–830 nm
GaAlAs or InGaAsP alloys	Semiconductor	760–2400 nm
Ti:sapphire	Solid state	700–1000 nm
Neodynium:YAG (Nd:YAG)	Solid state	1064 nm
Holium:YAG (Ho:YAG)	Solid state	2.10 μm
Ytterbium-doped fiber	Optical fiber	1030–1080 nm
Erbium-doped fiber	Optical fiber	1530–1620 nm
Thulium-doped fiber	Optical fiber	1750–2100 nm
Erbium:YAG (Er:YAG)	Solid state	2.94 μm
Free-electron laser (FEL)	Electromagnetic	5.4, 6.1, 6.45, 7.7 μm
Carbon-dioxide (CO_2)	Gas	10.6 μm

their optical radiation. The characteristics include the spectrum over which the source emits, the emitted power levels as a function of wavelength, the optical power per unit solid angle emitted in a given direction, the polarization of the light, and the coherence properties of the emission. In addition, depending on the physical operating principles of the light source, it can emit light in either a continuous mode or a pulsed mode.

4.1 Radiometry

Before examining the physical and functional characteristics of various optical sources, it will be helpful to define some radiometric terminology. Radiometry deals with measuring the power and the geometric nature of optical radiation [1–3].

4.1.1 Optical Flux and Power

A key radiometric parameter for characterizing a light source is the *flux* (also known as *optical flux* or *radiant flux*), which is the rate of energy flow with respect to time. Because energy per time is power, this parameter gives the *optical power* emitted by a source. Standard symbols that are recommended for power are Φ (upper case Greek letter phi) or P. The letter P will be used in this book. The flux or power typically is measured in units of watts (W).

4.1.2 Irradiance or Exposure Rate

When illuminating a tissue area for applications such as imaging, therapy, or surgery, the quantity of interest is the *irradiance*, which is designated by the symbol E. Equivalently the irradiance is called the *exposure rate* because it designates the energy incident on a tissue area as a function of time. This parameter is the power incident from all directions in a hemisphere, or the power emitted through an optical aperture, onto a target surface area as shown in Fig. 4.2. Thus, irradiance E is given by

$$E = dP/dA \qquad (4.1)$$

where dP is the incremental amount of power hitting the incremental area dA. Irradiance is also known as the *power density* (or *flux density*) and is measured in units of power per area, for example, W/m^2 or mW/cm^2.

> **Example 4.1** Consider an optical source that emits a highly collimated circular beam of light. Suppose the beam diameter is 5 mm and let the power level or radiant flux be 1 W. Neglecting any divergence of the beam, compute the irradiance.

Fig. 4.2 Illustration of irradiance: optical power falling on a tissue area

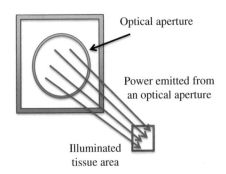

Optical aperture

Power emitted from an optical aperture

Illuminated tissue area

Solution: From Eq. (4.1) the irradiance is

$$E = \frac{\text{Power}}{\text{Area}} = \frac{1\text{ W}}{\pi (2.5 \times 10^{-3}\text{ m})^2} = 5.09 \times 10^4\text{ W/m}^2$$
$$= 5.09 \times 10^3\text{ mW/cm}^2$$

4.1.3   Radiant Intensity

Radiant intensity, or simply *intensity* I, is the power per unit solid angle Ω and is measured in watts per steradian (W/sr). Because intensity is the derivative of power with respect to solid angle Ω (that is, $dP/d\Omega$), then that the integral of the intensity over the solid angle is power. The angular intensity distribution is illustrated comparatively in Fig. 4.3 for a lambertian source and a highly directional laser diode in which the output beams have rotational symmetry. The power delivered at an angle θ measured relative to a normal to the emitting surface can be approximated by the expression

$$I(\theta) = I_0 \cos^{g-1}\theta \qquad (4.2)$$

Here I_0 is the intensity normal to the source surface and $g \geq 1$. For example

$g = 1$ for an isotropic source (emits the same intensity in all directions)
$g = 2$ for a lambertian source (the intensity varies as $\cos\theta$ because the projected area of the emitting surface varies as $\cos\theta$ with viewing direction)
$g \geq 30$ for a lensed LED lamp or a laser diode (for example, $g = 181$ in Fig. 4.3).

Fig. 4.3 Intensity patterns for a lambertian source and the lateral output of a highly directional laser diode. Both sources have I_0 normalized to unity

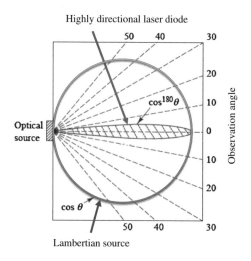

Example 4.2 (a) At what angle from the normal in a lambertian LED is the intensity level 50 % of the normal intensity? This is the half-power point. (b) At what angle from the normal axis does the light appear only 40 % as intense as it does when viewed down its centerline?

Solution:

(a) For a lambertian source the parameter $g = 2$ in Eq. (4.2). Using this equation, at the half-power point $I(\theta) = I_0 \cos \theta = 0.50 I_0$. Thus $\cos \theta = 0.50$ so that $\theta = 60°$.
(b) In this case $I(\theta) = I_0 \cos \theta = 0.40 I_0$, or $\cos \theta = 0.40$ so that $\theta = 67°$.

4.1.4 Radiant Exposure or Radiant Fluence

Radiant exposure or *radiant fluence* is the optical energy received by a surface per unit area. Equivalently radiant exposure can be viewed as the irradiance of a surface integrated over the time of irradiation. For example, for a light pulse the radiant exposure is the average power of the pulse multiplied by the pulse duration. A standard symbol for radiant exposure is H and the units are J/cm^2. The radiant exposure also is known as the *energy density*. For biophotonics applications H ranges from approximately 1 mJ/cm^2 to 1 kJ/cm^2.

4.1.5 Radiance

Radiance indicates how much of the optical power coming from an emitting or reflecting surface will fall within a given solid angle in a specified direction. This parameter is measured in units of watts per steradian per square meter ($W/sr/m^2$). Thus knowing the radiance and the angle at which a tissue target is illuminated will allow a determination of the total power incident on a tissue area. Radiance, designated by L, is defined by

$$L = \frac{d^2P}{dA d\Omega \cos \theta} \approx \frac{P}{A\Omega \cos \theta} \tag{4.3}$$

where
L   is the observed or measured radiance ($W/sr/m^2$) in the direction θ
d   is the mathematical differential operator
P   is the total power (W) emitted
θ is the angle between the surface normal and the observation direction
A   is the surface area (m^2) of the source
Ω is the solid angle (sr) subtended by the observation or measurement.

The approximation on the right-hand side of Eq. (4.3) only holds for small A and small Ω where $\cos \theta$ is approximately constant.

Often it is necessary to designate the above quantities as a function of wavelength in order to correlate the light characteristics to the responses of various tissue types to particular spectral bands. This designation yields the following parameters:

- *Spectral flux* is the radiant flux (power) per wavelength interval
- *Spectral irradiance* is the flux density (power density) per wavelength interval
- *Spectral radiant intensity* gives the radiant intensity as a function of wavelength
- *Spectral radiant exposure* gives the radiant exposure as a function of wavelength
- *Spectral radiance* specifies the radiance as a function of wavelength.

4.2   Arc Lamps

Lamps used for scientific and medical biophotonics applications are mainly high-pressure gas-discharge lamps or *arc lamps*. Such lamps are based on the creation of an intense light when an electrical discharge passes through a high-pressure gas or a vapor. When the lamp is in operation, the pressure is on the order of 10–40 MPa, where a pascal (Pa) unit is one newton per meter squared (N/m^2). For comparison purposes, the atmospheric pressure at sea level is about 101 kPa. Because the arc length usually is only a few millimeters, these lamps are known as *short-arc lamps* [4–6].

In general the discharge is sustained by the ionization of mercury (Hg) vapor or of inert gases like xenon (Xe), argon (Ar), and neon (Ne). These light sources come in many sizes with a wide range of output characteristics. The outputs can be sharp spectral lines, narrow spectral bands, or wide spectral bands ranging from the ultraviolet to the infrared regions. Some sources operate in a continuous wave (CW) mode and others are used in a pulsed mode. The lamps can be used for microscopy, endoscopy, minimally invasive surgery, and medical fiber optic applications.

High-pressure mercury arc lamps have a high-intensity line-type spectral output in the UV and visible regions as Fig. 4.4 shows. The spectral lines are mainly in the 300–600 nm region. The output powers can range from 50 to 500 W with nominal focused beam sizes of 5 mm. In general the mercury sources are useful when certain wavelengths in the line spectra are suitable for monochromatic irradiation for fluorescence spectroscopic applications.

High-pressure xenon arc lamps are popular and highly versatile radiation sources. As Fig. 4.5 shows, these lamps emit an intense quasi-continuous spectrum ranging from about 250 nm in the UV region to greater than 1000 nm in the near infrared. The output is relatively continuous from 300 to 700 nm and has a number of sharp lines near 475 nm and above 800 nm. The emission of Xe lamps is about 95 % similar to daylight. The lamps are made of two tungsten electrodes encapsulated in a

Fig. 4.4 Typical emission spectrum from a mercury arc lamp

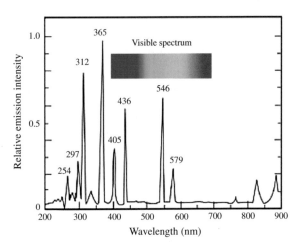

Fig. 4.5 Typical emission spectrum from a xenon arc lamp

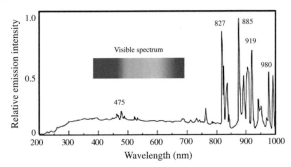

quartz envelope that contains a xenon gas under high pressure. The maximum optical power emission from a Xe lamp comes from an area approximately 0.3 mm by 0.5 mm in the arc region. Typically these lamps have an integrated parabolic reflector, which collects the light that has been emitted into a large solid angle and either images it onto a pinhole opening (with typical f numbers between 1.0 and 2.0) or onto an objective lens to produce a collimated output.

Pulsed xenon lamps or *flash lamps* produce microsecond to millisecond flashes of high intensity light containing a broadband spectrum. This light can be generated over a continuous spectrum from about 160 nm in the ultraviolet to 2000 nm in the infrared. Typical outputs are 15 W with arc beam diameters of 3 or 8 mm.

4.3 Light-Emitting Diodes

Owing to attractive features such as low power consumption, good reliability, long lifetimes, and availability of different emission colors in a broad range of wavelengths, light-emitting diodes are used worldwide for many categories of lighting [7–10]. The applications include spotlights, vehicle taillights, road lights, indicator lights on equipment, surgical lights, and medical therapy lights. Continued enhancements in spectral emission ranges, beam profiles, electrical power consumption, and optical output powers have enabled expansions of LED usage to life sciences research and biophotonics medical applications in the UV, visible, and near infrared spectral regions.

4.3.1   LED Operation and Structures

The basic operating principle of semiconductor LEDs is the recombining of electron-hole pairs, which results in the creation of photons in a *pn* junction region [11–13]. A is the interface between n-type*pn junction* and p-type semiconductor materials within a continuous crystal. These materials are created by adding a small percentage of foreign atoms (called *impurity atoms*) into the regular lattice structure of pure materials such as silicon. This process is called *doping*. Doping with impurity atoms that have five valence electrons produces an *n-doped semiconductor* by contributing extra electrons. The addition of impurity atoms with three valence electrons produces a *p-doped semiconductor* by creating an electron deficiency or a *hole*. Thus, in a p-doped region charge transport only occurs in the form of hole conduction, whereas electron conduction is the charge transport mechanism in an n-doped region.

To achieve a high radiance and a high efficiency, the LED structure must provide a means of confining the charge carriers and the stimulated optical emission to the active region of the pn junction where radiative electron-hole recombination takes place. *Carrier confinement* is used to achieve a high level of radiative recombination in the active region of the device, which yields a high efficiency. *Optical confinement* is of importance for preventing absorption of the emitted radiation by the material surrounding the pn junction and for guiding the emitted light out of the device, for example, into an optical fiber.

Normally a sandwich layering of slightly different semiconductor alloys is used to achieve these confinement objectives. This layered structure is referred to as a *double-heterostructure*. The two basic LED configurations are the surface emitter and the edge-emitter. In the *surface emitter* shown in Fig. 4.6, the plane of the active light-emitting region is oriented perpendicularly to the axis of the optical fiber into which the light is coupled. In this configuration, a well is etched through the substrate of the device, into which a fiber is then cemented in order to accept the emitted light. Typically the active region is limited to a circular section having an area compatible with the fiber-core end face.

For a surface emitter the beam pattern is *lambertian*, as Fig. 4.3 illustrates, which means that LEDs emit light into a hemisphere. Consequently, in setups that do not use an optical fiber for light delivery, some type of lens is needed for collection and distribution of the emitted light into a specific pattern for various applications. A frequently used LED lens design is the total internal reflection (TIR) lens. These lenses can produce output beams that are collimated or that give a uniform illumination, for example, of a circular, ring, square, or rectangular shape [14, 15].

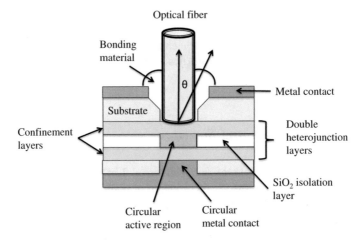

Fig. 4.6 Schematic (not to scale) of a high-radiance surface-emitting LED

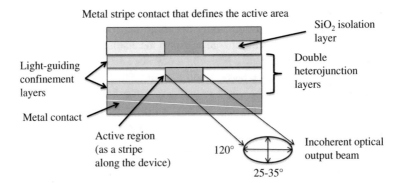

Fig. 4.7 Schematic (not to scale) of an edge-emitting double-heterojunction LED

The *edge emitter* depicted in Fig. 4.7 consists of an active junction region, which is the source of the incoherent light, and two guiding layers. This structure forms a waveguide channel that directs the optical radiation toward the edge of the device where it can be coupled to an optical fiber core. The emission pattern of the edge emitter is more directional than that of the surface emitter, as is illustrated in Fig. 4.7. In the plane parallel to the junction, where there is no waveguide effect, the emitted beam is lambertian with a half-power beam width of $\theta_\parallel = 120°$. In the plane perpendicular to the junction, the half-power beam can be on the order $\theta_\perp \approx 25\text{–}35°$, thereby forming an elliptical output beam.

Fig. 4.8 Energy band
concept for a semiconductor

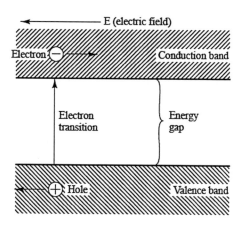

4.3.2 LED Wavelengths and Device Uses

The operation of semiconductor devices such as LEDs, laser diodes, and photodiodes can be understood by considering the *energy-band concept* shown in Fig. 4.8. In a semiconductor the valence electrons of the constituent material atoms occupy a band of energy levels called the *valence band*. This is the lowest band of allowed energy states. The next higher band of allowed energy levels for the electrons is called the *conduction band*. In a pure crystal at low temperatures, the conduction band is completely empty of electrons and the valence band is completely full. These two bands are separated by an *energy gap*, or *bandgap*, in which no energy levels exist. When electrons are energetically excited across the bandgap, a number of freely moving electrons appear in the conduction band. This process leaves an equal concentration of vacancies, or *holes*, in the valence band. Both the free electrons and the holes are mobile within the material and can contribute to electrical conductivity when an external electric field is applied. Light emission takes place when electrons drop from the conduction band and recombine with holes in the valence band.

The emission wavelengths of LEDs depend on the material composition in the pn junction because different materials have different bandgap energies, which determine the wavelength of the emitted light. Common materials include GaN, GaAs, InP, and various alloys of AlGaAs and of InGaAsP. Whereas the full-width half-maximum (FWHM) power spectral widths of LEDs in the 800-nm region are around 35 nm, the widths increase in longer-wavelength materials. For devices operating in the 1300–1600-nm region, the spectral widths vary from around 70–180 nm. Figure 4.9 shows an example for surface and edge-emitting devices with a peak wavelength at 1546 nm. The output spectral widths of surface-emitting LEDs tend to be broader than those of edge-emitting LEDs because of different internal absorption effects of the emitted light in the two device structures.

In biophotonics and medical disciplines, there is a growing use of LEDs in the UV spectrum ranging from 100 to 400 nm. Care must be exercised when using UV light because, as noted in Sect. 1.3, within the optical radiation spectrum,

Fig. 4.9 Typical spectral patterns for edge-emitting and surface-emitting LEDs at 1546 nm showing that the patterns are wider for surface emitters

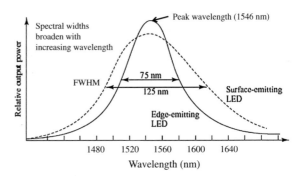

broad-spectrum ultraviolet radiation is the strongest and most damaging to living things. Ultraviolet radiation (UVR) is classified into three standard wavelength ranges identified as UV-A (315–400 nm), UV-B (280–315 nm), and UV-C (100–280 nm). Among the uses of UV radiation are the following:

- 230–400 nm: optical sensors and instrumentation
- 240–280 nm: sterilization of surface areas and water
- 250–405 nm: body fluid analysis
- 270–300 nm: protein analysis, drug discovery
- 300–320 nm: medical light therapy
- 315–400 nm: curing of dental bonding material
- 390–410 nm: cosmetic sterilization.

Uses of LEDs in the visible and near infrared spectral regions include photodynamic therapy (skin cancer and acne treatments), fluorescence spectroscopy, blood and oxygen sensors, endoscopic illumination of internal organs with simulated natural light, cosmetic treatments (e.g., wrinkle removal and hair growth), treatment of prenatal jaundice, promotion of wound healing, and optogenetic neural stimulation.

4.3.3 Modulation of an LED

A variety of biophotonics measurement techniques use pulsed or modulated light signals. These systems include optical spectroscopy, tomography, modulated Raman spectroscopy, and neuroimaging. Both LEDs and laser diodes are advantageous for such uses because of the ability to directly modulate the optical output with a time-varying electric current. The *response time* or *frequency response* of an optical source dictates how fast an electrical input drive signal can vary the light output level. The maximum possible modulation frequency depends largely on the *recombination lifetime* (also called *injection carrier lifetime*) τ_i of the electron-hole pairs in the recombination region (the active light-producing region) of the device. This time can vary widely between different LED types. If the drive current is modulated at a frequency $f = \omega/2\pi$ (where the ordinary frequency f is measured in hertz and the angular frequency ω is given in radians per second), the optical output power of the device will vary as

$$P(\omega) = P_0\left[1 + (\omega\tau_i)^2\right]^{-1/2} \qquad (4.4)$$

where P_0 is the power emitted at zero modulation frequency.

Example 4.3 A particular LED has a 5-ns injected carrier lifetime. When no modulation current is applied to the device, the optical output power is 0.250 mW for a specified dc bias. Assuming other factors that might affect the modulation speed are negligible, what are the optical outputs at modulation frequencies f of (a) 10 MHz and (b) 100 MHz?

Solution:

(a) From Eq. (4.4) the optical output at 10 MHz is

$$P(\omega) = \frac{0.250\,\mathrm{mW}}{\sqrt{1 + [2\pi(10 \times 10^6)(5 \times 10^{-9})]^2}} = 0.239\,\mathrm{mW} = 239\,\mu\mathrm{W}$$

(b) Similarly, the optical output at 100 MHz is

$$P(\omega) = \frac{0.250\,\mathrm{mW}}{\sqrt{1 + [2\pi(100 \times 10^6)(5 \times 10^{-9})]^2}} = 0.076\,\mathrm{mW} = 76\,\mu\mathrm{W}$$

Thus the output of this particular device decreases at higher modulation rates.

The modulation bandwidth of an LED can be defined in either electrical or optical terms. Normally, electrical terms are used because the bandwidth is determined via the associated electrical circuitry. Thus the *modulation bandwidth* is defined as the point where the electrical signal power, designated by $P_{\mathrm{elec}}(\omega)$, has dropped to half its constant value as the modulation frequency increases. This is the *electrical 3-dB point*; that is, the modulation bandwidth is the frequency at which the output electrical power is reduced by 3 dB with respect to the input electrical power.

In a semiconductor optical source the light output power varies linearly with the drive current. Thus the values of currents rather than voltages (which are used in electrical systems) are compared in optical systems. To find the relationship, first let $i(\omega) = i_{\mathrm{out}}$ be the output current at frequency ω and let $i(0) = i_{\mathrm{in}}$ be the input current at zero modulation. Then, because $P_{\mathrm{elec}}(\omega) = i^2(\omega)R$, with R being the electrical

Fig. 4.10 Frequency response of an optical source showing the electrical and optical 3-dB-bandwidth points

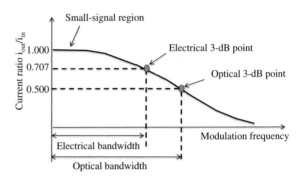

resistance, the ratio of the output electrical power at the frequency ω to the electric power at zero modulation is

$$\text{Ratio}_{\text{elec}} = 10 \log\left[\frac{P_{\text{elec}}(\omega)}{P_{\text{elec}}(0)}\right] = 10 \log\left[\frac{i^2(\omega)}{i^2(0)}\right] \tag{4.5}$$

where $i(\omega)$ is the electrical current in the detection circuitry. The electrical 3-dB point occurs at that frequency point where the detected electrical power $P_{\text{elec}}(\omega) = P_{\text{elec}}(0)/2$. This happens when

$$\frac{i^2(\omega)}{i^2(0)} = \frac{i_{\text{out}}^2}{i_{\text{in}}^2} = \frac{1}{2} \tag{4.6}$$

or at the point where $i(\omega)/i(0) = i_{\text{out}}/i_{\text{in}} = 1/\sqrt{2} = 0.707$, as is illustrated in Fig. 4.10.

Sometimes, the modulation bandwidth of an LED is given in terms of the 3-dB bandwidth of the modulated optical power $P(\omega)$; that is, it is specified at the frequency where $P(\omega) = P_0/2$ where P_0 is the optical power at zero modulation. In this case, the 3-dB bandwidth is determined from the ratio of the optical power at frequency ω to the unmodulated value of the optical power. Since the detected current is directly proportional to the optical power P, this ratio is

$$\text{Ratio}_{\text{optical}} = 10\log\left[\frac{P(\omega)}{P_0}\right] = 10\log\left[\frac{i(\omega)}{i(0)}\right] = 10\log\left[\frac{i_{\text{out}}}{i_{\text{in}}}\right] \tag{4.7}$$

The *optical 3-dB point* occurs at that frequency where the ratio of the currents is equal to 1/2. As shown in Fig. 4.10, this corresponds to an electrical power attenuation of 6 dB.

Example 4.4 Consider the particular LED described in Example 4.3, which has a 5-ns injected carrier lifetime.

(a) What is the 3-dB optical bandwidth of this device?

(b) What is the 3-dB electrical bandwidth of this device?

Solution:

(a) The 3-dB optical bandwidth occurs at the modulation frequency for which $P(\omega) = 0.5P_0$. Using Eq. (4.4) this condition happens when

$$\frac{1}{\sqrt{1 + (\omega\tau_i)^2}} = 0.5$$

so that $1 + (\omega\tau_i)^2 = 4$ or $\omega\tau_i = \sqrt{3}$. Solving this expression for the frequency $f = \omega/2\pi$ yields

$$f = \frac{\sqrt{3}}{2\pi\tau_i} = \frac{\sqrt{3}}{2\pi(5 \times 10^{-9})} = 55.1 \, \text{MHz}$$

(b) The 3-dB electrical bandwidth is $f/\sqrt{2} = (55.1 \, \text{MHz}) \times 0.707 = 39.0 \, \text{MHz}$.

4.4 Lasers for Biophotonics

During the past several decades, various categories of lasers have made a significant impact in biophotonics [16–21]. Among the many application areas are cardiology, dentistry, dermatology, gastroenterology, gynecology, microscopy, microsurgery, neurosurgery, ophthalmology, orthopedics, otolaryngology, spectroscopy, and urology. The advantages of lasers over other light sources are the following:

- Lasers can have monochromatic (single-wavelength) outputs, so that the device wavelengths can be selected to match the absorption band of the material to be analyzed
- The output beam can be highly collimated, which means that the laser light can be directed precisely to a specific spot
- Depending on the laser, the outputs can range from mW to kW
- Short pulse durations are possible (e.g., a few femtoseconds) for use in applications such as fluorescence spectroscopy where the pulse width needs to be shorter than the desired time-resolution measurement.

This section first explains the basic principles of laser construction and operation and then gives examples of laser diodes, solid-state lasers, gas lasers, and optical fiber lasers.

Fig. 4.11 Pumping of
electrons to higher energy
levels to produce a population
inversion and the ensuing
light emission

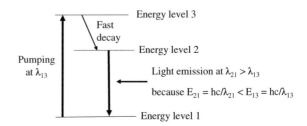

4.4.1 Basic Laser Actions

A lasing medium can be a gas, a liquid, a solid-state crystal, a semiconductor, or an optical fiber. Despite their material and physical differences, the basic principle of operation is the same for each type of laser. Laser action is the result of three key processes: boosting electrons (or molecules) to a higher energy level by a method such as photon absorption, a fast decay process of the excited electron to a slightly lower metastable energy level, and stimulated emission. The simple energy-level diagram in Fig. 4.11 represents these three processes. Here E_1 is the ground-state energy and E_2 and E_3 are two excited-state energies. According to Planck's law as described in Sect. 2.3, a transition of an electron between the ground state and an excited state can occur through the absorption of a photon. Vice versa, the transition from an excited state to a lower state results in the emission of a photon.

For lasing action the lasing material must first absorb energy from some external source. The external energy raises electrons in the material atoms from a ground-state energy level E_1 to a higher energy level E_3. Because E_3 is an unstable state, through a fast decay transition the electrons drop quickly (in the order of fs) to a lower metastable energy level E_2. This process thus can produce a *population inversion* between levels 1 and 2, which means that the electron population of the excited states is greater than in the ground state. The process of raising electrons to higher energy levels is called *pumping*. Pumping can be done either electronically or optically.

From the E_2 state the electron can return to the ground state by emitting a photon of energy $h\nu_{12} = E_2 - E_1$. This process can occur without any external stimulation and is called *spontaneous emission*. The electron also can be induced to make a downward transition from the excited level to the ground-state level by an external stimulation such as an incoming photon. If a photon of energy $h\nu_{12}$ impinges on the system while the electron is still in its excited E_2 state, the electron is immediately stimulated to drop to the ground state and gives off a photon of energy $h\nu_{12} = E_2 - E_1$. This emitted photon is in phase with the incident photon, and the resultant emission is known as *stimulated emission*. Now the incident and emitted photons together can induce further stimulated emissions, which can lead to a lasing action.

Fig. 4.12 Schematic of a simple laser design and the allowed lasing modes

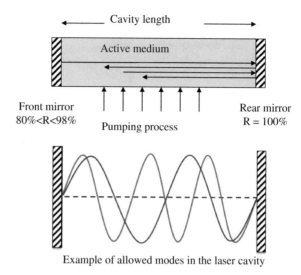

Example of allowed modes in the laser cavity

A simple laser design consists of the following:

- An *active medium* (or *gain medium*) in which atoms or molecules can be excited to higher energies
- An electrical or optical energy *pump* source
- Two reflectors (high-reflection *mirrors*)

The active medium and the two mirrors form a *resonator* or a *laser cavity*, as shown in Fig. 4.12. The lasing action takes place when light builds up in amplitude as it is reflected by the end mirrors and travels back and forth in the cavity. The *cavity length* determines the *emitted light wavelength*. Usually the cavity length is fixed so that the laser emits at a number of specific wavelengths that satisfy the condition of standing wave patterns as shown on the bottom of Fig. 4.12. Each of these standing waves forms a *lasing mode* of the cavity, which gives a specific wavelength. In some devices the cavity length can be changed. This structure results in a *tunable laser*, which emits at a selection of different wavelengths depending on the cavity length.

Consider the case when the lasing medium is a semiconductor material, which is the basis of a laser diode. In this case the pumping is carried out by electron injection by means of an external bias current. The relationship between optical output power and laser diode drive current is shown in Fig. 4.13. At low diode currents only spontaneous radiation is emitted. Both the spectral range and the beam width of this emission are broad like that from an LED. A dramatic and sharply defined increase in the optical power output occurs at a current level I_{th} known as the *lasing threshold*. As this transition point is approached from lower drive current values, the spectral range and the beam width of the emitted light both become significantly narrower with increasing drive current. The emission is then a

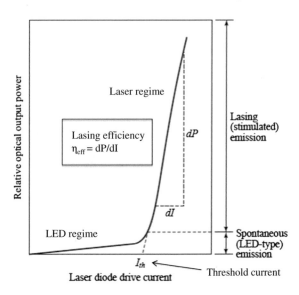

Fig. 4.13 Relationship between optical output power and laser diode drive current

lasing condition that is dominated by stimulated emission. The lasing efficiency is given by

$$\eta_{\text{eff}} = \frac{dP}{dI} \tag{4.8}$$

where dP is the change in optical output for a change dI in the diode drive current.

Depending on the device material and structure, the light from a laser can be a *continuous wave* (CW) output or the light can be modulated or pulsed. Very short-duration high-energy pulses can be created with either a solid-state Q-switched laser or a mode-locked laser, which can be a laser diode, a solid-state laser, or a fiber laser. The quality (or Q) of the laser cavity can be controlled in a *Q-switched laser*. For example, an optical element can change the optical transmission gain of the cavity from very low to very high. This feature acts like a shutter to allow light to leave the device or not. Pulse durations as low as nanoseconds (10^{-9} s) are possible with a Q-switched laser. In a *mode-locked laser* the different longitudinal modes (running along the cavity) are *locked in phase*. This process can produce a series of very short pulses with pulse durations ranging from picoseconds (10^{-12} s) to femtoseconds (10^{-15} s).

4.4.2   Laser Diodes

Various categories of semiconductor laser diodes are available for biophotonics uses with wavelengths commonly ranging from 375 nm to about 3.0 µm, with some devices having emission wavelengths up to 10 µm. The advantages of laser

Fig. 4.14 A digital input in the lasing region produces a pulsed optical output

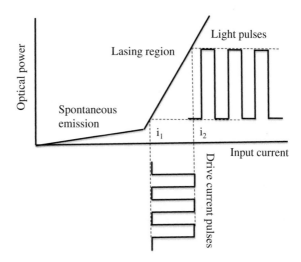

diode modules include small size and weight (e.g., 1 mm across and weighing a few grams), low electrical power requirements of typically a few mW, and high coherent optical output up to tens of mW. The optical output of laser diodes can be varied directly with electrical pumping, which can be an electrical information-carrying signal. As is shown in Fig. 4.14 for a digital signal, this operation is carried out in the linear lasing region by switching between two drive currents i_1 and i_2 and is known as *direct modulation*. Up to a certain device-dependent modulation limit the optical output power varies directly with an electrical drive signal. This limit is reached for pulse rates around 2.5 Gb/s. External modulation techniques are needed for data rates beyond this limit [13].

Two basic laser diode types are the Fabry-Perot laser and the double-heterostructure configuration. In a *Fabry-Perot laser* the light is generated within a semiconductor lasing cavity such as that shown in Fig. 4.12, which is known as a *Fabry-Perot cavity*. The mirror facets are constructed by making two parallel clefts along natural cleavage planes of the semiconductor crystal. The purpose of the mirrors is to establish a strong optical feedback in the longitudinal direction. This feedback mechanism converts the device into an oscillator (and hence a light emitter) with a gain mechanism that compensates for optical losses in the cavity at certain resonant optical frequencies. Each resonant frequency results in optical emission at a certain wavelength, which yields the multimode Fabry-Perot emission pattern shown in Fig. 4.15. The output has a Gaussian shaped envelope. The spacing between the modes is given by

$$\Delta\lambda = \frac{\lambda^2}{2nL} \tag{4.9}$$

Fig. 4.15 Typical emission spectrum from a Fabry-Perot laser diode

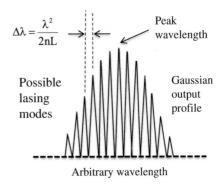

$$\Delta\lambda = \frac{\lambda^2}{2nL}$$

Peak wavelength

Possible lasing modes

Gaussian output profile

Arbitrary wavelength

where n is the refractive index of the lasing cavity and L is the cavity length. This can be related to the frequency spacing Δv through the relationship $\Delta v/v = \Delta\lambda/\lambda$ to yield

$$\Delta v = \frac{c}{2nL} \tag{4.10}$$

Example 4.5 A GaAs Fabry-Perot laser operating at 850 nm has a 500-µm length and a refractive index n = 3.7. What are the frequency spacing and the wavelength spacing?

Solution: From Eq. (4.10) the frequency spacing is

$$\Delta v = \frac{3 \times 10^8 \, \text{m/s}}{2(3.7)(500 \times 10^{-6} \, \text{m})} = 81 \, \text{GHz}$$

From Eq. (4.9) the wavelength spacing is

$$\Delta\lambda = \frac{(850 \times 10^{-9} \, \text{m})^2}{2(3.7)(500 \times 10^{-6} \, \text{m})} = 0.195 \, \text{nm}$$

In order to create a single-mode optical output, one of a variety of double-heterostructure configurations can be used. Two popular versions are the *distributed feedback laser* or *DFB laser* and the *vertical cavity surface-emitting laser* (VCSEL), which are similar to the LED structures shown in Figs. 4.7 and 4.6, respectively. The DFB laser is used widely by the telecom industry and has a high degree of reliability. The laser materials consist of compound semiconductor materials such as GaAlAs or InGaAsP alloys with emission wavelengths ranging from 760 to 2400 nm depending on the particular alloy used. A typical single-mode DFB output is shown in Fig. 4.16 for a 1557.3 nm peak wavelength.

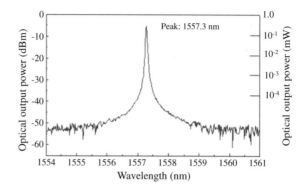

Fig. 4.16 Typical DFB laser output spectrum (power shown in mW and dBm)

The VCSEL is a single-mode laser, which is made mainly for emissions in the 750–980-nm range [21]. A common peak wavelength is 850 nm from a GaAs/AlGaAs material. VCSEL devices based on InP alloys are available for operation at 1300, 1550, and 2360 nm. In a VCSEL the emission occurs perpendicular to the active layer. This structure allows a high density of emitters to be produced over a small area, and thus enables VCSELs to be configured as closely spaced arrays of lasers. Individual and arrays of VCSELs have been applied to areas such as absorption spectroscopy, brain imaging, and cell sorting in microfluidic channels.

A fourth class of semiconductor laser diodes is the *quantum cascade laser* (QCL) that emits in the mid- to far-infrared spectrum [22, 23]. The devices can cover the biophotonics spectral range of 2.75 μm to beyond 10 μm. The QCL devices are available in either a broadband Fabry-Perot structure or as a single-wavelength DFB laser. The QCL operates in a continuous-wave (CW) mode and can have optical outputs up to 500 mW.

4.4.3   Solid-State Lasers

Solid-state lasers use lasing media such as crystals or glasses that are doped with rare earth ions or transition metal ions [24–27]. To achieve the population inversion needed for a lasing condition, solid-state lasers are optically pumped with flash lamps, arc lamps, or laser diodes. The advantages of *diode-pumped solid-state lasers* (DPSS lasers) are compact designs, long lifetimes, and very good beam qualities.

The characteristics of several solid-state lasers that are commonly used in biophotonics are listed below.

- **Ruby lasers** use a synthetic ruby crystal as the gain medium to produce pulses of deep red 694.3-nm wavelength light with a linewidth of 0.53 nm. This laser is energized by optically pumping it with a xenon flashtube. Ruby lasers initially were used for tattoo and hair removal, but now usually the alexandrite and Nd: YAG lasers described below are carrying out these functions.

- **Neodymium:yttrium aluminum garnet (Nd:YAG) lasers** use a neodymium-doped yttrium aluminum garnet ($Y_3Al_5O_{12}$) synthetic crystal, which is known by the acronym YAG. The Nd:YAG laser can be pumped by a laser diode or by a lamp. The most commonly used wavelength from Nd:YAG lasers is 1064 nm. Other emission lines are at 946, 1123, 1319, 1338, 1415, and 1444 nm. In continuous operation versions the output power can be a few milliwatts for miniature setups to several watts for larger models. Pulsed versions can generate pulses of a few nanoseconds in duration with microjoule pulse energies. Applications of these lasers are in the areas of ophthalmology, oncology, urology, pulmonology, hair removal, prostate surgery, and toenail infections.
- **Erbium:YAG (Er:YAG) lasers** are erbium-doped YAG crystals that lase at 2940 nm. This wavelength is close to an absorption line of water. Thus, because water strongly absorbs the light, the laser light interaction with tissue results in evaporation and ablation, such as the smoothing or resurfacing of skin (for example, wrinkle removal). Osseous minerals also absorb this wavelength strongly, so an Er:YAG laser is advantageous for sawing and drilling in bone and teeth.
- **Holmium:YAG (Ho:YAG) lasers** are holmium-doped YAG crystals that lase at 2100 nm. Its use in ophthalmology includes correcting mild to moderate cases of farsightedness and some types of astigmatism. Other applications include tissue ablation, kidney stone removal, and intraoral soft tissue surgery.
- **Titanium sapphire lasers** (also called Ti:Al_2O_3 lasers or Ti:sapphire lasers) are tunable lasers that emit light in the 650–1100 nm range (red to near infrared). As its name implies, the lasing medium is a sapphire (Al_2O_3) crystal that is doped with titanium ions. This laser usually is pumped with another laser that emits in the 514–532-nm region, such as an argon laser. A Ti:sapphire laser can generate ultra-short pulses, which can be as short as 6 femtoseconds in duration with an average output power of 0.5–1.5 W. The pulse repetition frequency nominally is between 70 and 90 MHz. A major application is for multiphoton microscopy, which is a key noninvasive laboratory tool for studying underlying biological phenomena.
- **Alexandrite lasers** are tunable in the 700–830-nm range. The average output power can be up to 100 W. Light in the emission spectrum of these lasers is absorbed well by melanin and dyes but not by blood. Therefore it is useful for destroying tissue elements that contain melanin (e.g., moles and hair roots), for removing tattoos, and for breaking up kidney stones (lithotripsy).

4.4.4 Gas Lasers

A wide selection of gas lasers is available for many different types of commercial, hobby, educational, and medical applications. Generally a *gas laser* produces a coherent light beam when an electric current is discharged through a gas medium. The emitted light typically is a very high quality, single-mode beam with a long coherence length. Common gas lasers used in biophotonics have helium-neon mixtures, argon, krypton, or carbon dioxide gases. The gas type used in the laser determines or influences the emission wavelength, the device efficiency, and the optical output power. Gas lasers are employed in a wide range of life sciences and medical disciplines including flow cytometry, cell sorting, microscopy, DNA sequencing, retinal scanning, ophthalmology, dermatology, hair and tattoo removal, and surgical and dental procedures.

The generic characteristics of some gas lasers are given below.

- **Helium–neon (He–Ne) lasers** consist of a gas cell with a length of around 20 cm and generate output powers from about 0.8–20 mW. The nominal beam width is a few millimeters. A standard wavelength is 632.8 nm (red light), but a He–Ne laser can emit at many other wavelengths by adjusting the characteristics of the lasing cavity. Other common wavelengths are 543.5 nm (green), 594.1 nm (yellow), 611.9 nm (orange), 1152.3 nm, or 3391.2 nm. Biophotonics applications include confocal microscopy, DNA sequencing, flow cytometry, and hematology.
- **Carbon dioxide (CO_2) lasers** use a mixture of CO_2, He, and N_2 for generating laser radiation mostly at 10.6 µm. In general the coated hollow-core fibers described in Sect. 3.8 are used to deliver this wavelength of light to a specific pinpoint location. Depending on the application a CO_2 laser can operate in either a continuous or a pulsed mode. These lasers can be made to emit up to several kilowatts, but optical powers of 10–20 W are sufficient for most medical applications such as surgery, ophthalmology, and cosmetics.
- **Argon (Ar) ion lasers** need a nominally 1-m long tube and provide a single-frequency narrow light beam (1.5 mm). These lasers emit 13 spectral lines in the UV through visible region from 351 to 528-nm. The commonly used wavelengths are 488.0 and 514.5 nm (green light) with output powers of up to 20 W in the green region. Because these two wavelengths are in the visible region, the light can be transmitted through optical fibers. The emission wavelengths are absorbed strongly by numerous tissue chromophores and thus can be used for coagulation applications and for vaporization of pigmented lesions in the skin and retina.
- **Krypton (Kr) ion lasers** are similar in both structure and beam properties to Ar ion lasers. These lasers emit at 11 wavelengths, the main ones being 530.9, 568.2, and 676.4 nm with output powers up to 10 W. The applications are similar to Ar lasers.

Example 4.6 In a He–Ne gas laser an electrical discharge pumps helium atoms to a 20.61-eV excited state. This excited helium level is very close to a 20.66-eV level in neon, so that upon collision of a helium atom and a neon atom, the energy can be transferred from the He to the Ne atom. From this excited state, the neon atom then drops to a lower 18.70-eV state thereby emitting a photon. What is the wavelength of this photon?

Solution: Using the relationship given in Eq. (2.21) it follows that

$$\lambda = \frac{hc}{E} = \frac{(6.626 \times 10^{-34}\,\text{J} \cdot \text{s})(2.998 \times 10^8\,\text{m/s})}{(20.66 - 18.70\,\text{eV})(1.602 \times 10^{-19}\,\text{J/eV})} = 633\,\text{nm}$$

4.4.5 Optical Fiber Lasers

In an *optical fiber laser*, a length of optical fiber acts as the resonator lasing cavity [28–31]. These devices have a single-mode fiber core doped with erbium, ytterbium, or thulium. First energy from a solid-state optical source is coupled into the cladding of the fiber. The injected light then moves into the fiber core and pumps the dopant. Desirable medical wavelength ranges for optical fiber lasers are the 1300-nm range for imaging applications and the 1550-nm to 4-μm region for surgical use. Output powers from fiber lasers can vary from milliwatts to over 100 W. This light can be directly coupled into an optical fiber transmission line for delivery to a tissue treatment or diagnosis site.

Thulium-doped and erbium-doped fiber lasers are two commonly used devices. Thulium-doped fiber lasers emit in the 2–4-μm spectrum with output powers up to 100 W. These lasers can be used for soft tissue surgery in otolaryngology, urology (e.g., prostate treatment), ophthalmology, and cardiology. Erbium-doped fiber lasers emit in the 1530–1610-nm spectrum for applications in optical coherence tomography, microsurgery, and skin resurfacing.

4.5 Superluminescent Diodes

In applications such as optical coherence tomography, it is desirable to have a light source that emits over a broad spectral band. One such device is a *superluminescent diode* (SLD). This device is an edge-emitting semiconductor light source that has the high output power and intensity features of a laser diode and the low coherence characteristics of a light-emitting diode. Typical spectral emission widths range from 5 nm to over 100 nm with center wavelengths that can be selected in the general range of 830–1550 nm. Optical output powers can vary from a few milliwatts to several tens of milliwatts.

4.6 Summary

The biophotonics field covers a wide spectrum ranging from about 190 nm in the UV to over 10 μm in the infrared. The numerous diverse biomedical and life sciences applications that are used within this spectral range necessitate implementation of a broad selection of light sources with different sizes, shapes, operational configurations, light output powers, and modulation capabilities. The source types described in this chapter include the following:

- Arc lamps create an intense light when an electrical discharge passes through a high-pressure gas or a vapor such as Hg, Ar, Xe, or Ne.
- Light-emitting diodes are highly reliable devices that can be used in the UV, visible, and near infrared spectral regions depending on the material.
- Excimer lasers contain a mixture of a noble gas (such as Ar, Kr, or Xe) and a reactive gas (such as fluorine or chlorine) and emit in the UV range from 193 to 350 nm.
- Gas lasers produce a coherent light beam when an electric current is discharged through a gas medium (He–Ne, Ar, Kr, or CO_2). Gas lasers are employed in a wide range of life sciences and medical disciplines.
- Solid-state crystal lasers use lasing media such as crystals or glasses that are doped with rare earth ions or transition metal ions. The advantages of these devices are compact designs, long lifetimes, and very good beam qualities.
- Semiconductor lasers are highly reliable devices that include Fabry-Perot lasers, DFB lasers, VCSELs, and quantum cascade lasers. Depending on the device material, they are used in all segments of the biophotonics spectrum.
- Optical fiber lasers are being implemented in the 1300-nm range for imaging applications and in the 1550-nm to 4-μm region for surgical use.

4.7 Problems

4.1 Consider a laser that emits a highly collimated beam of light. Suppose the beam diameter is 2 mm and let the power level or radiant flux, be 100 mW. Neglecting any divergence of the beam, show that the irradiance is 3.18 W/cm^2.

4.2 For a certain laser diode consider a modified lambertian approximation to the emission pattern of the form $I(\theta) = I_0 \cos^m \theta$. Suppose that for this laser diode the half-power level occurs at an angle of 15° from the normal to the emitting surface. Show that the value of m is 20.

4.3 Analogous to Fig. 4.3, write a MatLab program to plot and compare the emission patterns from a lambertian source and a source with an emission pattern given by $I(\theta) = I_0 \cos^3 \theta$. Assume both sources have the same peak intensity I_0, which is normalized to unity in each case.

4.4   Excimer lasers generally contain a mixture of a noble gas (such as argon, krypton, or xenon) and a reactive gas (such as fluorine or chlorine) and emit in the UV range from 193 to 350 nm. Using Web resources, find the possible pulse repetition rates and output powers for the five excimer lasers listed in Table 4.1. Describe the uses of such lasers for eye surgery.

4.5   Using Web resources, find some characteristics of alexandrite and Nd:YAG lasers for removal of unwanted hair. Consider different physical and operational factors such as irradiance, wavelength, pulse duration, and spot size diameter.

4.6   An important parameter of a GaAs light source when coupling it to an optical fiber is the value of its refractive index as a function of wavelength. For the range $\lambda = 0.89\text{--}4.1$ μm the refractive index is given by

$$n^2 = 7.10 + \frac{3.78\lambda^2}{\lambda^2 - 0.2767}$$

where λ is given in micrometers. Compare the refractive indices of GaAs at 810 and 900 nm. [Answer: n = 3.69 at 810 nm and 3.58 at 900 nm.]

4.7   By differentiating the expression $E = hc/\lambda$, show why the full-width half-maximum (FWHM) power spectral width of LEDs becomes wider at longer wavelengths.

4.8   A particular LED has a 5-ns injected carrier lifetime. When no modulation current is applied to the device, the optical output power is 0.250 mW for a specified dc bias. Assuming other factors that might affect the modulation speed are negligible, show that the optical outputs at modulation frequencies f of (a) 25 MHz is 197 μW and (b) 150 MHz is 52 μW.

4.9   Consider a particular LED that has a 4.5-ns injected carrier lifetime. (a) Show that the 3-dB optical bandwidth of this device is 61.2 MHz (b) Show that the 3-dB electrical bandwidth of this device is 43.3 MHz.

4.10 A GaAs Fabry-Perot laser operating at 860 nm has a 550-μm length and a refractive index n = 3.7. Show that the frequency spacing and the wavelength spacing are 73.7 GHz and 0.182 nm, respectively.

4.11 In order for a mode to be supported in a lasing cavity, the condition $N = 2L/\lambda$ must be satisfied, where L is the cavity length, λ is the wavelength of the allowed mode, and N is an integer. Suppose a lasing cavity has 0.400-mm length. Show that 1500 nm is not an allowed wavelength, but that 1498.13 and 1500.94 nm correspond to lasing modes.

4.12 Assume that the cleaved mirror end faces of a GaAs laser are uncoated and that the outside medium is air, which has a refractive index of 1.00. If the refractive index of GaAs is 3.60, show that the reflectance for light incident normal to the air-laser end face is 0.32. Recall the expression for reflectance given in Eq. (2.35).

4.13 A Nd:YAG laser can be pumped by a laser diode or by a lamp to several different energy levels. One possibility is to use a wavelength of 808.7 nm to raise the energy level to 1.534 eV. From there the excited state quickly drops to an energy level of about 1.430 eV. From this slightly lower excited state a common transition is to an energy level of 0.264 eV. Show that the wavelength of the emitted photon for this particular transition is 1064 nm.

4.14 Thulium optical fiber lasers are of growing interest to various medical applications, such as laser surgery and photo dermatology. (a) Using resources from the Web, list some characteristics of these lasers, such as emission wavelengths, optical power output, modulation capabilities, and beam characteristics. (b) In a few sentences, describe two medical applications of thulium fiber lasers.

References

1. E.F. Zalewski, Radiometry and photometry, chap. 34, in *Handbook of Optics: Design, Fabrication and Testing, Sources and Detectors, Radiometry and Photometry*, eds. by M. Bass, C. DeCusatis, J. Enoch, V. Lakshminaravanan, G. Li, C. MacDonald, V. Mahajan, E. Van Stryland, vol. 2, 3rd edn. (McGraw-Hill, New York, 2010)
2. I. Moreno, LED intensity distribution, paper TuD6, in *OSA International Optical Design Conference*, Vancouver, CA, June 2006
3. J.M. Palmer, Radiometry and photometry: units and conversions, chap. 7, in *Handbook of Optics: Classical Optics, Vision Optics, X-Ray Optics*, eds. by M. Bass, J. Enoch, E. Van Stryland, W.L. Wolfe, vol. 3, 2nd edn. (McGraw-Hill, New York, 2000)
4. D. Nakar, A. Malul, D. Feuermann, J.M. Gordon, Radiometric characterization of ultrahigh radiance xenon short-arc discharge lamps. Appl. Opt. **47**(2), 224–229 (2008)
5. Newport Corp., *DC-Arc-Lamps*. www.newport.com, July 2015
6. Hamamatsu Photonics, *Super-Quiet Xenon Flash Lamp Series*. www.hamamatsu.com, Sept 2013
7. W. Henry, MicroLEDs enabling new generation of fluorescence instruments. Biophotonics **20** (3), 25–28 (2013)
8. M.-H. Chang, D. Das, P.V. Varde, M. Pecht, Light emitting diodes reliability review. Microelecton. Reliab. **52**, 762–782 (2012)
9. W.D. van Driel, X.J. Fan, eds., *Solid State Lighting Reliability: Components to Systems* (Springer, Berlin, 2013)
10. R.-H. Horng, S.-H. Chuang, C.-H. Tien, S.-C. Lin, D.-S. Wuu, High performance GaN-based flip-chip LEDs with different electrode patterns, Optics Express **22**(S3), A941–A946 (2014)
11. D.A. Neaman, *Semiconductor Physics and Devices*, 4th edn. (McGraw-Hill, New York, 2012)
12. S.O. Kasap, *Optoelectronics and Photonics: Principles and Practices*, 2nd edn. (Pearson, Upper Saddle River, NJ, 2013)
13. G. Keiser, *Optical Fiber Communications*, 4th US edn., 2011; 5th international edn., 2015 (McGraw-Hill, New York)
14. H. Ries, J. Muschaweck, Tailored freeform optical surfaces. J. Opt. Soc. America A **19**, 590–595 (2002)
15. L.-T. Chen, G. Keiser, Y.-R. Huang, S.-L. Lee, A simple design approach of a Fresnel lens for creating uniform light-emitting diode light distribution patterns. Fiber Integ. Optics **33**(5–6), 360–382 (2014)
16. Q. Peng, A. Juzeniene, J. Chen, L.O. Svaasand, T. Warloe, K.-E. Giercksky, J. Moan, Lasers in medicine, Rep. Prog. Phys. **71**, 056701 (2008)
17. B. Kemper, G. von Bally, *Coherent laser measurement techniques for medical diagnostics*, chap. 9, in *Biophotonics*, eds. by L. Pavesi, P.M. Fauchet (Springer, Berlin, 2008)

18. R. Riesenberg, A. Wutting, Optical sources, chap. 4, pp. 263–295, in *Handbook of Biophotonics*, eds. by J. Popp, V.V. Tuchin, A. Chiou, S.H. Heinemann, vol. 1: Basics and Techniques (Wiley, New York, 2011)
19. A. Müller, S. Marschall, O.B. Jensen, J. Fricke, H. Wenzel, B. Sumpf, P.E. Andersen, Diode laser based light sources for biomedical applications. Laser Photonics Rev. **7**(5), 605–627 (2013)
20. H. Tu, S.A. Boppart, Coherent fiber supercontinuum for biophotonics. Laser Photonics Rev. **7** (5), 628–645 (2013)
21. R. Michalzik, *VCSELs: Fundamentals, Technology and Applications of Vertical-Cavity Surface-Emitting Lasers* (Springer, Berlin, 2013)
22. M. Razeghi, High-performance InP-based mid-IR quantum cascade lasers. IEEE J. Sel. Topics Quantum Electron. **15**(3), 941–951 (2009)
23. S. Liakat, K.A. Bors, L. Xu, C.M. Woods, J. Doyle, C.F. Gmachi, Noninvasive in vivo glucose sensing on human subjects using mid-infrared light. Biomed. Opt. Express **5**, 2397–2404 (2014)
24. T. Watanabe, K. Iwai, T. Katagiri, Y. Matsuura, Synchronous radiation with Er:YAG and Ho:YAG lasers for efficient ablation of hard tissues. Biomed. Opt. Express **1**, 337–346 (2010)
25. J. Kozub, B. Ivanov, A. Jayasinghe, R. Prasad, J. Shen, M. Klosner, D. Heller, M. Mendenhall, D.W. Piston, K. Joos, M.S. Hutson, Raman-shifted alexandrite laser for soft tissue ablation in the 6- to 7-µm wavelength range. Biomed. Opt. Express **2**, 1275–1281 (2011)
26. G. Deka, K. Okano, F.-J. Kao, Dynamic photopatterning of cells *in situ* by Q-switched neodymium-doped yttrium ortho-vanadate laser, J. Biomed. Optics **19**, 011012 (2014)
27. K. Baek, W. Deibel, D. Marinov, M. Griessen, M. Dard, A. Bruno, H.-F. Zeilhofer, P. Cattin, P. Juergens, A comparative investigation of bone surface after cutting with mechanical tools and Er:YAG laser. Lasers Surg. Med. **47**, 426–432 (2015)
28. P.F. Moulton, G.A. Rines, E.V. Slobodtchikov, K.F. Wall, G. Frith, B. Samson, A.L.G. Carter, Tm-doped fiber lasers: fundamentals and power scaling. IEEE J. Sel. Topics Quantum Electron. **15**, 85–92 (2009)
29. H. Ahmad, A.Z. Zulkifli, K. Thambiratnam, S.W. Harun, 2.0-µm Q-switched thulium-doped fiber laser with graphene oxide saturable absorber. IEEE Photonics J. **5**(4), 1501108 (2013)
30. W. Shi, Q. Fang, X. Zhu, R.A. Norwood, N. Peyghambarian, Fiber lasers and their applications. Appl. Opt. **53**(28), 6554–6568 (2014)
31. L.J. Mortensen, C. Alt, R. Turcotte, M. Masek, T.-M. Liu, D.C. Cote, C. Xu, G. Intini, C.P. Lin, Femtosecond bone ablation with a high repetition rate fiber laser source. Biomed. Opt. Express **6**, 32–42 (2015)

Chapter 5
Fundamentals of Optical Detectors

Abstract The photodetection devices used in biophotonics disciplines are semiconductor-based *pin* and avalanche photodiodes, photomultiplier tubes, and optical detector arrays. The photodetectors can be either single-channel elements or multichannel devices. With a single-channel element only one spectral channel in a biophotonics setup can be monitored at any instance in time. However, multichannel devices can measure multiple spectral channels simultaneously or observe spatial channels individually in different time sequences. Associated with photodetection setups is the need for optical filters, optical couplers, and optical circulators. Optical filters selectively transmit light in one or more specific bands of wavelengths. Optical couplers split or combine two or more light streams, tap off a small portion of optical power for monitoring purposes, or transfer a selective range of optical power. An optical circulator is a non-reciprocal multi-port device that directs light sequentially from port to port in only one direction.

A photodetection device is an essential and crucial element in most biophotonics disciplines that involve light sensing and analysis. Among the photodetection devices covered in this chapter are semiconductor-based *pin* and avalanche photodiodes, photomultiplier tubes, and detector arrays based on complementary metal-oxide-semiconductor (CMOS) and charge-coupled device (CCD) technologies. The photodetectors can be either single-channel elements or multichannel devices consisting of one-dimensional rows or two-dimensional arrays of detectors. With a single-channel element only one spectral channel can be monitored at any instance in time. However, multichannel devices can measure multiple spectral channels or examine spatial channels individually in different time sequences. Multichannel devices enable the simultaneous monitoring of a wide spectrum of multiple individual wavelengths or narrow spectral channels.

© Springer Science+Business Media Singapore 2016
G. Keiser, *Biophotonics*, Graduate Texts in Physics,
DOI 10.1007/978-981-10-0945-7_5

5.1 The *pin* Photodetector

A common semiconductor photodetector is the *pin photodiode*, shown schematically in Fig. 5.1 [1–8]. The device structure consists of a *pn junction*, which is the interface between p-doped and n-doped materials within a continuous crystal. As a result of free electron diffusion across the pn junction, an insulating zone or *intrinsic region* (*i region*) that contains no free *charge carriers* (electrons or holes) is built up between the p and n regions. This i region results because free electrons from the n region will diffuse across the pn junction and be captured by the holes in the p region, and similarly free holes from the p region will cross the junction interface and be captured by electrons in the n region. In normal operation a sufficiently large reverse-bias voltage is applied across the device through a load resistor R_L so that the intrinsic region is fully depleted of carriers. That is, the intrinsic n and p carrier concentrations are negligibly small in comparison with the impurity concentration in this region. Thus the i region is referred to as the *depletion region*.

As a photon flux penetrates into a photodiode, it will be absorbed as it progresses through the material. Suppose P_{in} is the optical power level falling on the photodetector at the device surface where $x = 0$ and let $P(x)$ be the power level at a distance x into the material. Then the incremental change $dP(x)$ in the optical power level as this photon flux passes through an incremental distance dx in the semiconductor is given by $dP(x) = -\alpha_s(\lambda)P(x)dx$, where $\alpha_s(\lambda)$ is the *photon absorption coefficient* at a wavelength λ. Integrating this relationship gives the power level at a distance x into the material as

$$P(x) = P_{in}\exp(-\alpha_s x) \tag{5.1}$$

The lower part of Fig. 5.1 gives an example of the power level as a function of the penetration depth into the intrinsic region, which has a width w. The width of the p region typically is very thin so that little radiation is absorbed there.

Example 5.1 Consider an InGaAs photodiode in which the absorption coefficient is 0.8 μm^{-1} at 1550 nm. What is the penetration depth x at which $P(x)/P_{in} = 1/e = 0.368$?

Solution: From Eq. (5.1) the power ratio is

$$\frac{P(x)}{P_{in}} = \exp(-\alpha_s x) = \exp[(-0.8)x] = 0.368$$

Therefore

$$-0.8x = \ln 0.368 = -0.9997$$

which yields x = 1.25 μm.

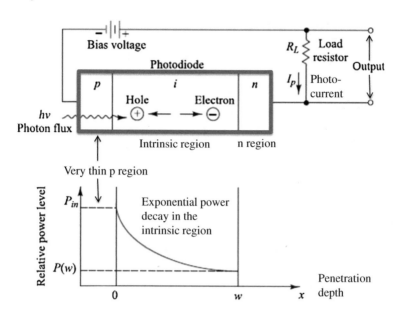

Fig. 5.1 *Top* Representation of a *pin* photodiode circuit with an applied reverse bias; *Bottom* An incident optical power level decays exponentially inside the device

Example 5.2 A high-speed InGaAs *pin* photodetector is made with a 0.15-μm thick depletion region. What percent of incident photons are absorbed in this photodetector at 1310 nm if the absorption coefficient is 1.5 μm^{-1} at this wavelength?

Solution: From Eq. (4.1), the optical power level at x = 0.15 μm relative to the incident power level is

$$\frac{P(0.15)}{P_{in}} = \exp(-\alpha_s x) = \exp[(-1.5)0.15] = 0.80$$

Therefore only 20 % of the incident photons are absorbed.

When the energy of an incident photon is greater than or equal to the bandgap energy E_g of the semiconductor material, the photon can give up its energy and excite an electron from the valence band to the conduction band. This absorption process generates mobile electron–hole pairs, as Fig. 5.2 shows. These electrons and holes are known as *photocarriers*, since they are photon-generated charge carriers that are available to produce a current flow when a bias voltage is applied across the device. The concentration level of impurity elements added to the material controls the number of charge carriers. Normally these carriers are generated mainly in the depletion region where most of the incident light is absorbed.

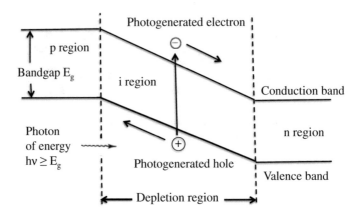

Fig. 5.2 Simple energy-band diagram for a *pin* photodiode. Photons with energies greater than or equal to E_g can generate free electron–hole pairs that become photocurrent carriers

The high electric field present in the depletion region causes the carriers to separate and be collected across the reverse-biased junction. This gives rise to a current flow in an external circuit, with one electron flowing for every carrier pair generated. This current flow is known as the *photocurrent*.

As the charge carriers flow through the material, some electron–hole pairs will recombine before they can be collected by the external circuit and hence will disappear. On the average, the charge carriers move a distance L_n or L_p for electrons and holes, respectively. This distance is known as the *diffusion length*. The time it takes for an electron or hole to recombine is known as the *carrier lifetime* and is represented by τ_n and τ_p, respectively. The lifetimes and the diffusion lengths are related by the expressions

$$L_n = (D_n \tau_n)^{1/2} \tag{5.2}$$

and

$$L_p = (D_p \tau_p)^{1/2} \tag{5.3}$$

where D_n and D_p are the *electron diffusion coefficient* and *hole diffusion coefficient*, respectively, which are expressed in units of centimeters squared per second.

The dependence of the optical absorption coefficient on wavelength is shown in Fig. 5.3 for several photodiode materials. As the curves show, α_s depends strongly on the wavelength. Thus a particular semiconductor material can be used only over a limited wavelength range. The upper wavelength cutoff λ_c is determined by the bandgap energy E_g of the material. If E_g is expressed in units of electron volts (eV), then λ_c is given in units of micrometers (μm) by

Fig. 5.3 Optical absorption coefficient as a function of wavelength of several different photodetector materials

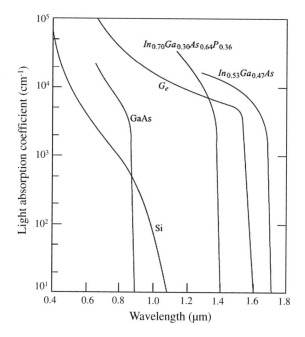

$$\lambda_c(\mu m) = \frac{hc}{E_g} = \frac{1.240}{E_g(eV)} \tag{5.4}$$

The cutoff wavelength is about 1.06 μm for Si and 1.6 μm for Ge. For longer wavelengths the photon energy is not sufficient to excite an electron from the valence to the conduction band in these two materials. Thus other semiconductor alloys have been developed for long wavelength operation.

Example 5.3 Consider a GaAs photodiode that has a bandgap energy of 1.43 eV at 300 K. What is the cutoff wavelength of this device?

Solution: From Eq. (5.2) the long-wavelength cutoff is

$$\lambda_c = \frac{hc}{E_g} = \frac{(6.625 \times 10^{-34}\,\text{J} \cdot \text{s})(3 \times 10^8\,\text{m/s})}{(1.43\,\text{eV})(1.6 \times 10^{-19}\,\text{J/eV})} = 869\,\text{nm}$$

This GaAs photodiode will not respond to photons that have a wavelength greater than 869 nm.

At the lower-wavelength end, the photon response cuts off as a result of the very large values of α_s at the shorter wavelengths. In this case, the photons are absorbed very close to the photodetector surface, where the recombination time of the generated electron–hole pairs is very short. The generated carriers thus recombine before they can be collected by the photodetector circuitry.

If the depletion region has a width w, then, from Eq. (5.1), the total power absorbed in the distance w is

$$P_{absorbed}(w) = \int_0^w \alpha_s P_{in} \exp(-\alpha_s x) dx = P_{in}(1 - e^{-\alpha_s w}) \qquad (5.5)$$

When taking into account a reflectivity R_f at the entrance face of the photodiode, then the *primary photocurrent* i_p resulting from the power absorption of Eq. (5.5) is given by

$$i_p = \frac{q}{h\nu} P_{in}(1 - e^{-\alpha_s w})(1 - R_f) \qquad (5.6)$$

where P_{in} is the optical power incident on the photodetector, q is the electron charge, and hv is the photon energy.

Two important characteristics of a photodetector are its quantum efficiency and its response speed. These parameters depend on the material bandgap, the operating wavelength, and the doping and thicknesses of the p, i, and n regions of the device. The *quantum efficiency* η is the number of the electron–hole carrier pairs generated per incident–absorbed photon of energy hv and is given by

$$\eta = \frac{\text{number of electron–hole pairs generated}}{\text{number of incident absorbed photons}} = \frac{i_p/q}{P_{in}/h\nu} \qquad (5.7)$$

Here, i_p is the photocurrent generated by a steady-state optical power P_{in} incident on the photodetector.

In a practical photodiode, 100 photons will create between 30 and 95 electron–hole pairs, thus giving a detector quantum efficiency ranging from 30 to 95 %. To achieve a high quantum efficiency, the depletion layer must be thick enough to permit a large fraction of the incident light to be absorbed. However, the thicker the depletion layer, the longer it takes for the photon-generated carriers to drift across the reverse-biased junction. Because the carrier drift time determines the response speed of the photodiode, a compromise has to be made between response speed and quantum efficiency.

Example 5.4 Consider the case when in a 100-ns pulse there are 6×10^6 photons at a 1300-nm wavelength that fall on an InGaAs *pin* photodetector. If an average of 5.4×10^6 electron–hole pairs are generated, what is the quantum efficiency?

Solution: From Eq. (5.7) the long-wavelength cutoff is

$$\eta = \frac{\text{number of electron} - \text{hole pairs generated}}{\text{number of incident absorbed photons}} = \frac{5.4 \times 10^6}{6 \times 10^6} = 0.90$$

Thus for this device the quantum efficiency at 1300 nm is 90 %.

The performance of a photodiode is often characterized by the *responsivity* R. This is related to the quantum efficiency by

$$R = \frac{i_p}{P_{in}} = \frac{\eta q}{h\nu} \tag{5.8}$$

The responsivity parameter is quite useful because it specifies the photocurrent generated per unit of optical power.

Example 5.5 Consider the case in which photons of energy 1.53×10^{-19} J are incident on a photodiode that has a responsivity of 0.65 A/W. If the input optical power level is 10 µW, what is the photocurrent?

Solution: From Eq. (5.8) the photocurrent generated is

$$i_p = RP_{in} = (0.65\,\text{A}/\text{W})(10\,\mu\text{W}) = 6.5\,\mu\text{A}$$

Typical *pin* photodiode responsivity values as a function of wavelength are shown in Fig. 5.4. Representative values are 0.65 A/W for silicon at 900 nm and 0.45 A/W for germanium at 1.3 µm. For InGaAs, typical values are 0.9 A/W at 1.3 µm and 1.0 A/W at 1.55 µm.

In most photodiodes the quantum efficiency is independent of the power level falling on the detector at a given photon energy. Thus the responsivity is a linear function of the optical power. That is, the photocurrent i_p is directly proportional to the optical power P_{in} incident upon the photodetector, so that the responsivity is constant at a given wavelength (a given value of energy hν). Note, however, that the quantum efficiency is not a constant at all wavelengths because it varies according to the photon energy. Consequently, the responsivity is a function of the wavelength and of the photodiode material (since different materials have different bandgap energies). For a given material, as the wavelength of the incident photon becomes longer, the photon energy becomes less than that required to excite an electron from the valence band to the conduction band. The responsivity thus falls off rapidly beyond the cutoff wavelength, as can be seen in Fig. 5.4.

Fig. 5.4 Comparison of the responsivity and quantum efficiency as a function of wavelength for *pin* photodiodes constructed of three different materials

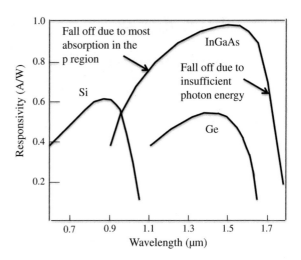

Example 5.6

(a) The quantum efficiency for InGaAs is around 90 % for the wavelength range 1300 nm $< \lambda <$ 1600 nm. Show that in this wavelength range the responsivity $R = 7.25 \times 10^5\,\lambda$.

(b) Show that at 1300 nm the responsivity $R = 0.92$ A/W.

(c) At wavelengths higher than 1600 nm the photon energy is not sufficient to excite an electron from the valence band to the conduction band. If the material $In_{0.53}Ga_{0.47}As$ has an energy gap $E_g = 0.73$ eV, what is the cutoff wavelength?

(d) Why does the responsivity drop off at low wavelengths?

Solution:

(a) From Eq. (5.8)

$$R = \frac{\eta q}{h\nu} = \frac{\eta q \lambda}{hc} = \frac{(0.90)(1.6 \times 10^{-19}\,C)\lambda}{(6.625 \times 10^{-34}\,J \cdot s)(3 \times 10^8\,m/s)} = 7.25 \times 10^5\lambda$$

(b) From the above equation, at 1300 nm

$$R = [7.25 \times 10^5\,(A/W)/m](1.30 \times 10^{-6}\,m) = 0.92\,A/W$$

(c) From Eq. (5.4)

$$\lambda_c(\mu m) = \frac{hc}{E_g} = \frac{1.240}{0.73} = 1.70\,\mu m$$

(d) At wavelengths less than 1100 nm, the photons are absorbed very close
     to the photodetector surface, where the recombination rate of the gen-
     erated electron–hole pairs is very short. The responsivity thus decreases
     rapidly for smaller wavelengths, since many of the generated carriers
     recombine rapidly and thus do not contribute to the photocurrent.

5.2   Avalanche Photodiodes

Avalanche photodiodes (APDs) internally multiply the primary signal photocurrent
before it enters the input circuitry of the following amplifier. This action increases
receiver sensitivity because the photocurrent is multiplied before encountering the
thermal noise associated with the receiver circuit. As shown in Fig. 5.5, *photocurrent multiplication* takes place when the photon-generated carriers traverse a
region where a very high electric field is present. In this high-field region, a
photon-generated electron or hole can gain enough energy so that it ionizes bound
electrons in the valence band upon colliding with them. This carrier multiplication
mechanism is known as *impact ionization*. The newly created carriers are also
accelerated by the high electric field, thus gaining enough energy to cause further
impact ionization. This phenomenon is the avalanche effect.

Fig. 5.5 Concept of
photocurrent multiplication in
an APD

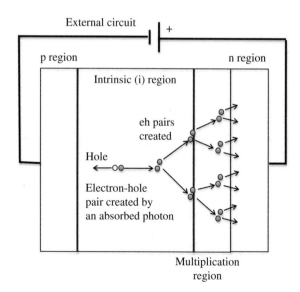

The multiplication M for all carriers generated in the photodiode is defined by

$$M = \frac{i_M}{i_p} \tag{5.9}$$

where i_M is the average value of the total multiplied output current and i_p is the primary unmultiplied photocurrent defined in Eq. (5.6). In practice, the avalanche mechanism is a statistical process, since not every carrier pair generated in the diode experiences the same multiplication. Thus, the measured value of M is expressed as an average quantity.

Analogous to the *pin* photodiode, the performance of an APD is characterized by its responsivity R_{APD}, which is given by

$$R_{APD} = \frac{\eta q}{h\nu} M = RM \tag{5.10}$$

where R is the unity gain responsivity.

Example 5.7 A given silicon avalanche photodiode has a quantum efficiency of 65 % at a wavelength of 900 nm. Suppose 0.5 μW of optical power produces a multiplied photocurrent of 10 μA. What is the multiplication M?

Solution: First from Eq. (5.8) the primary photocurrent generated is

$$i_p = RP_{in} = \frac{\eta q \lambda}{hc} P_{in} = \frac{(0.65)(1.6 \times 10^{-19}\,\text{C})(9 \times 10^{-7}\,\text{m})}{(6.625 \times 10^{-34}\,\text{J} \cdot \text{s})(3 \times 10^{8}\,\text{m/s})} (5 \times 10^{-7}\,\text{W})$$
$$= 0.235\,\mu\text{A}$$

Then from Eq. (5.9) the multiplication is

$$M = \frac{i_M}{i_p} = \frac{10\,\mu\text{A}}{0.235\,\mu\text{A}} = 43$$

Thus the primary photocurrent is multiplied by a factor of 43.

5.3 Photodiode Noises

5.3.1 Signal-to-Noise Ratio

When detecting a weak optical signal, the photodetector and its following amplification circuitry need to be designed so that a desired signal-to-noise ratio is

Fig. 5.6 Simple circuit model of a photodetector receiver

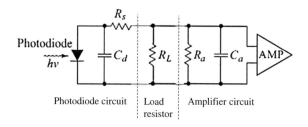

maintained for proper signal interpretation. The power *signal-to-noise ratio* (designated by SNR) at the output of an optical receiver is defined by

$$\text{SNR} = \frac{\text{signal power from photocurrent}}{\text{photodetector noise power} + \text{amplifier noise power}}$$
$$= \frac{\text{mean square signal current}}{\sum \text{mean square noise currents}} \tag{5.11}$$

The noise currents in the receiver arise from various photodetector noises and the thermal noise associated with the combined resistance of the photodiode and the amplifier circuitry.

To see the interrelationship of the different types of noises affecting the signal-to-noise ratio, consider the circuit of a simple receiver model shown in Fig. 5.6. The photodiode has a small series resistance R_s, a total capacitance C_d consisting of junction and packaging capacitances, and a bias (or load) resistor R_L. The amplifier following the photodiode has an input capacitance C_a and a resistance R_a. For practical purposes, R_s typically is much smaller than the load resistance R_L and can be neglected.

5.3.2 Noise Sources

The mean square value of the signal current i_s is given by

$$\langle i_s^2 \rangle = \langle i_p^2(t) \rangle M^2 \tag{5.12}$$

where $i_p(t)$ is the primary time varying current resulting from a time varying optical power $P_{in}(t)$ falling on the photodetector and M is the avalanche gain (M = 1 for a *pin* photodiode).

The principal noise sources associated with photodetectors are shot noise (also called *quantum noise*) and dark-current noise generated in the photodiode material. The *shot noise* arises from the statistical nature of the production and collection of photoelectrons when an optical signal is incident on a photodetector. These statistics follow a Poisson process. Because the fluctuations in the number of photocarriers created are a fundamental property of the photodetection process, the shot noise sets the lower limit on the receiver sensitivity when all other conditions are optimized. The *shot noise current* i_{shot} has a mean-square value in a receiver electrical bandwidth B_e that is proportional to the average value of the photocurrent i_p, that is,

$$\langle i_{shot}^2 \rangle = 2qi_p B_e M^2 F(M) \tag{5.13}$$

where F(M) is a *noise figure* associated with the random nature of the avalanche process. For an APD the noise figure is typically 3–6 dB. For *pin* photodiodes M and F(M) are unity.

The photodiode *dark current* is the current i_D that continues to flow through the bias circuit of the device when no light is incident on the photodiode. This is a combination of bulk and surface dark currents, but in general the surface dark current is negligible. The *bulk dark current* i_{DB} arises from electrons and/or holes that are thermally generated in the pn junction of the photodiode. In an APD, these liberated carriers also get accelerated by the high electric field present at the pn junction, and are therefore multiplied by the avalanche gain mechanism. The mean-square value of this dark current is given by

$$\langle i_{DB}^2 \rangle = 2qi_D M^2 F(M)B_e \tag{5.14}$$

where i_D is the primary (unmultiplied) detector bulk dark current, which is listed on component data sheets.

To simplify the analysis of the receiver circuitry, one can assume that the amplifier input impedance is much greater than the load resistance R_L, so that the thermal noise from R_a is much smaller than that of R_L. The photodetector load resistor then dominates and contributes a mean-square *thermal noise current*

$$\langle i_T^2 \rangle = \frac{4k_B T}{R_L}B_e \tag{5.15}$$

where k_B is Boltzmann's constant and T is the absolute temperature. Using a load resistor that is large but still consistent with the receiver bandwidth requirements can reduce this noise.

Example 5.8 An InGaAs *pin* photodiode has the following parameters at a wavelength of 1300 nm: i_D = 4 nA, η = 0.90, R_L = 1000 Ω, and the surface leakage current is negligible. Assume the incident optical power is 300 nW (−35 dBm), the temperature is 293 K, and the receiver bandwidth is 20 MHz. Find (a) The primary photocurrent; (b) The mean-square shot noise current;

(c) The mean-square dark current noise; and (d) The mean-square thermal noise current.

Solution:

(a) From Eq. (5.8) the primary photocurrent is

$$i_p = RP_{in} = \frac{\eta q \lambda}{hc} P_{in}$$
$$= \frac{(0.90)(1.6 \times 10^{-19}\,\text{C})(1.3 \times 10^{-6}\,\text{m})}{(6.625 \times 10^{-34}\,\text{J} \cdot \text{s})(3 \times 10^8\,\text{m/s})} (3 \times 10^{-7}\,\text{W}) = 0.282\,\mu\text{A}$$

(b) From Eq. (5.13) the mean-square shot noise current for a *pin* photodiode is

$$\langle i_{shot}^2 \rangle = 2qi_p B_e = 2(1.6 \times 10^{-19}\,\text{C})(0.282 \times 10^{-6}\,\text{A})(20 \times 10^6\,\text{Hz}) = 1.80 \times 10^{-18}\,\text{A}^2$$

or $\langle i_{shot}^2 \rangle^{1/2} = 1.34\,\text{nA}$

(c) From Eq. (5.14) the mean-square dark current is

$$\langle i_{DB}^2 \rangle = 2qi_D B_e = 2(1.6 \times 10^{-19}\,\text{C})(4 \times 10^{-9}\,\text{A})(20 \times 10^6\,\text{Hz}) = 2.56 \times 10^{-20}\,\text{A}^2$$

or $\langle i_{DB}^2 \rangle^{1/2} = 0.16\,\text{nA}$

(d) From Eq. (5.15) the mean-square thermal noise current for the receiver is

$$\langle i_T^2 \rangle = \frac{4k_B T}{R_L} B_e = \frac{4(1.38 \times 10^{-23}\,\text{J/K})(293\,\text{K})}{1000\,\Omega} (20 \times 10^6\,\text{Hz}) = 323 \times 10^{-18}\,\text{A}^2$$

or $\langle i_T^2 \rangle^{1/2} = 18\,\text{nA}$

Thus for this receiver the rms thermal noise current is about 14 times greater than the rms shot noise current and about 100 times greater than the rms dark current.

5.3.3 *Noise-Equivalent Power and Detectivity*

The sensitivity of a photodetector is describable in terms of the *minimum detectable optical power*. This is the optical power necessary to produce a photocurrent of the same magnitude as the root-mean-square (rms) of the total noise current, or equivalently, to yield a signal-to-noise ratio of 1. This optical signal power is referred to as the *noise equivalent power* or NEP, which is designated in units of $W/\sqrt{Hz}$.

As an example, consider the thermal-noise-limited case for a *pin* photodiode. A thermal-limited SNR occurs when the optical signal power is low so that thermal noise dominates over shot noise. Then the SNR becomes

$$\text{SNR} = R^2 P^2 / (4k_B T B_e / R_L) \qquad (5.16)$$

To find the NEP, set the SNR equal to 1 and solve for P to obtain

$$\text{NEP} = \frac{P_{min}}{\sqrt{B_e}} = \sqrt{4k_B T / R_L} / R \qquad (5.17)$$

Example 5.9 Let the responsivity $R = 0.90$ A/W for an InGaAs photodetector operating at 1550 nm. What is the NEP in the thermal-noise-limited case if the load resistor $R_L = 1000\ \Omega$ and T = 300 K?

Solution: From (5.17) the value for NEP is

$$\text{NEP} = [4(1.38 \times 10^{-23}\ \text{J/K})(300\ \text{K})/1000\ \Omega]^{1/2} / (0.90\ \text{A/W})$$
$$= 4.52 \times 10^{-12}\ \text{W}/\sqrt{\text{Hz}}$$

The parameter *detectivity*, or D*, is a figure of merit for a photodetector used to characterize its performance. The detectivity is equal to the reciprocal of NEP normalized per unit area A.

$$D^* = A^{1/2} / \text{NEP} \qquad (5.18)$$

Its units commonly are expressed in cm $\sqrt{\text{Hz}}/\text{W}$.

5.3.4 Comparisons of Photodiodes

Tables 5.1 and 5.2 list some generic operating characteristics for GaP, Si, Ge, and InGaAs *pin* photodiodes and avalanche photodiodes, respectively. The values were derived from various vendor data sheets and from performance numbers reported in the literature. The performance values are given as guidelines for comparison purposes. Detailed values on specific devices for particular applications can be obtained from suppliers of photodetector and receiver modules.

Table 5.1 Generic operating parameters of GaP, Si, Ge, and InGaAs *pin* photodiodes

Parameter	Symbol	GaP	Si	Ge	InGaAs
Wavelength range (nm)	λ	250–380	400–1100	800–1650	1100–1700
Responsivity (A/W)	R	0.07–0.1	0.4–0.6	0.4–0.5	0.75–0.95
Dark current (nA)	I_D	0.01	1–10	50–500	0.5–2.0
Rise time (ns)	τ_r	500	0.5–1	0.1–0.5	0.05–0.5
Modulation bandwidth (GHz)	B_m	0.001	0.3–0.7	0.5–3	1–2
Bias voltage (V)	V_B	5	5	5–10	5

Table 5.2 Generic operating parameters of Si, Ge, and InGaAs avalanche photodiodes

Parameter	Symbol	Si	Ge	InGaAs
Wavelength range (nm)	λ	400–1100	800–1650	1100–1700
Avalanche gain	M	20–400	50–200	10–40
Dark current (nA)	I_D	0.1–1	50–500	10–50 @M = 10
Rise time (ns)	τ_r	0.1–2	0.5–0.8	0.1–0.5
Gain • bandwidth (GHz)	$M \cdot B_m$	100–400	2–10	20–250
Bias voltage (V)	V_B	150–400	20–40	20–30

Note Si can be enhanced to extend its operation to the 200–400-nm spectrum, but with a lower responsivity of about 0.1 A/W for a *pin* photodiode at 200 nm compared to 0.4–0.6 A/W in the 750–950-nm region

5.4 Multichannel Detectors

Only one spectral channel can be monitored at any instance in time with a single-channel photodetection element. However, multichannel devices can measure multiple spectral or spatial channels simultaneously in different time sequences [9–11]. Two common multichannel devices are based on *complementary metal-oxide-semiconductor* (CMOS) and *charge-coupled device* (CCD) technologies. Both CCD and CMOS devices consist of one-dimensional or two-dimensional arrays of light-sensitive *pixels*. These pixels are individual photodetection elements that convert incoming photons to electrons and collect them in a potential well. After a certain time period the electrons in the well are counted and converted to an electrical signal, which corresponds to the optical power flow onto the device.

5.4.1 CCD Array Technology

A number of architectural variations exist for a CCD. A simple device model for an individual pixel is shown in Fig. 5.7. Each pixel is a metal-oxide-semiconductor (MOS) capacitor with an optically transparent gate contact, an opaque transfer gate,

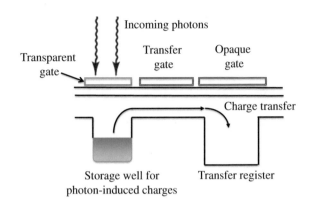

Fig. 5.7 Operational structure of one stage of a CCD array

and a charge transfer register situated under another opaque gate. During device operation the light-sensing portion is biased to produce a potential well under it for storing electric charge. The transfer gate is biased at a voltage V_1 to establish a potential barrier that confines the photogenerated electrons in the potential well. When the pixel is illuminated, the charge in the well increases with time and is proportional to the light intensity. After a specific illumination time (or *light integration time*) the bias V_1 is pulsed low and the transfer gate voltage V_2 is biased high. This action shifts the accumulated charge to a transfer register potential well under the transfer gate and from there it is transferred to a readout device that performs charge-to-voltage conversion. After the charges have been read out, the bias voltages V_1 and V_2 return to their original values and the cycle starts over. This process produces a measurable time-dependent output current, which is a proportional to the light intensity falling the pixel.

An example of a 4×4 CCD array is shown in Fig. 5.8, which is a 16-pixel image sensor. Following the specified light integration time, all the photosensing wells are emptied simultaneously into their associated CCD transfer register (or *shift register*) wells. The charges in the shift register wells then are transferred row by row to the output CCD register array. Thereby each row is sent sequentially to the output for signal interpretation. This process results in a time-dependent output current, which is a measure of the light intensity falling on various areas of the CCD array. Note that in an actual CCD image sensor or camera, the CCD array typically consists of millions of pixels.

5.4.2 CMOS Array Technology

In contrast to a CCD array in which the contents of the wells are read out row by row for measurement by an external circuit, in a CMOS array the charge-to-voltage conversion process occurs inside each pixel. The pixel area is thus more complex. As Fig. 5.9 shows, the pixel contains a photodiode for detecting light and electronic

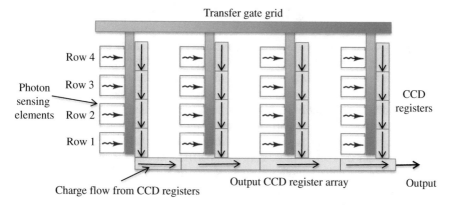

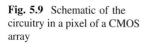

Fig. 5.8 Illustrative example of a CCD imager using a 4 × 4 pixel array

Fig. 5.9 Schematic of the
circuitry in a pixel of a CMOS
array

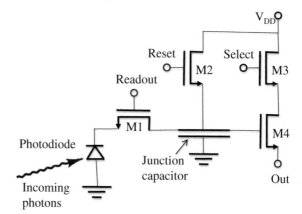

conversion and amplification circuitry consisting of four metal-oxide-semiconductor field-effect transistors (MOSFETs) plus a junction capacitor. Because each pixel has its own built-in amplifier, the pixel is referred to as an *active pixel*. When light is incident on a pixel, the charges that are built up from the photodetection process are converted to volts, buffered by a source-follower transistor, and then are transferred to thin metal output buses through transistor switches that reside inside each pixel to generate an output video stream.

   When light is being detected, the MOSFETs M1, M2, and M3 are turned off and the capacitor gets charged to nominally 3.3 V. The voltage on the photodiode depends on the incoming light intensity and the integration time (the time during which light is measured). Then M1 is turned on to determine the charge on the photodiode. This action causes electrons to flow from the photodiode through M1, thereby discharging the capacitor. This process takes about 1 μs. After this procedure, M1 is turned off and the photodiode again detects light. The voltage on the capacitor represents the light signal, which is measured by turning on M3 and activating M4. The resulting output current is related to the capacitor voltage, which is a measure of the light intensity.

The CMOS device architecture and operating process have several advantages. The operating voltages required for a CMOS array and the power consumption are lower than for a CCD array. Digital timing and control signals along with other digital image processing functions can be located directly on the chip. Because each pixel is individually selectable, functions such as random access of specific pixels, data windowing (taking a small subset of a large data set), and having a variable frame rate within a chosen set of pixels can all be supported. Having additional circuitry within a pixel element and the operation of this circuitry increases the complexity of the chip, but allows cameras containing CMOS arrays to be smaller than cameras that use CCD arrays. However, devices containing CMOS arrays tend to have lower image quality because more of the chip area is covered with circuitry.

5.5 Photomultiplier Tubes

A *photomultiplier tube* (PMT) is an optical detection device that uses an external photoelectric effect, in contrast to semiconductor-based photodiodes that use an internal photoelectric effect. In an external photoelectric process, when light strikes a semiconductor-coated metal or glass plate that is located in a vacuum, electrons (referred to as *photoelectrons*) are emitted from the material surface into the vacuum. Because PMTs have fast response speeds and superior sensitivity, they are used widely in biophotonics research, medical equipment, and analytical instruments, particularly when very low light intensities need to be measured.

As Fig. 5.10 shows, a generic PMT consists of a photocathode (also called a photoemissive cathode), a series of dynodes, and an anode, all of which are encapsulated in an evacuated glass enclosure. A *dynode* is an electrode in a vacuum tube that serves as an electron multiplier through a secondary emission process, which occurs when primary electrons of sufficient energy hit the dynode material.

Fig. 5.10 Basic structure of a PMT with a reflection mode photocathode

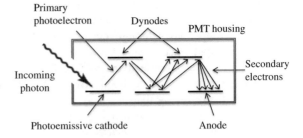

Photocathodes can be configured to operate in either a reflection mode or a transmission mode. A *reflection mode photocathode* usually consists of a compound semiconductor formed on a metal plate so that the photoelectrons travel in the opposite direction as the incident light. In the *transmission mode photocathode* a compound semiconductor is deposited as a thin film on an optically transparent glass plate. In this case the photoelectrons travel in the same direction as the incident light.

During the PMT operation, first photons enter the PMT through a window or faceplate in the vacuum tube and hit the photocathode. This process excites electrons in the photocathode material thereby causing the emission of photoelectrons into the vacuum. The photoelectrons then are accelerated and directed to the first dynode by means of a focusing electrode that follows the photocathode. Upon hitting the first dynode the photoelectrons are multiplied by means of secondary emission. This secondary emission process then is repeated at each successive dynode to generate additional electrons. At the end of the PMT an anode collects and measures the secondary electrons emitted by the last dynode. This cascading effect creates from 10^5 to 10^7 electrons for each photon hitting the first cathode. If a PMT contains n dynode stages and each dynode has a secondary electron multiplication factor of δ, then the gain of the PMT is given by

$$G = \delta^n \tag{5.19}$$

Numerous PMT designs are available for a wide range of applications. The PMT operation depends on the entrance window and photocathode materials, the number of dynodes, the electrode configurations and their arrangement, and the voltages applied to the electrodes. In general a photocathode is made from compound semiconductors, which are materials made from two or more elements from different groups of the periodic table. Examples of popular materials for reflection mode PMT photocathodes are the following:

- Cesium iodide (Cs-I) with a MgF_2 window for operation in the 115–200 nm range
- Bialkali (antimony-rubidium-cesium or Sb-Ru-Cs and antimony-potassium-cesium or Sb-K-Cs) with a quartz or UV-transmitting glass window for the 185–750 nm spectral range
- Multialkali (Na-K-Sb-Cs) with a UV-transmitting glass window for the 185–900 nm spectral range
- InGaAsP-Cs (InGaAsP activated with cesium) for the 950–1400 nm spectral range.
- InGaAs-Cs (InGaAs activated with cesium) for the 950–1700 nm spectral range.

5.6 Optical Filters

Optical filters are devices that selectively transmit light in one or more specific bands of wavelengths while absorbing or reflecting the remainder [12–16]. The absorption or reflection properties give rise to the two broad categories of absorptive filters and dichroic filters, respectively. In an *absorptive filter*, the unwanted light spectrum is blocked by being absorbed and contained within the filter. In a *dichroic filter* one spectral slice of light is reflected and another spectral range is transmitted through the device. A characteristic of an absorptive filter is that light can impinge on the filter from a wide range of angles and the filter will maintain its transmission and absorption properties. In contrast, a dichroic filter is highly angle sensitive. In general, an absorptive filter is placed perpendicular to the optical beam, whereas a dichroic filter is placed at a 45° angle relative to the optical beam.

Historically optical filters were designed to pass a single designated wavelength or a single narrow spectral band. This characteristic imposed a time-consuming limitation in areas such as fluorescence imaging when trying to analyze events occurring at multiple wavelengths. In that case, different sets of emission and excitation filters had to be inserted into the optical path one at a time based on the specific wavelength-dependent process that was being examined. Thus the ability to evaluate live specimens or time-sensitive events was difficult.

The implementation of the worldwide telecom network brought a demand for highly reliable dichroic optical filters with stringent spectral performance that could be reproduced and controlled precisely during manufacture. This fabrication technology involves depositing a series of thin optical interference coatings on a flat glass surface to create durable dichroic filters that are resilient to temperature and humidity effects. This durability is important for both telecom and biomedical applications. These devices often are called *thin film filters* (TFF). Such TFF devices can be made to pass several different wavelength bands simultaneously and thus allow the use of a series of color-dependent biomarkers to label and analyze multiple time-sensitive events in the same specimen.

Some key parameters of dichroic filters are illustrated in Fig. 5.11. Several categories of optical filters are of interest in biophotonics applications. *Longpass filters* allow only light above a specified wavelength to be transmitted, with all other wavelengths being reflected or absorbed. The *cut-on wavelength* or central wavelength of a longpass filter is specified at 50 % of the peak transmission. Conversely, *shortpass filters* allow only light below a specified wavelength to be transmitted. The *cut-off wavelength* of a shortpass filter is specified at 50 % of the peak transmission. Longpass and shortpass optical filters can have a very sharp wavelength-separation slope, in which case they sometimes are referred to as *edge filters*. *Bandpass filters* are used to transmit one or more extremely narrow (<2–10 nm) or broad (e.g., 50 nm or 80 nm) bands of wavelengths while blocking (or

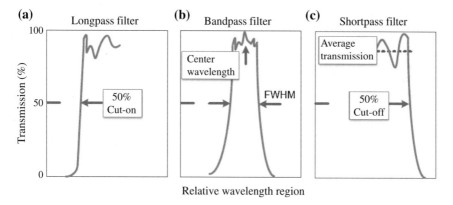

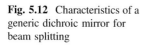

Fig. 5.11 Three categories of dichroic filters and some terminology

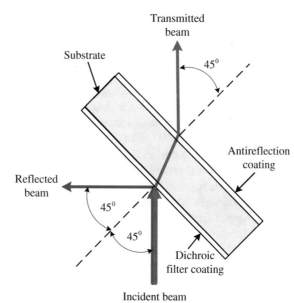

Fig. 5.12 Characteristics of a generic dichroic mirror for beam splitting

reflecting) adjacent wavelengths over extended spectral ranges. The wavelength bandwidth often is described in terms of the *full-width half-maximum* (FWHM) response of a filter. As shown in Fig. 5.11, the FWHM is the width of the spectral transmission window at the half-power point.

A basic application of a dichroic filter (or a TFF) is to separate a beam of light into two beams with differing spectral regions, as Fig. 5.12 illustrates. Reciprocally, by reversing the arrows in Fig. 5.12 the dichroic filter can combine two optical beams from different spectral bands into a single optical beam. Such a device

Fig. 5.13 Example of a longpass dichroic mirror designed with a 905-nm separation wavelength, which has a 400–875-nm reflection band and a 935–1300-nm transmission band

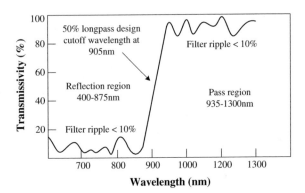

commonly is referred to as a *dichroic mirror*, *beam splitter*, or *beam combiner*. Typically a dichroic mirror is placed at a 45° angle to the incoming light beam, as shown in Fig. 5.12.

Applications of dichroic mirrors are in areas such as confocal microscopes, spectroscopy, laser beam delivery mechanisms, flow cytometry, and laser traps. Figure 5.13 shows an example of the transmission and reflection performance for a generic dichroic mirror, which is designed so that the spectral separation occurs at 905 nm. Below the design wavelength is a spectral band over which the device will reflect more than 90 % of the incident light, that is, the average reflection $R_{ave} > 90$ %. Above the design wavelength is a spectral band for which the device will transmit more than 90 % of the incident light, that is, the average transmission $T_{ave} > 90$ %. Such a design is called a *longpass mirror*.

The design is reversed for *shortpass mirrors*. In this case, the device will transmit a finite spectral band below the design wavelength and will reflect another longer-wavelength band above the design wavelength. Depending on the device quality, actual filters have varying degrees of cutoff sharpness between the two spectral bands and exhibit some ripple ranging up to 10 % in the transmission or reflection characteristics. The key dichroic mirror performance parameters are the reflection band, the transmission band, and the 50 % design cutoff wavelength. Table 5.3 lists typical values of key operating parameters for three commercial longpass and shortpass dichroic mirrors.

Standard thin film filters in the visible region are available as blue, green, yellow, cyan, magenta, and red types. For example, a blue filter passes 495 ± 15 nm, a

Table 5.3 Transmission and reflection bands for three example commercial dichroic mirrors

Parameter	505 nm Longpass	638 nm Longpass	805 nm Shortpass
Reflection band ($R_{ave} > 90$ %)	380–490 nm	580–621 nm	820–1300 nm
Transmission band ($T_{ave} > 90$ %)	520–700 nm	655–700 nm	400–790 nm
50 % Cutoff wavelength	505 nm	638 nm	805 nm

green filter passes 505 ± 15 nm, and a red filter passes 610 ± 15 nm. In the near-IR region, thin-film filters are available in a wide range of passbands varying from 50 to 800 GHz, or equivalently, spectral widths ranging from 0.3 to 4.5 nm in the 1300-nm wavelength region or from 0.4 to 6.4 nm in the 1550-nm wavelength region. Many different standard and custom TFF designs are available commercially for other wavelength regions.

5.7   Optical Couplers and Optical Circulators

Two optical elements that are widely used in optical fiber-based equipment setups are optical couplers and optical circulators. The concept of an *optical coupler* or *beam splitter* encompasses a variety of functions, including splitting a light signal into two or more streams, combining two or more light streams, tapping off a small portion of optical power for monitoring purposes, or transferring a selective range of optical power between fibers. When discussing couplers it is customary to designate couplers in terms of the number of input ports and output ports on the device. In general, an N × M coupler has N ≥ 1 input ports and M ≥ 2 output ports. For example, a coupler with two inputs and two outputs is a 2 × 2 coupler. A 1 × N device (one input and N outputs) is popularly known as a *star coupler*. These devices can be made in the form of planar waveguides or they can be made from optical fibers that are fused together along a specific coupling length. The latter device often is called a *fused-fiber coupler*. Biophotonics applications of fused-fiber couplers include laser surgery, optical coherence tomography, endoscopy, optical power level monitoring in light therapy, spectroscopy, and biosensors.

A 2 × 2 optical fiber coupler, shown in Fig. 5.14, takes a fraction α of the power from input 1 and places it on output 1 and the remaining fraction 1-α exits from output 2. Similarly, a fraction 1-α of the power from input 2 is directed to output 1 and the remaining power goes to output 2. The parameter α is called the *coupling ratio*. As an example of nomenclature, if the parameter α = 50 %, the device is called a 50/50 coupler or a *3-dB coupler*. For light level monitoring purposes, a *tap coupler* might have α be a few percent (e.g., 1–5 %). A coupler can be designed to be either wavelength-selective or wavelength-independent (sometimes called wavelength-flat) over a wide usable range. In a wavelength-independent device, α is independent of the wavelength; in a wavelength-selective device, α depends on the wavelength. Table 5.4 lists some common specifications for optical couplers.

Fig. 5.14 Operational concept of an optical fiber coupler or splitter

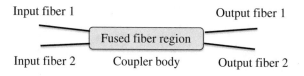

Table 5.4 Common specifications for optical fiber couplers

Coupler parameter	Specification
Coupling ratio	1/99–50/50
Directivity	≥50 dB (1 × 2)
	≥60 dB (2 × 2)
Reflectance	≥ −55 dB
Polarization dependent loss (PDL)	≥0.1 dB (≥0.3 dB for taps of ≥15 %)
Operating bandpass	±40 nm
Operating temperature:	−40 to +85 °C
Standard lead length	1 m

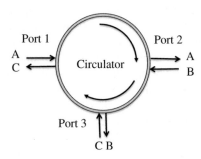

Fig. 5.15 Operational concept of a three-port optical circulator

An *optical circulator* is a non-reciprocal multi-port passive device that directs light sequentially from port to port in only one direction. As illustrated in Fig. 5.15 for a 3-port circulator, an optical input signal A on port 1 is sent out on port 2, an input B on port 2 is sent out on port 3, and an input C on port 3 is sent out on port 1. Available configurations include jacketed optical fiber attachments and fibers with connector terminations. Table 5.5 lists some average values of key operating

Table 5.5 Typical values of key operating parameters for commercial 3-port and 4-port optical circulators (*WDL* wavelength dependent loss; *PDL* polarization dependent loss)

Parameter	3-Port circulator	4-Port circulator
Center wavelength	1310 or 1550 nm (±20 nm)	1290–1330, 1525–1565, or 1570–1610 nm
Isolation	50 dB	50 dB
Insertion loss	0.8 dB	0.6 dB
WDL	0.20 dB	0.15 dB
PDL	0.20 dB	0.10 dB
Power handling	500 mW	500 mW
Size	5.5 mm × 60 mm	5.5 mm × 60 mm
Operating temperature	0 to +70 °C	0 to +70 °C

parameters for commercial 3-port and 4-port optical circulators. In the biophotonics field this device is used in conjunction with thin-film filters or FBGs to combine or separate a series of narrow wavelength bands for applications such as laser-scanning microscopy, optical coherence tomography, imaging techniques (for example, monitoring tissue changes induced by laser thermal therapy), spectroscopic diagnostics, and processing of biosensor signals.

5.8   Summary

Photodetection or light measuring devices for biophotonics applications are needed in the spectral range from about 190 nm to 10 μm, which covers the ultraviolet, visible, and infrared regions. Depending on the particular discipline in which the device is used, the performance characteristics can include sensing very low light levels, responding to light pulses as fast as femtoseconds, and possessing low internal noise characteristics. The devices used in biophotonics include semiconductor-based *pin* and avalanche photodiodes, photomultiplier tubes, and detector arrays based on complementary metal-oxide-semiconductor (CMOS) and charge-coupled device (CCD) technologies.

The photodetectors can be either single-channel elements or multichannel devices consisting of one-dimensional rows or two-dimensional arrays of detectors. Multichannel devices can measure multiple spectral or spatial channels simultaneously in different time sequences. Two common multichannel devices used extensively in cameras are based on CMOS and CCD technologies.

For many biophotonics applications it may be necessary to select a narrow spectral band or to reject light above or below a particular wavelength. Optical filters are devices that selectively transmit light in one or more specific bands of wavelengths while absorbing or reflecting the remainder. The absorption or reflection properties give rise to the two broad categories of absorptive filters and dichroic filters, respectively.

5.9   Problems

5.1   An InGaAs pin photodetector has an absorption coefficient of 1.0 μm^{-1} at 1550 nm. Show that the penetration depth at which 50 % of the photons are absorbed is 0.69 μm.

5.2   If the absorption coefficient of silicon is 0.05 μm^{-1} at 860 nm, show that the penetration depth at which $P(x)/P_{in} = 1/e = 0.368$ is x = 20 μm.

5.3   A particular InGaAs pin photodiode has a bandgap energy of 0.74 eV. Show that the cutoff wavelength of this device is 1678 nm.

5.4   If an optical power level P_{in} is incident on a photodiode, the electron–hole generation rate G(x) is given by

$$G(x) = \Phi_0 \alpha_s \exp(-\alpha_s x)$$

Here Φ_0 is the incident photon flux per unit area given by

$$\Phi_0 = \frac{P_{in}(1 - R_f)}{Ah\nu}$$

where A is the detector area, R_f is the surface reflectance, and hν is the photon energy. From this, show that the primary photocurrent in the depletion region of width w is given by Eq. (5.6).

5.6   Consider a sinusoidally modulated optical signal P(t) of frequency ω, modulation index m, and average power P_0 given by

$$P(t) = P_0(1 + m\cos \omega t)^2$$

Show that when this optical signal falls on a photodetector, the mean-square signal current $\langle i_s^2 \rangle$ generated consists of the following sum of a dc (average) component i_{dc} and a signal current i_p

$$\langle i_s^2 \rangle = i_{dc}^2 + \langle i_p^2 \rangle = (RP_0)^2 + \frac{1}{2}(mRP_0)^2$$

where the responsivity R is given by Eq. (5.8).

5.7   A given InGaAs pin photodiode has the following parameters at 1550 nm: i_D = 1.0 nA, η = 0.95, R_L = 500 Ω, and the surface leakage current is negligible. Assume the incident optical power is 500 nW (-33 dBm) and the receiver bandwidth is 150 MHz. Show that the primary current and the mean-square noise currents are as follows: (a) i_p = 0.593 μA; (b) $\langle i_{shot}^2 \rangle$ = 2.84 $\times$ 10^{-17} A^2; (c) $\langle i_{DB}^2 \rangle$ = 4.81 $\times$ 10^{-20} A^2; (d) $\langle i_T^2 \rangle$ = 4.85 $\times$ 10^{-15} A^2.

5.8   A given InGaAs avalanche photodiode has a quantum efficiency of 90 % at a wavelength of 1310 nm. Suppose 0.5 μW of optical power produces a multiplied photocurrent of 8 μA. Show that the multiplication M = 16.

5.9   A biophotonics engineer has a choice between two cameras that are based on a CCD array or a CMOS array, respectively. Using resources from the Web or from vendor catalogs, list some of the tradeoffs between selecting one or the other of these instruments. Consider factors such as speed of response, image quality, camera power consumption, camera size, and the capability to enhance certain areas of the viewed image.

5.10 A certain PMT has 12 dynode stages. In the first five stages, each primary electron can stimulate four secondary electrons. In the next seven dynodes, each primary electron can stimulate five secondary electrons. Show that the gain of this PMT is 8 $\times$ 10^7.

5.11 A PMT has 10 dynodes. At each dynode one incoming primary electron creates four secondary electrons. The PMT is used to detect a faint laser beam that is operating at 632.8 nm. The quantum efficiency of the cathode at this wavelength is 25 %. The laser power incident on the PMT is 1nW.

(a) Show that the number of incident photons per second at the cathode is 3.19×10^9 photons/s. Recall that power is equal to photon energy times the number of arriving photons per second.

(b) Show that 7.97×10^8 photoelectrons are generated at the cathode per second.

(c) Show that the gain of this PMT is 1048576.

(d) The anode current is the cathode current multiplied by the gain of the PMT. Show that the anode current is 133.6 μA. Recall that $q = 1.6 \times 10^{-19}$ C.

5.12 Many types of spectroscopic systems require the capability of a measurement system to view several narrow spectral bands simultaneously. Using resources from the Web, identify vendors that supply multichannel bandpass optical filters and list some characteristics of some typical filters in the visible spectrum. Consider parameters such as channel spacing, operating spectral bandwidth, channel isolation, and insertion loss.

References

1. S. Donati, *Photodetectors: Devices, Circuits, and Applications* (Prentice-Hall, New York, 2000)
2. M. Johnson, *Photodetection and Measurement* (McGraw-Hill, New York, 2003)
3. B.L. Anderson, R.L. Anderson, *Fundamentals of Semiconductor Devices* (McGraw-Hill, New York, 2005)
4. A. Beling, J.C. Campbell, InP-based high-speed photodetectors: tutorial. J. Lightw. Technol. **27**(3), 343–355 (2009)
5. R. Riesenberg, A. Wutting, Optical detectors, chap. 5, in *Handbook of Biophotonics*, vol. 1: Basics and Techniques, eds. by J. Popp, V.V. Tuchin, A. Chiou, S.H. Heinemann (Wiley, New York, 2011), pp. 297–343
6. C. Jagadish, S. Gunapala, D. Rhiger, eds., *Advances in Infrared Photodetectors*, vol. 84 (Academic Press, Cambridge, 2011)
7. M.J. Deen, P.K. Basu, *Silicon Photonics: Fundamentals and Devices* (Wiley, Hoboken, NJ, 2012)
8. G. Keiser, *Optical Fiber Communications*, chap. 6, 4th US edn., 2011; 5th international edn., 2015 (McGraw-Hill)
9. G.C. Holst, T.S. Lomheim, *CMOS/CCD Sensors and Camera Systems* (SPIE Press, Bellingham, WA, 2007)
10. Hamamatsu, *Photomultiplier Tube Handbook*, 3rd edn. (2007). https://www.hamamatsu.com/resources/pdf/etd/PMT_handbook_v3aE.pdf
11. W. Becker, *The bh TCSPC Handbook* (Becker & Hickl, Berlin, Germany, 2014)
12. V. Kochergin, *Omnidirectional Optical Filters* (Springer, New York, 2003)

13. J. Jiang, J.J. Pan, Y. H. Guo, G. Keiser, Model for analyzing manufacturing-induced internal stresses in 50-GHz DWDM multilayer thin film filters and evaluation of their effects on optical performances. *J. Lightwave Technol.* **23**(2), 495–503 (2005)
14. R.R. Willey, *Field Guide to Optical Thin Films* (SPIE Press, Bellingham, WA, 2006)
15. H.A. Macleod, *Thin-Film Optical Filters*, 4th edn. (Taylor & Francis, London, UK, 2010)
16. M. Ohring, D. Gall, S.P. Baker, *Materials Science of Thin Films: Deposition and Structure*, 3rd edn. (Academic Press, Cambridge, 2015)

Chapter 6
Light-Tissue Interactions

Abstract A fundamental challenge in biophotonics is to understand the interaction of light with multilayered, multicomponent, and optically inhomogeneous biological tissues. The effects of light-tissue interactions include reflection and refraction when light encounters different tissue types, absorption of photon energy, and multiple scattering of photons. Light absorption determines how far light can penetrate into a specific tissue. It depends strongly on wavelength and is important in the diagnosis and therapy of abnormal tissue conditions. Scattering of photons in tissue is another significant factor in light-tissue interactions. Together, absorption and multiple scattering of photons cause light beams to broaden and decay as photons travel through tissue. Light can interact with biological tissue through many different mechanisms, including photobiomodulation, photochemical interactions, thermal interactions (e.g., coagulation and vaporization), photoablation, plasma-induced ablation, and photodisruption. Two key phenomena used in tissue analyses are random interference patterns or speckle fields and the principles of fluorescence.

The interaction of light with biological tissues is a complex process because the constituent tissue materials are multilayered, multicomponent, and optically inhomogeneous. As shown in Fig. 6.1, the basic effects of light-tissue interactions include reflection at a material interface, refraction when light enters a tissue structure that has a different refractive index, absorption of photon energy by the material, and multiple scattering of photons in the material. Owing to the fact that diverse intermingled biological tissue components have different optical properties, then along some path through a given tissue volume, various physical parameters (for example, the refractive index, the absorption coefficients, and the scattering coefficients) can vary continuously or undergo abrupt changes at material boundaries, for example, at interfaces between soft biological tissue and organs or bone tissue.

In this chapter first Sect. 6.1 describes some basic reflection and refraction phenomena that occur when light impinges on tissue. This description expands on the discussion in Sect. 2.4 by considering applications of reflection and refraction in biophotonics. Following this discussion, Sect. 6.2 addresses light absorption, which is a highly important factor in the diagnosis of tissue properties and in the therapy of

© Springer Science+Business Media Singapore 2016
G. Keiser, *Biophotonics*, Graduate Texts in Physics,
DOI 10.1007/978-981-10-0945-7_6

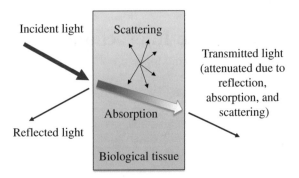

Fig. 6.1 Basic effects of light-tissue interactions

abnormal tissue conditions based on the interaction of light with tissue. Absorbed light can be converted into heat, be radiated in a fluorescent process, or be consumed in photochemical reactions. The strength of the absorption coefficients for different tissue components determines how far light can penetrate into a specific tissue at a particular wavelength and also determines how much energy a specific tissue absorbs from a particular optical source. The degree of absorption depends on the tissue type and in many cases is a strong function of wavelength. Light is readily absorbed in the UV (<400 nm) and mid-IR (>2 μm) regions. Thus light cannot penetrate deeply into tissue in these spectral ranges and there is little attenuation due to scattering. As Sect. 6.2 describes, there is a lower attenuation window in the visible and near-IR spectral range running from about 400 nm to 2 μm, which allows deeper penetration of light into tissue.

Scattering of photons in tissue is another significant factor in the behavior of light-tissue interactions, as Sect. 6.3 describes. Together, absorption and multiple scattering of photons cause light beams to broaden and decay as photons travel through tissue. Scattering dominates over absorption in the 600–1600-nm spectral range, and both forward scattering and backscattering of incident light within tissue are used in a variety of biophotonics applications. In a scattering process, light interacts with tissue components so that some photons in a light beam are diverted from their original path and sent in different directions. Typically light can penetrate to a depth of several centimeters into a tissue. However, strong scattering of light generally prevents observers from getting a clear image of tissue characteristics beyond a few millimeters in depth. Scattering of photons can be either an elastic process or an inelastic process. In the case of elastic scattering, the incident and scattered photons have the same energy, which also means that the wavelengths do not change during the scattering event. In this chapter, first Sect. 6.3 introduces the basic concepts of elastic scattering. This material includes a discussion of Rayleigh scattering, which is a particular category of elastic scattering. The effects of elastic scattering are used in many biophotonics applications such as optical coherence tomography, confocal microscopy, and elastic scattering spectroscopy.

Inelastic scattering occurs at a much lower probability than elastic scattering (by a factor of about 10^{-6}) and involves an exchange of energy between the photon and the scattering element. Raman scattering is the major inelastic scattering process used in biophotonics. This process is the basis for procedures such as Raman vibrational spectroscopy and surface-enhanced Raman scattering (SERS). Raman scattering has a rapidly growing number of applications used for the study of the structures and dynamic functions of biological molecules. The process also is of importance for diagnosing and monitoring the progress of diseases such as cataract formations, precancerous and cancerous lesions in human soft tissue, artherosclerotic lesions in coronary arteries, and bone and dental pathologies.

The effects of absorption and scattering of light in tissue are considered separately in Sects. 6.2 and 6.3 to illustrate their basic effects. However, in actual tissues both of these effects are present simultaneously. Their combined effect is described in Sect. 6.4.

LED and laser light can interact with biological tissue through many different mechanisms [1–11]. This chapter classifies light-tissue interactions into six categories that are commonly used to describe therapeutic and surgical applications. As is discussed in Sect. 6.5, these interactions can be categorized as photobiomodulation (also called low-level light therapy), photochemical interactions, thermal interactions (e.g., coagulation and vaporization), photoablation, plasma-induced ablation, and photodisruption. The degree of light-tissue interaction depends on tissue characteristics (such as the coefficients of reflection, absorption, and scattering) and the parameters of the irradiating light. The light related parameters include wavelength, pulse width and amplitude, pulse rate or the duration of a continuous-wave exposure, and the focal spot size of the light beam. It is important to note that very high optical intensities can be achieved through the combination of small spot sizes and short pulse durations. This condition holds even for moderate pulse energies.

Next, Sect. 6.6 introduces the phenomenon of random interference patterns, or speckle fields, in relation to the scattering of laser light from weakly ordered media such as tissue. The appearance of speckles arises from coherence effects in light-tissue interactions. Among the applications areas of this effect are the study of tissue structures and cell flow monitoring (see Chap. 10).

Finally, Sect. 6.7 presents the basic concepts of fluorescence, which is the property of certain atoms and molecules to absorb light at a particular wavelength and, subsequently, to emit light of a longer wavelength after a short interaction time. This physical phenomenon is widely used in a variety of biophotonics sensing, spectroscopic, and imaging modalities.

Following the descriptions in this chapter on light-tissue interactions, subsequent chapters discuss further applications such as biosensing, microscopy, spectroscopy, imaging, and the use of light in microsystems, nanophotonics, and neurophotonics.

6.1   Reflection and Refraction Applications

The characteristics of reflection and refraction are discussed in Chap. 2. These two factors are related by the Fresnel equations described in Sect. 2.4. This section describes some uses in biophotonics of reflection and refraction effects.

6.1.1 Refraction in Ophthalmology

Refraction is of biophotonics interest when dealing with transparent tissues such as certain ophthalmological material (e.g., eye materials of the cornea, lens, vitreous humor, and aqueous humor) [12–14]. As Fig. 6.2 shows, the human eye has a series of refractive media that focus the image of objects onto the retina. The cornea is the major optical medium in the eye because, in addition to having a large radius of curvature in the central area, it has a high refractive index of 1.376 compared to 1.00 for air. The crystalline lens has a refractive index of 1.406 at its center, and can change its shape dynamically to allow the eye to focus on objects located at various distances. Eye treatments involving considerations of the refractive index include using excimer (UV) lasers for surgical remodeling of the cornea to correct vision disorders and for cataract surgery to replace clouded lenses. The goals of refractive eye surgery are to treat degenerative eye disorders like keratoconus (a thinning condition of the cornea that causes visual distortion) and to alleviate or cure common vision disorders such as myopia (nearsightedness), hyperopia (farsightedness), and astigmatism.

> **Example 6.1** Consider a visible light beam of parallel light rays that impinges on a cornea, as shown in Fig. 6.2. Suppose the outer ray hits the cornea curvature at an angle $\theta_1 = 20°$ with respect to the surface normal. If the indices of refraction of air and the cornea are 1.000 and 1.376, respectively, what is the angle θ_2 at which the light from this ray enters the aqueous humor area behind the cornea?
>
> **Solution**: From Snell's law as given by Eq. (2.22) the refracted angle is
>
> $$\theta_2 = \sin^{-1}[(1.000 \sin 20°)/1.376] = 14.4°$$
>
> Thus the ray is bent toward the normal and then travels to the lens where it is focused onto the retina along with the other entering light rays.

The discussions in Sect. 6.5 on photoablation, plasma-induced ablation, and photodisruption light-tissue interactions describe further procedures and examples of eye surgery applications.

6.1.2 Specular Reflection

The amount of light that is reflected by an object depends strongly on the texture of the reflecting surface. Reflection from smooth surfaces such as mirrors, polished

Fig. 6.2 The major components of the human eye and their refractive indices

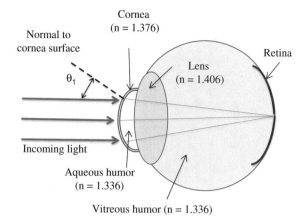

Vitreous humor (n = 1.336)

glass, or crystalline materials is known as *specular reflection*. In specular reflection the surface imperfections are smaller than the wavelength of the incident light, and the incident light is reflected at a definite angle following Snell's law. However, two characteristics of the reflected light need to be explained. First, except at normal incidence, when light interacts with a medium the intensities of the refracted and reflected beams do not add up to the intensity of the incident beam. The reason for this is that the intensity is defined as the power per unit area. Because the cross-sectional area of the refracted beam is different from the incident and reflected beams, the intensities of the refracted and reflected beams are not complementary. Only the total energy in the beams is conserved, if there are no optical power losses in the reflection and refraction processes.

A second factor related to reflection, as can be seen from Eqs. (2.25) and (2.26), is that the perpendicular reflectance and the parallel reflectance have different values and depend on the angle of incidence. Examples of the perpendicular and parallel reflectances for light traveling in air being reflected at an air-water interface are shown in Fig. 6.3 as a function of the incident angle. The reflectance for circularly polarized or unpolarized light is also shown in Fig. 6.3. Here $n_{air} = 1.00$ and $n_{water} = 1.33$. Although typically tissue surfaces are not smooth, the curves in Fig. 6.3 can be used to describe specular reflection from wet tissue surfaces. The perpendicular reflectance shown in Fig. 6.3 increases monotonically as the angle of incidence increases. On the other hand, the parallel reflectance first decreases gradually until it reaches a value of zero at a specific angle and then it increases again. The angle for zero parallel reflectance is called the *Brewster angle*. For an air-water interface the parallel reflectance reaches a value of zero at a 53° angle of incidence. Note that Eqs. (2.25) and (2.26) show that the reflectances depend on the refractive indices of the adjoining materials. Thus the value of the Brewster angle depends on the value of the refractive index and changes for interfaces between other materials with different refractive indices. For increasingly larger values of the angle of incidence beyond the Brewster angle, the parallel reflectance increases

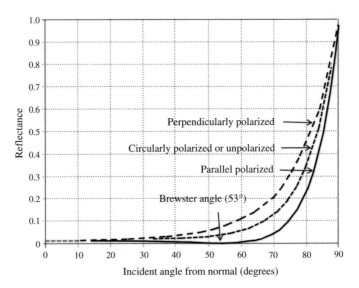

Fig. 6.3 Perpendicular, average (circular or unpolarized), and parallel reflectances for an air-water interface

monotonically and reaches a value of unity at 90°. This condition of unity reflectance means that the wave is traveling parallel to the material interface so that there is no reflected wave anymore.

Example 6.2 Consider an optical fiber that illuminates a tissue area as shown in Fig. 6.4. Let the fiber axis make a 60° angle with respect to the normal to the tissue surface. Suppose the fiber NA is such that the light emerging from the fiber illuminates an oval area with the near and far edges of the beam making angles of 50° and 70°, respectively. What are the perpendicular reflectances for points at each of these angles? Let $n_{air} = 1.000$ and $n_{tissue} = 1.400$.

Solution: Consider the perpendicular reflectance coefficient expression given in Prob. 2.10. In this problem $n_{21} = n_2/n_1 = n_{tissue}/n_{air} = 1.400$.

(a) For $\theta_1 = 50°$ and with $n_{21}^2 = 1.960$ the perpendicular reflectance is

$$R_\perp = (r_\perp)^2 = \left[\frac{\cos\theta_1 - \left(n_{21}^2 - \sin^2\theta_1\right)^{1/2}}{\cos\theta_1 + \left(n_{21}^2 - \sin^2\theta_1\right)^{1/2}}\right]^2$$

$$= \left[\frac{\cos 50° - \left(1.960 - \sin^2 50°\right)^{1/2}}{\cos 50° + \left(1.960 - \sin^2 50°\right)^{1/2}}\right]^2 = 0.085 = 8.5\,\%$$

(b) Similarly, for $\theta_1 = 70°$ the perpendicular reflectance is 25.4 %.

Fig. 6.4 Angular
dependence of reflectance of
light from a tissue with
n = 1.400

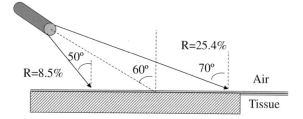

6.1.3 Diffuse Reflection

Diffuse reflection results from the reflection of light off of microscopically uneven surfaces for which the sizes of the surface imperfections are comparable or larger than the wavelength of the incident light. This type of reflection tends to send light in all directions, as Fig. 6.5 shows. Because most surfaces in the real world are not smooth, most often incident light undergoes diffuse reflection. Note however that the path followed by each individual light ray impinging on an uneven surface still obeys Snell's law as it hits an incremental surface area. Diffuse reflection also is the main mechanism that results in scattering of light within biological tissue, which is a turbid or random medium with many different types of intermingled materials, such as cell nuclei, organelles, and collagen, that reflect light in all directions. Such diffusely scattered light can be used to probe and get images of spatial variations in macroscopic optical properties of biological tissues. This is the basis of elastic scattering spectroscopy, also known as diffuse reflectance spectroscopy, which is a non-invasive imaging technique for detecting changes in the physical properties of cells in biological tissues (see Chap. 10).

When a beam of photons illuminates a diffusely scattering tissue sample, the *diffuse reflectance* is defined as the photon remission per unit surface area from the tissue. Monte Carlo simulation methods can predict the diffuse reflectance value accurately, but this is a computationally intensive process [15]. A faster but still mathematically involved computational approach is to use *diffusion theory*. Details of this method are presented in the literature [16–18].

Fig. 6.5 Illustration of
diffuse reflection from an
uneven surface

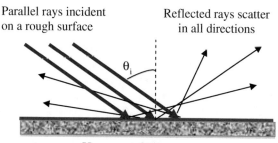

6.2   Absorption

Light absorption in tissue depends on the molecular composition. Molecules will absorb light when the photon energy equals an interval between quantum mechanical energy levels of the molecules. In biological matter the constituent atoms of molecules are principally elements such as H, C, N, O, and S. After light has entered a tissue, interactions between the electromagnetic fields of the incoming photons and the molecules of the medium will cause absorption of some of the light energy. This absorption results from the conversion of a portion of the light energy into thermal motion or an increase in molecular vibrations in the material. That is, photons will give up their energy to move the vibrational state of a molecule to a higher vibrational state. The degree of the absorption depends on factors such as the electronic structure of the atoms and molecules in the material, the wavelength of the light, the thickness of the absorbing layer, and the temperature.

6.2.1   Absorption Characteristics

According to the theory of quantum mechanics, the vibrational energy of a molecule is quantized into discrete energy levels. For an ideal system, the molecular motion can be described by a harmonic oscillator, which obeys *Hooke's Law*. This law assumes that when a system is displaced from equilibrium by a distance d, then there is a restoring force F that is proportional to the displacement by a factor k, that is, F = kd. The discrete energy levels for a harmonic oscillator are evenly spaced. However, this idealized situation does not hold in a real system, because the restoring force is no longer proportional to the displacement. In this case the molecular bonding energy can be described in terms of an *anharmonic oscillator* or a *Morse potential*, as shown in Fig. 6.6 [19–21]. In this description, the energy separation between vibrational levels decreases with increasing energy until a continuum state is reached where molecular breakup (dissociation) takes place (see Sect. 6.5.5 for the results of this effect).

In Fig. 6.6, at a specific vibrational level the displacement d is the oscillating variation of the distance between the nuclei of the constituent atoms of the molecule in an excited state and the equilibrium bond distance r_e. For example, at the vibrational level $v = j$ the interatomic distance d vibrates between $d_{j,min}$ and $d_{j,max}$. Thus as the vibrational energy level approaches the *dissociation energy* D_e, the increasingly large oscillating distances between the atoms eventually causes the molecule to break apart. In Fig. 6.6, the dissociation energy D_e is defined relative to the dissociated atoms and D_0 is the true energy required for dissociation of the molecule.

Light absorption is quantified by a parameter called the *absorption cross section* σ_a, which is the absorbing strength of an object. This parameter gives the proportionality between the intensity $I_0 = P_0/A$ of an incident light beam of power P_0

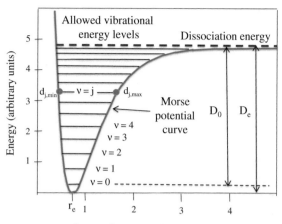

Fig. 6.6 Illustration of molecular vibrational energy levels described by a Morse potential

falling on the object of cross-sectional area A and the amount of power P_{abs} absorbed from the incident beam. That is, the absorption cross section has units of area and is defined by

$$\sigma_a = \frac{P_{abs}}{I_0} = \frac{P_{abs}}{P_0/A} \qquad (6.1)$$

The absorption cross section further defines a parameter μ_a, which is the *absorption coefficient* of the material. In a tissue that has a uniform distribution of identical absorbers with a number density ρ, the absorption coefficient is

$$\mu_a = \rho \sigma_a \qquad (6.2)$$

The absorption coefficient is defined as the probability per unit path length that a photon is absorbed in a particular material. It is measured in units of inverse length, such as cm^{-1} or mm^{-1}.

As noted in Sect. 5.1 in relation to photon absorption in semiconductor photo-diodes, in biological tissues there is an analogous behavior of photon absorption as a function of the distance x that the light penetrates into a tissue. This behavior states that in a tissue layer the intensity of a collimated light beam is attenuated due to absorption in accordance with the exponential *Beer-Lambert Law*

$$I(x) = (1 - R)I_0 \exp(-\mu_a x) \qquad (6.3)$$

where $I(x)$ is the intensity at a distance x into the tissue, I_0 is the intensity of the incident light, μ_a is the absorption coefficient of the material, and R is the coefficient of Fresnel reflection at the tissue surface at a normal beam incidence, as given in Eq. (2.35). The absorption-induced intensity decreases as a light beam travels through tissue is shown in Fig. 6.7 assuming there is no surface reflection. For example, Fig. 6.7 shows the remaining intensity $I(d)$ after the light has traveled to a

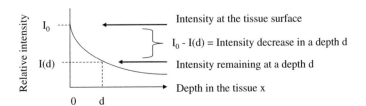

Fig. 6.7 Photon absorption as a function of the distance x that the light penetrates into a tissue

depth d into the tissue. Note that the Beer-Lambert Law sometimes is called the Lambert-Beer Law, Lambert's Law, or Beer's Law.

In most cases the absorption coefficient is wavelength dependent, particularly when the absorption coefficient is expressed explicitly in terms of the material properties. That is, suppose there are N different types of absorbing molecules in a medium. Let the concentration of the nth molecule type be c_n (measured in moles/liter) and let its *molar absorption coefficient* (also known as the *molar extinction coefficient*) be $\varepsilon_n(\lambda)$, which is measured in M^{-1} cm^{-1} where M is given in moles/liter. Then the absorption coefficient is given by

$$\mu_a = \sum_{n=1}^{N} c_n \varepsilon_n(\lambda) \tag{6.4}$$

This is the sum of the molar concentration of various absorbers c_n present in the material multiplied by the molar absorption coefficient of each absorber ε_n, which typically is wavelength dependent and is a measure of how strongly the nth molecule type absorbs light.

A parameter used in relation to absorption is the *optical density* $D(\lambda)$, which also is known as *absorbance*. For absorption over a distance d, using Eqs. (6.3) and (6.4), this parameter is defined as

$$D(\lambda) = ln\left(\frac{I_0}{I_a}\right) = \sum_{n=1}^{N} c_n \varepsilon_n(\lambda) d \tag{6.5}$$

For a larger optical density, more optical power gets attenuated as a beam propagates through a medium.

The inverse of the absorption coefficient is called the *penetration depth* or the *absorption length* L_a. Thus

$$L_a = 1/\mu_a \tag{6.6}$$

The absorption length is the light penetration distance x into the tissue for which the intensity has dropped to 1/e of its initial value, that is, when $x = L_a = 1/\mu_a$.

Example 6.3 Consider an epidermis skin tissue in which the absorption coefficient is 300 cm^{-1} at 308 nm and 35 cm^{-1} at 633 nm. Using Eq. (6.3), what are the absorption lengths for each of these wavelengths at which $I(x)/I_0 = 1/e = 0.368$?

Solution: (a) From Eq. (6.3) at 308 nm the intensity ratio is

$$\frac{I(x)}{I_0} = \exp(-\mu_a x) = \exp[(-300)x] = 0.368$$

Therefore

$$-(300\,cm^{-1})x = \ln 0.368 = -0.9997$$

which yields $x = 3.33 \times 10^{-3}$ cm = 0.033 mm.
(b) Similarly, at 633 nm

$$-(35\,cm^{-1})x = \ln 0.368 = -0.9997$$

which yields $x = 2.86 \times 10^{-2}$ cm = 0.286 mm.

6.2.2 Absorption in Biological Tissues

In general, absorption in biological tissues is due largely to water molecules and macromolecules such as proteins and chromophores [7, 22–25]. By definition, a *chromophore* is the part of an organic molecule that absorbs certain wavelengths and transmits or reflects other wavelengths. The reflection and transmission process causes the molecule or compound to have a specific color. As Fig. 6.8 shows, water is the main absorbing substance in the infrared region. Because water is the principal constituent of most tissues, its absorption characteristic is of importance in light-tissue interactions.

In the UV and visible regions the important absorbing substances include the aromatic amino acids (such as tryptophan, tyrosine, and phenylalanine), proteins, melanins, DNA, and porphyrins (which include hemoglobin, myoglobin, vitamin B12, and cytochrome). The aromatic amino acids exhibit large absorption peaks in the 200–230-nm spectrum and have local peaks in the 250–280 nm range. The absorption coefficients of three of these amino acids at their local peak absorption wavelengths are listed in Table 6.1. Proteins typically have a peak absorption value at about 280 nm, which depends on the concentrations of amino acids, such as tryptophan.

Two important absorbing substances in tissue are the chromophore melanin and the protein hemoglobin. *Melanin* is the basic chromophore that gives human skin,

Fig. 6.8 Absorption coefficients of select biological tissues and some common biological laser wavelengths

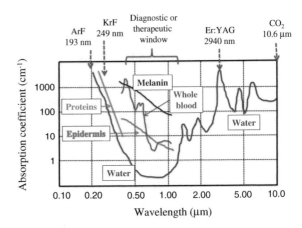

Table 6.1 Absorption coefficients of three amino acids at their local peak absorption wavelengths

Amino acid	Peak absorption wavelength (nm)	Absorption coefficient (cm^{-1})
Tryptophan	280	5400
Tyrosine	274	1400
Phenylalanine	257	200

hair, and eyes their color. As shown in Fig. 6.8, the absorption coefficient of melanin increases monotonically when moving from the visible toward the UV spectrum. Melanin is an effective absorber of light and is able to dissipate over 99.9 % of absorbed UV radiation. Because of this property, melanin can help protect skin cells from UV-B radiation damage.

Hemoglobin is the fundamental protein that transports oxygen from the lungs to the body and also functions as an antioxidant and regulator of iron metabolism. Each hemoglobin molecule within the red blood cells has four iron-containing heme groups plus the protein globin. Oxygen atoms readily bind to the iron atoms with the result that hemoglobin molecules carry 97 % of the oxygen in the blood. *Hemoglobin saturation* describes the extent to which hemoglobin is loaded with oxygen molecules. The hemoglobin is fully saturated when oxygen is attached to all four heme groups. Hemoglobin with no oxygen atoms bound to it is called *deoxyhemoglobin* and is denoted by Hb. When oxygen binds with the iron atoms, the resulting hemoglobin molecule changes its structure and now is called *oxyhemoglobin* (denoted by HbO_2). This structural change results in a change of the absorption spectrum. The spectra of deoxyhemoglobin and oxyhemoglobin are shown in Fig. 6.9 in terms of the molar extinction coefficient [26]. Both hemoglobin forms absorb strongly up to 600 nm and have absorption peaks around 280, 420, 540, and 580 nm, which are used for analyses of blood health. At the 600-nm wavelength the HbO_2 spectrum drops sharply by almost two orders of magnitude and then stays near the new low level for longer wavelengths. The absorption of the

Fig. 6.9 Spectra of deoxyhemoglobin and oxyhemoglobin in terms of the molar extinction coefficient (The hemoglobin data are compiled in Prahl [26])

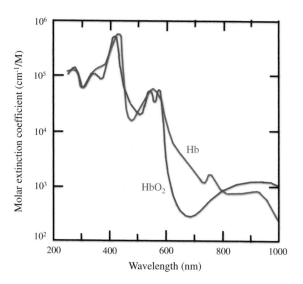

Hb spectrum decreases monotonically with increasing wavelength and eventually intersects the HbO_2 spectrum at around 800 nm.

Example 6.4 The comparison of the absorption coefficients of blood with and without oxygen binding at two different wavelengths forms the basis of pulse oximetry, which measures the oxygen content in blood. Consider an aqueous solution that contains a mixture of deoxyhemoglobin and oxyhemoglobin. In critical clinical settings (e.g., emergency rooms, intensive care units, and operating rooms), for such a solution the *oxygen saturation* S is a characteristic of interest for determining the cardio-respiratory status of a patient. This parameter is the percentage of hemoglobin saturated with oxygen. The task of the measurement system is to find S if the extinction coefficients and the sample thickness d are known.

Solution: One method for measuring S is through the comparison of the optical densities at two different wavelengths where there is a large difference between the extinction coefficients for deoxyhemoglobin and oxyhemoglobin. Let $\varepsilon_o(\lambda_i)$ and $\varepsilon_d(\lambda_i)$ be the extinction coefficients at wavelength λ_i for HbO_2 and Hb, respectively, and let c be the total concentration of hemoglobin molecules. If S is the oxygen saturation, then the optical densities $D(\lambda_1)$ and $D(\lambda_2)$ at wavelengths λ_1 and λ_2, respectively, are

$$D(\lambda_1) = S\varepsilon_o(\lambda_1)cd + (1-S)\varepsilon_d(\lambda_1)cd \qquad (6.7)$$

$$D(\lambda_2) = S\varepsilon_o(\lambda_2)cd + (1-S)\varepsilon_d(\lambda_2)cd \qquad (6.8)$$

By eliminating the factor cd and letting $D_{12} = D(\lambda_1)/D(\lambda_2)$, then solving these two equations for S yields

$$S = \frac{\varepsilon_d(\lambda_1) - \varepsilon_d(\lambda_2)D_{12}}{[\varepsilon_o(\lambda_2) - \varepsilon_d(\lambda_2)]D_{12} - [\varepsilon_o(\lambda_1) - \varepsilon_d(\lambda_1)]} \qquad (6.9)$$

where $D(\lambda_1)$ and $D(\lambda_2)$ are the measured optical densities at λ_1 and λ_2, respectively.

6.3 Scattering

Scattering of light occurs when photons encounter a compact object having a refractive index that is different from the surrounding material. The scattering process will redirect the photons to follow diverted paths. From a physics point of view, scattering in tissue is the same process as the scattering of sunlight by the atmosphere, which gives the sky a blue appearance and makes clouds appear white. However, scattering in biological media is much more involved due to the complex intermingled structure of tissues. Because most human tissues are heterogeneous materials that have spatial variations in their optical properties, they are considered to be cloudy or opaque materials known as *turbid media*. The consequence of these spatial density and optical variations is that tissues strongly scatter light. If the decrease in optical power due to absorption is small compared to scattering losses, then a large fraction of the photons in a light beam entering a tissue are scattered multiple times. This process, which results in a diffuse distribution of the light beam, is being used for imaging methodologies such as those described in Chap. 10.

Light scattering in tissue depends on the wavelength of the light together with the sizes, structures, and refractive indices of tissue components that are responsible for the scattering, for example, cell membranes, collagen fibers, and organelles. These scattering processes are widely used in both diagnostic and therapeutic biomedical photonics applications. In diagnostic applications such as imaging, disease-induced changes in the tissue can affect scattering properties, thereby giving diagnostic indicators for distinguishing healthy tissue from diseased tissue. In therapeutic applications using the principles of laser-tissue interactions (see Sect. 6.5), photon scattering data is useful for functions such as determining light dosage levels and for assessing progress during a light therapy session. Chaps. 9 and 10 have more details on these applications.

The light scattering elements are contained within the four basic tissue types, which are epithelium tissue (a sheet of cells that covers a body surface or lines a cavity inside the body), connective tissue, nervous tissue, and muscle tissue. Organs in the body are composed of a layered combination of these basic tissues. As an example, within the esophagus the first layer of the esophageal wall contains epithelial cells, as Fig. 6.10 shows. Covering this epithelium layer is connective tissue and several layers of smooth muscle cells. Glands, nerve fibers, and blood vessels run through the connective and muscle tissues.

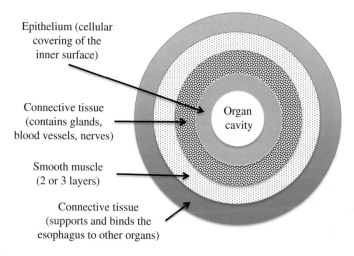

Fig. 6.10 Layered combination of basic organ tissues

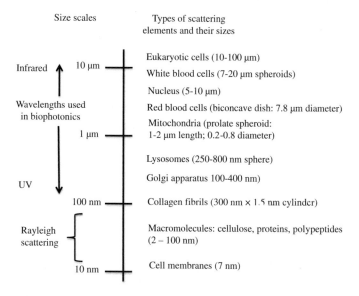

Fig. 6.11 Sizes of scattering elements in biological tissue

Table 6.2 Refractive indices
of some common tissue
components

Tissue component	Refractive index
Cytoplasm	1.350–1.375
Epidermis	1.34–1.43
Extracellular fluids	1.35–1.36
Human liver	1.367–1.380
Melanin	1.60–1.70
Mitochondria	1.38–1.41
Tooth enamel	1.62–1.73
Whole blood	1.355–1.398

As indicated in Fig. 6.11, the sizes of scattering elements in biological tissue range from about 0.01 to 20 μm. This corresponds to the sizes of biological tissue elements ranging from cell membranes to whole cells. As shown in Fig. 6.11, eukaryotic cell sizes generally range from 10 μm and higher. For example, red blood cells have the shape of a biconcave disk with a 7.8-μm diameter and 2.6-μm thickness, white blood cells are spheroids ranging in size from 7 to 20 μm, and platelets are oblate spheroids that are approximately 3.6 μm in diameter and 0.9 μm in thickness.

Photons are scattered most strongly by particles that have a size approximately equal to a wavelength used in diagnostic processes (500–1600 nm) and with a refractive index similar to the surrounding medium in which the particles are embedded. Thus two primary strong scattering elements in a cell are cell nuclei and mitochondria. The average refractive index of biological tissue ranges from 1.34 to 1.62, which is higher than the 1.33 value of water. Table 6.2 lists the refractive indices of some common tissue components.

Scattering can be either an elastic or inelastic process. In *elastic scattering* the incident and scattered photons have the same energy. This type of scattering describes most light-tissue interactions. The uses of this phenomenon in biophotonics techniques include elastic scattering spectroscopy and diffuse correlation spectroscopy (see Chap. 9). *Inelastic scattering* involves the exchange of energy between a photon and the molecule responsible for the scattering. As Sect. 6.3.2 describes, during inelastic scattering, energy can be transferred either from the photon to the molecule or the photon can gain energy from the molecule. This phenomenon is used in biophotonics techniques such as fluorescent scattering-based microscopy, multiphoton florescence scattering, Raman scattering microscopy, and coherent anti-Stokes Raman scattering.

6.3.1 Elastic Scattering

For a spherical particle of any size, light scattering can be analyzed exactly by the Mie scattering theory [16, 27]. Note that even though the shapes of some scattering elements in biological tissue are not necessarily spherical, their scattering behavior can be modeled fairly well by Mie theory applied to spheres of comparable size and refractive index. The Mie theory is somewhat mathematically involved and will not be addressed here. However, Mie scattering is of importance when examining scattering from blood cells, which have sizes ranging from about 2.5 to 20 μm depending on the blood cell type. If the size of the tissue particle is much smaller than the wavelength of the incident light (i.e., less than about 100 nm), the simpler Rayleigh scattering theory can be used.

Scattering is quantified by a parameter called the *scattering cross section* σ_s. This parameter indicates the scattering strength of an object and gives the proportionality between the intensity $I_0 = P_0/A$ of a light beam of power P_0 incident on the object of geometrical cross-sectional area A and the amount of power P_s scattered from it. That is, the scattering cross section is defined by

$$\sigma_s = \frac{P_s}{I_0} \qquad (6.10)$$

which has units of area. As depicted in Fig. 6.12, basically the scattering cross section describes the area that a scattering object removes from the cross section of an incident light beam during the process of diverting the amount of scattered power P_s from the beam. It is important to note that the scattering cross section is not the projected geometric area of an object, because different identically sized objects can have different scattering cross sections.

Similar to absorption, the non-scattered intensity component I_s remaining after light has traveled a distance x in a medium can be described by a Beer-Lambert law as

$$I_s(x) = I_0 exp(-\mu_s x) \qquad (6.11)$$

Here μ_s is the *scattering coefficient* of the material. The scattering coefficient is defined as the probability per unit path length that a photon is scattered in a particular material and is measured in units of cm^{-1} or mm^{-1}. In a tissue that has a uniform distribution of identical scatters with a number density ρ, the scattering coefficient is

$$\mu_s = \rho\sigma_s \qquad (6.12)$$

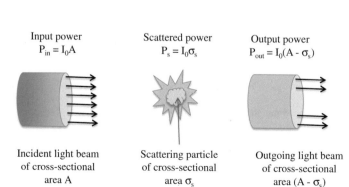

Input power
$P_{in} = I_0 A$

Scattered power
$P_s = I_0\sigma_s$

Output power
$P_{out} = I_0(A - \sigma_s)$

Incident light beam
of cross-sectional
area A

Scattering particle
of cross-sectional
area σ_s

Outgoing light beam
of cross-sectional
area $(A - \sigma_s)$

Fig. 6.12 Concept of the scattering cross section

The inverse of the scattering coefficient is called the *scattering mean free path* L_s. Thus

$$L_s = 1/\mu_s \tag{6.13}$$

which represents the average distance that a photon travels between scattering events. Figure 6.13 illustrates the photon scattering process and the concept of L_s.

Example 6.5 Consider an epidermis skin tissue in which the scattering coefficient is 1400 cm^{-1} at 308 nm and 450 cm^{-1} at 633 nm. What is the scattering mean free path for each of these wavelengths at which $I(x)/I_0 = 1/e = 0.368$?

Solution: (a) From (6.11) at 308 nm the intensity ratio is

$$\frac{I(x)}{I_0} = \exp(-\mu_a x) = \exp[(-1400)x] = 0.368$$

Therefore

$$-(1400\,\text{cm}^{-1})x = \ln 0.368 = -0.9997$$

which yields $x = 7.14 \times 10^{-4}$ cm = 7.14 μm.
(b) Similarly, at 633 nm

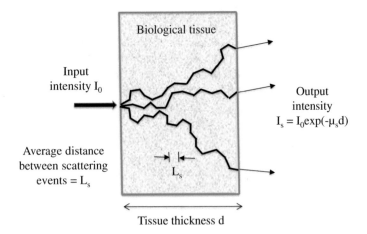

Fig. 6.13 The photon scattering process and the concept of L_s

$$-\left(450\,\mathrm{cm}^{-1}\right)x = \ln 0.368 = -0.9997$$

which yields $x = 2.22 \times 10^{-3}$ cm = 0.022 mm.

6.3.2 Rayleigh Scattering

Rayleigh scattering is a category of elastic scattering. In this particular process the scattering particles are smaller than the wavelength of the light impinging on the tissue [3, 5, 16, 27]. If the incoming light is unpolarized, a superposition of lightwaves with all possible polarizations yields the following intensity distribution due to Rayleigh scattering

$$I_s = I_0 \frac{8\pi^2\alpha^2}{r^2\lambda^4}\left(1+\cos^2\theta\right) \tag{6.14}$$

Here r is the distance from an observation point to the scattering molecule, θ is the scattering angle measured relative to the original propagation direction of the photon, and α is the *molecular polarizability*, which is proportional to the dipole moment induced in the molecule by the electric field of the light (see Sect. 2.7). The molecular polarizability of a sphere with radius a is given by [16]

$$\alpha = \frac{n_{rel}^2 - 1}{n_{rel}^2 + 2}a^3 \tag{6.15}$$

where $n_{rel} = n_s/n_b$ is the relative refractive index of the sphere of index n_s embedded in a background material of index n_b. The Rayleigh scattering cross section is given by

$$\sigma_s = \frac{8\pi k^4 \alpha^2}{3} \tag{6.16}$$

where $k = 2\pi n_b/\lambda$ is the propagation constant.

Example 6.6 Consider Eq. (6.14) for Rayleigh scattering. Show the effect that the λ^{-4} behavior has on the relative intensity I_s/I_0 in the range 300–800 nm.

Solution: Because only the λ^{-4} behavior is of interest in this question, let the parameter defined by

Fig. 6.14 Comparison of
Rayleigh scattering in the UV,
visible, and IR regions

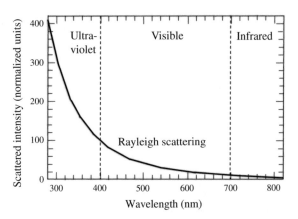

$$A \equiv \frac{8\pi^2\alpha^2}{r^2}\left(1+\cos^2\theta\right)$$

be a fixed quantity. Then the relative intensity I_s/I_0 is shown in Fig. 6.14 as a
function of λ. The plot is given in normalized units such that $A/(300\ \text{nm})^4 = 100$. Note that even in the range of the visible spectrum, the
Rayleigh scattering effect is significantly smaller in the red region (around
680 nm) compared to the blue region (around 450 nm).

Example 6.7 Consider a spherical particle that has a radius $a = 10$ nm and a
refractive index $n_s = 1.57$. (a) If the sphere is in a background material of
refractive index $n_b = 1.33$, what is the scattering cross section at a wavelength
$\lambda = 400$ nm? (b) How does this compare to the geometric cross section of the
sphere?

Solution: (a) First from Eq. (6.15) with $n_{rel} = n_s/n_b = 1.57/1.33 = 1.18$ and
with $k = 2\pi n_b/\lambda$, it follows from Eq. (6.16) that the scattering cross section is

$$\sigma_s = \frac{8\pi k^4\alpha^2}{3} = 2.15 \times 10^{-20}\text{m}^2$$

(b) The geometric cross section of the sphere is $\pi a^2 = 3.14 \times 10^{-18}\ \text{m}^2$

The *scattering efficiency* Q_s is another parameter used in describing scattering.
This parameter is defined as [16]

$$Q_s = \frac{8x^4}{3}\left(\frac{n_{rel}^2 - 1}{n_{rel}^2 + 2}\right)^2 \tag{6.17}$$

Fig. 6.15 Illustration of angular scattering

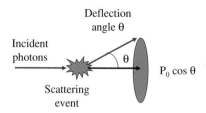

where x = $2\pi n_b a/\lambda$. The scattering efficiency gives the ratio of the energy scattered by a particle to the total energy in the incident beam that is intercepted by the geometric cross section of the particle.

6.3.3 Anisotropy Factor

Observations of scattering in biological tissues have shown that photons tend to be scattered in the forward direction. In practice it is useful to define a probability function $p(\theta)$ to predict the scattering direction of a photon at an angle θ, as shown in Fig. 6.15. If $p(\theta)$ does not depend on θ, then the scattering is isotropic (equal intensity in all directions). If $p(\theta)$ does depend on θ, then the scattering is anisotropic (having different values in different directions).

Anisotropy is a measure of the number of scattered photons continuing in the forward direction after a single scattering event. A commonly used approximation for $p(\theta)$ that fits well with experimental data is the Henyey-Greenstein function [3, 16]

$$p(\theta) = \frac{1 - g^2}{(1 + g^2 - 2g \cos \theta)^{3/2}} \tag{6.18}$$

The *scattering anisotropy factor* g is the mean cosine of the scattering angle θ. The value of g varies from −1 to +1; g = −1 denotes purely backward scattering, g = 0 indicates isotropic scattering, and g = +1 denotes total forward scattering, which describes Mie scattering for large particles.

Example 6.8 From Eq. (6.18) plot the value of $p(\theta)$ for values of g = 0.8 and 0.9 for scattering angles ranging from 0 to 40°.

Solution: The curves from Eq. (6.18) are given in Fig. 6.16. In most biological tissues scattering occurs predominantly in the forward direction, that is, for g approaching the value 1.

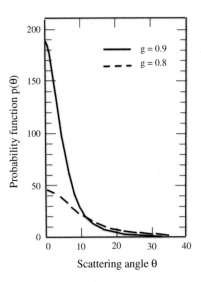

Fig. 6.16 Henyey-Greenstein function for two coefficients of anisotropic scattering

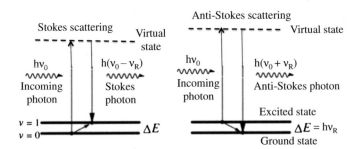

Fig. 6.17 Illustration of Stokes scattering (*left*) and anti-Stokes scattering (*right*); the thick horizontal lines represent the vibrational energy levels of the material (e.g., $v = 0$ and $v = 1$)

6.3.4   Inelastic (Raman) Scattering

A major type of inelastic scattering is *Raman scattering* of photons by their interaction with a molecule [28–30]. This effect is used in Raman spectroscopy and is complementary to infrared absorption spectroscopy methods to obtain information about molecular compositions, structures, and interactions in biological tissue samples. As noted in Fig. 6.6, molecules directly absorb photons during an infrared absorption process. This effect excites the molecules to higher vibrational states. In contrast, in a Raman inelastic scattering event two possible reactions can occur during the interaction of a photon with a molecule. Either a small amount of energy is transferred from the photon to the molecule, or the molecule can transfer some energy to the photon.

The molecular transitions occurring in these two processes are shown in Fig. 6.17. Here $h\nu_0$ is the energy of the incident photon and $h\nu_R$ is the energy gained or given up by the photon through the Raman scattering event. In the process of inelastic collisions between photons and vibrating molecules, an extremely short-lived transition state or *virtual state* is created. This transition state brings the collective quantum energy state of the molecule and the photon to a high energy level, which is called the *virtual level*. From this virtual level the molecule can then relax to an energetically excited molecular level (e.g., $v = 1$) if the original molecular state was the ground state, or to an energy level (e.g., $v = 0$) below its initial state if the original molecular state was already in a higher excited state (e.g., $v = 1$). The first process (from its lowest vibrational energy level to a higher vibrational state) is called *Stokes scattering*. If the incoming photon interacts with a molecule that already is in a higher vibrational state, the molecule can transfer some of its energy to the photon during the scattering event. This process is called *anti-Stokes scattering*.

6.4 Scattering with Absorption

The previous sections addressed absorption and scattering independently. However, in actual tissues these two processes are present simultaneously. In this case the *total extinction coefficient* μ_t (also referred to as the *total interaction coefficient*) for the combined absorption coefficient and scattering coefficient is given by

$$\mu_t = \mu_a + \mu_s \tag{6.19}$$

which is the probability of photon interaction with a medium per infinitesimal path length. This leads to the *total mean free path* L_t of an incident photon

$$L_t = \frac{1}{\mu_t} = \frac{1}{\mu_a + \mu_s} \tag{6.20}$$

In some spectral regions either μ_a or μ_s can be a dominant attenuation factor in tissue. Both mechanisms are always present, but generally occur in variable ratios.

Another factor of interest when considering both absorption and scattering is the *optical albedo a*. This parameter is defined as the ratio of the scattering coefficient to the total attenuation coefficient, that is,

$$a = \frac{\mu_s}{\mu_t} = \frac{\mu_s}{\mu_a + \mu_s} \tag{6.21}$$

For attenuation that is exclusively due to absorption $a = 0$, whereas $a = 1$ when only scattering occurs. For the case $a = 0.5$, the coefficients of absorption and scattering are of equal magnitude, that is, $\mu_a = \mu_s$.

Similar to the scattering efficiency given by Eq. (6.17), the *absorption efficiency* Q_a is defined as the fraction of the incident light beam absorbed per unit geometrical cross-sectional area of the absorbing particle. Combining the absorption and scattering efficiencies then gives the *total extinction efficiency* Q_{ext}, which is the sum of the absorbed and scattered energy of the light beam

$$Q_{ext} = Q_a + Q_s \qquad (6.22)$$

Example 6.9 Suppose that a stomach muscle tissue sample has an absorption coefficient of 3.3 cm^{-1} and a scattering coefficient of 30 cm^{-1} for a Nd:YAG laser emitting at a wavelength of 1064 nm. What is the optical albedo of the tissue?

Solution: From Eq. (6.21) the optical albedo is

$$a = \frac{\mu_s}{\mu_t} = \frac{\mu_s}{\mu_a + \mu_s} = \frac{30}{3.3 + 30} = 0.90.$$

6.5 Light-Tissue Interaction Mechanisms

Interactions between natural light and biological tissue are important for life, but also can cause harmful effects. On the beneficial side, plants require sunlight for the photosynthesis process needed for their survival, and humans depend on UV light to produce essential vitamin D in their skin. However, too much exposure to natural light can produce skin inflammations, can lead to skin cancer, and can cause wrinkles and age spots.

On the other hand, interactions between artificial light and biological tissue is a major aspect of biophotonics that is used extensively in medical and cosmetic therapy for procedures such as treatments of vascular lesions, removal of pigmented lesions and tattoos, speeding of wound healing, resurfacing of skin to remove scars and cosmetic defects, and for cancer detection and treatment. In addition, numerous laser-based imaging techniques using a variety of light-tissue interaction processes are deployed worldwide.

Depending on the characteristics of a specific light source and the properties of the irradiated tissue, a wide range of light interactions with biological tissue can occur. The main tissue properties involved in these interactions are the coefficients of reflection, absorption, and scattering together with heat conduction and heat capacity of the tissue. Because the absorption and scattering coefficients of tissue

Fig. 6.18 Tissue penetration depth of some common medical lasers

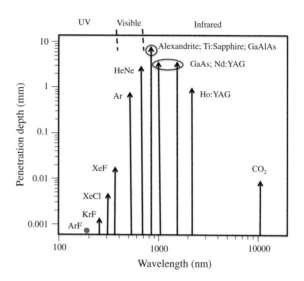

components depend strongly on wavelength, there are significant variations in the penetration depth of the light from lasers operating at different wavelengths. In Fig. 6.18 examples of the light penetration depths are shown for different lasers across the 190-nm to 10-μm spectrum. For example, light from a 193-nm ArFl laser or a 2.96-μm Er:YAG laser is totally absorbed in the first few micrometers of tissue owing to high absorption from amino acids in the UV and water absorption in the IR. In contrast, because of an absence of strongly absorbing chromophores, collimated light in the 600–1200-nm region can penetrate several centimeters into tissue and the associated scattered light can travel several millimeters through the tissue. As the collimated beam passes through tissue, it is exponentially attenuated by absorption and generates a heating effect in the tissue. Scattering also attenuates the collimated beam and the scattered light in the tissue forms a diffuse volume around the collimated beam as the light travels through the tissue.

Note that some terminology confusion can occur because the terms *intensity* and *irradiance* both are used in the literature to describe the power level falling on a specific tissue area [1–8, 31]. This terminology variation often is traditional and depends on the particular topic. In this section, the term *irradiance* will be used to designate the light exposure rate given in $J/s/cm^2$ or W/cm^2. The important parameters of tissue exposure include the following:

- The wavelength (or the photon energy) of the incident light
- The optical power emitted by the source
- The irradiance (the power per unit area usually measured in W/cm^2)
- The spot size (the area irradiated on the tissue)
- The spatial profile of the spot (the irradiance variation across the beam)
- The irradiation time for a CW laser

Fig. 6.19 Categories of light-tissue interaction modes and their irradiance and exposure characteristics

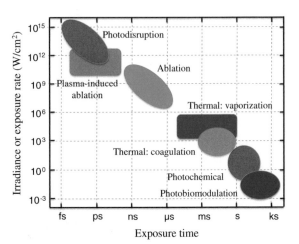

- The pulse duration for a pulsed laser
- The temporal profile of the pulse duration (the irradiance variation with time during the pulse)
- The spectral profile of the light beam (the variation in irradiance as a function of wavelength)
- The polarization state of the light

The key parameters of CW light sources include the optical power, irradiation time, and spot size of the light beam. For pulsed lasers the key parameters are the energy per pulse, irradiation time, spot size, pulse repetition rate (how often the optical power delivery takes place), and number of pulses administered. For CW light an important parameter is irradiance (given in W/cm^2), which is a function of the power delivered and the laser spot size on the tissue.

The interactions of light with biological tissue can be categorized according to the irradiance (or equivalently, the exposure rate) measured in W/cm^2 (or $J/s/cm^2$) and the exposure time, as shown in Fig. 6.19. The light-tissue interaction conditions can range from very high irradiances over extremely short times (10^{10}–10^{15} W/cm^2 over corresponding time periods of 10^{-9}–10^{-15} s, that is, nanoseconds to femtoseconds) to the situation of low irradiances over long time periods (1–100 mW/cm^2 over corresponding time periods of hours to minutes). Thus the irradiances span a range of nineteen orders of magnitude and the light exposure time ranges over more than sixteen orders of magnitude. Note that the circles, ovals, and rectangles shown in Fig. 6.19 merely give approximate ranges of the irradiances and exposure times for the generic light-tissue interaction modes described below.

In the various modes of interaction the *radiant exposure* or *energy density* ranges from approximately 1 mJ/cm^2–1 kJ/cm^2. The energy density sometimes is used as a parameter to describe light doses, but this can lead to incorrect applications, because energy is defined as power multiplied by time, that is,

$$\text{energy(J)} = \text{power(W)} \times \text{time(s)} \tag{6.23}$$

Thus, if the power is doubled and the time is halved, the same energy is delivered to the tissue but a different biological effect may be observed. A more precise dosage parameter definition therefore is to prescribe a given irradiance during a specific illumination time.

Example 6.10 Consider two tissue samples that have areas of 0.1 and 1 cm^2. If the samples each are illuminated for 1 s with an irradiance of 10 mW/cm^2, what is the energy deposited in each of these tissue samples?

Solution: The energy is given by
Energy = Irradiance $(W/cm^2) \times$ pulse time(s) $\times$ area (cm^2)

(a) $\text{Energy}_1 = (10 \ mJ/s/cm^2) \times 1 \ s \times 0.1 \ cm^2 = 1 \ mJ$
(b) $\text{Energy}_2 = (10 \ mJ/s/cm^2) \times 1 \ s \times 1 \ cm^2 = 10 \ mJ$

The fundamental light-tissue interaction modes for therapeutic and surgical uses can be broadly categorized into the following six classes:

- **Photobiomodulation**: This interaction class [also called *biostimulation* or *low-level light therapy* (LLLT)] deals with the stimulation of biological systems by means of low irradiance levels. In contrast to other interaction modes, here the light levels do not result in tissue ablation, heating effects, or cell destruction. Instead the irradiances induce a photochemical effect to stimulate tissue to heal and recover from inflammation and other effects. Its clinical applications include alleviating acute and chronic pain, treating sprains and strains from sports injuries, speeding up wound healing, treating nerve and brain injuries, and promoting tissue and bone healing.
- **Photochemical reaction**: An increase in the irradiance beyond that used for photobiomodulation can cause an electron in a molecule to absorb a high-energy photon. As a result, molecules with an electron in a higher energy state can undergo a chemical reaction by exchanging or sharing electrons with other molecules. Such a photochemical light-tissue interaction can destroy unwanted, cancerous, or diseased cells. This mode can be used to treat certain skin diseases and to kill cancer cells, for example, in processes such as photodynamic therapy.
- **Thermal effects**: Coagulation and vaporization are thermal effects that are based on localized heating of a selective region using a wavelength at which the tissue exhibits high absorption. The conversion of photon energy into heat energy occurs via increased molecular vibrations due to photon absorption and from collisions of photons with molecules. The localized coagulation and vaporization effects are used in disciplines such as tissue cutting in laser surgery, tissue ablation for scar removal and facial contouring, and tissue bonding in ophthalmology.

- **Photoablation**: When electrons in a molecule absorb high-energy UV photons, the electrons can be raised to such high non-bonding orbitals that the molecule breaks up immediately. This molecular breakup causes a rapid expansion of the irradiated tissue volume with an accompanying ejection of tissue molecules. The tissue is removed in a very clean and precise fashion without the tissue damage that can arise from thermal effects in coagulation and vaporization processes. Note that this photoablation process differs from the ablation due to a thermal vaporization effect.
- **Plasma-Induced Photoablation**: When intense irradiation levels of greater than 10^{11} W/cm^2 impinge on biological tissue in pulses of less than one nanosecond duration, molecules are torn apart and form a plasma. As the plasma expands it ablates tissue material in a clean and precise fashion. One application is in the treatment of cataracts.
- **Photodisruption**: Similar to plasma-induced photoablation, in photodisruption very high irradiance levels result in plasma formation. In this case, mechanical effects such as shock waves, jetting of material, and bubble formation followed by cavitation, accompany the plasma-creation process. Photodisruption is widely used for minimally invasive surgery, for breaking up kidney stones or gallstones, and in ophthalmology for drilling holes in the cornea or lenses.

The following subsections give some details on these light-tissue interaction categories. Table 6.3 summarizes their characteristics.

6.5.1   Photobiomodulation

Photobiomodulation [also called *biostimulation* or *low-level light therapy* (LLLT)] refers to the stimulation of biological systems by means of low light levels with the aim to provide a therapeutic, healing, or cosmetic effect [32–38]. In photobiomodulation, cells or tissue are exposed to low levels of red to near-IR light from low-power lasers or high-radiance LEDs to either stimulate or (in some occasions) to inhibit cellular functions. This procedure is used in numerous medical areas to reduce cell and tissue death, relive acute and chronic pain and inflammation, accelerate wound healing, treat inflamed and ulcerated oral tissues, alleviate tinnitus and other nerve injury conditions, and treat edema (swellings), indolent (non-healing) ulcers, burns, and dermatitis.

Light sources commonly employed for photobiomodulation emit at wavelengths between 600 and 1070 nm. Normally optical power levels range from 1 to 1000 mW with irradiance levels varying from 1 mW/cm^2 to 5 W/cm^2. The light sources typically run in a pulsed mode, but sometimes continuous-wave beams are delivered. Treatment time ranges are 30–60 s per treatment point a few times a week for up to several weeks. The skin layers that are relevant to photobiomodulation consists of the upper epidermis and lower dermis. These layers are approximately 100–150 μm and 1500–3000 μm thick, respectively, and are separated by a

Table 6.3 Summary for six generic laser-tissue interaction effects

Interaction mode	Tissue effect	Typical sources	Exposure time	Irradiance levels	Applications
Photobiomodulation	Photochemical effects	600–1100 nm lasers and LEDs	30 s–10^3 s treatments with pulsed or CW light	1 mW/cm^2 to 1 W/cm^2	Heal tissue, nerve injuries, wounds; reduce inflammation and pain
Photochemical	Photochemical effects	Red LEDs or lasers	500 ms–600 s	100 mW/cm^2 to 80 W/cm^2	Photodynamic therapy
Thermal	Local changes in tissue temperature produce various effects	CO_2, Nd:YAG, Er:YAG	50 µs–10 s	10–10^5 W/cm^2	Coagulation, vaporization
Photoablation	UV photons used to break molecular bonds; precise ablation	ArF, KrF, XeCl	1 ns–1 µs	10^7–10^{11} W/cm^2	Ophthalmology
Plasma-induced ablation	Very clean ablation from an ionized plasma	Nd:YAG	100 fs to 500 ps	10^{10}–10^{13} W/cm^2	Ophthalmology, caries therapy
Photodisruption	Cutting and fragmentation of tissue by mechanical forces	Nd:YAG, Ti-sapphire	10 fs to 200 ps	10^{11}–10^{16} W/cm^2	Lithotripsy, ophthalmology

Fig. 6.20 A general concept
of the perceived underlying
mechanism for
photobiomodulation

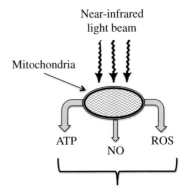

basement membrane (a thin connective tissue). The photons involved in photo-biomodulation usually are attenuated before they reach the deeper fatty layer of the skin.

In contrast to other laser-tissue interactions that use higher optical power levels to generate ablation, heating, or cell destruction effects, photobiomodulation acts by inducing a photochemical therapeutic reaction in the cell. The *first law of photo-biology* states that in order for low-power visible light to generate any reaction in biological tissue, the electronic absorption bands associated with molecular pho-toacceptors or chromophores in the tissue must absorb the photons [33–38]. Thus photobiomodulation is based on the principle that, when light impinges on certain chromophore molecules, the photon energy causes electrons to jump from low-energy levels to higher energy levels. In nature, the energy that is thereby stored in the higher energy levels in the chromophore molecules can be used by the biological system to carry out various cellular tasks, such as photosynthesis and photomorphogenesis (the control of plant development by light). Examples of chromophores include chlorophyll in plants, bacteriochlorophyll in bluegreen algae, flavoproteins or respiratory enzymes in cells, and hemoglobin in red blood cells.

A general concept of the perceived underlying mechanism for photobiomodu-lation is illustrated in Fig. 6.20. Because mitochondria play an important role in energy generation and metabolism, any proposed mechanism of photobiomodula-tion involves some type of interaction of light with mitochondria. It has been found that mitochondrial components absorb red to near-infrared light, which results in the transformation of light energy to metabolic energy thereby producing a mod-ulation of the biological function of cells [36]. Investigations have shown that illumination by laser light increases both the intracellular reactive-oxygen species (ROS) and the synthesis of adenosine triphosphate (ATP) in mouse embryonic fibroblasts (cells that maintain the structural integrity of connective tissue) and other cell types in culture. In addition, the release of nitric oxide (NO) has been observed.

- ROS are chemically active molecules (e.g., superoxide and hydrogen peroxide) that are involved in cell signaling, regulation of cell cycle progression, the activation of enzymes, and the synthesis of nucleic acid and protein.
- ATP is stored in the cytoplasm and nucleoplasm of every cell, and directly supplies the energy for the operation of essentially all physiological mechanisms.
- NO is a biological agent that dilates blood vessels. The NO molecules prevent the muscles in artery walls from tightening, thereby preventing the walls from narrowing. As a result, blood flows more easily through the arteries, so that blood pressure is reduced and the heart does not have to pump as hard.

The results of these processes are cell proliferation, growth factor production, and an enhancement of cell motility (which describes the movement and energy consumption of cells).

Among the clinical applications that have been examined for photobiomodulation are the following [32–38]:

- Physiotherapy and rehabilitation clinics: alleviate acute or chronic pain, reduce neck and lower-back pain, improve mobility
- Dermatology: treat edema (swelling of body tissues), promote burn healing, and treat dermatitis
- Surgery: promote wound and tissue healing, alleviate post-operative pain
- Neurology: treat conditions related to stroke, traumatic brain injury, degenerative brain disease, spinal cord injury, and nerve conditions
- Rheumatology: treat arthritis and other rheumatic diseases
- Dentistry: treat inflamed oral tissues, promote ulceration healing, improve bone formation and healing
- Sports medicine: treat sports injuries such as sprains and strains, muscle soreness and fatigue, and tendon injuries
- Veterinary medicine (e.g., dogs and racehorses): reduce swelling, increase mobility after leg injuries, promote wound and tissue healing.

6.5.2 Photochemical Interaction

Whereas photobiomodulation processes are based on a photochemical effect that uses low irradiances to promote tissue healing and cell growth, an increase in the irradiance results in a different photochemical light-tissue interaction methodology that is used to destroy unwanted, cancerous, or diseased cells. For example, in processes such as *photodynamic therapy* (PDT) the interaction of light with photosensitive drugs produces reactive oxygen species that cause necrosis (cell death) or apoptosis (programmed cell death). PDT is a treatment modality for a wide range of diseases and skin conditions that involve fast-growing or pathological cells or

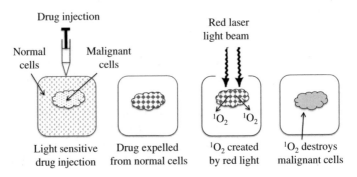

Fig. 6.21 Concept of photodynamic therapy procedure

bacteria [39–42]. The applications of PDT include treatment of acne, age-related macular degeneration, arteriosclerosis, and cancer.

As shown in Fig. 6.21, the PDT process uses a specific light wavelength in conjunction with a light-sensitive drug that is administered before exposure to the light. The drug can be mixed into a drink, injected into the body, or administered topically on a diseased skin area or a tumor. The selected wavelength (typically from a laser emitting in the red spectral region) corresponds to the absorption peaks of the photosensitive drug. The photosensitive agent accumulates strongly in malignant cells but not in healthy tissue. The role of the drug is to absorb the light energy and transfer it to the diseased cells. The process of the photosensitive agent absorbing energy from the light results in the formation of singlet oxygen ($^1O_2$). The singlet oxygen is a highly toxic reactant that causes an irreversible oxidation of cell structures. The reactant has a short lifetime (<0.04 µs) in biological tissue and thus has a short radius of interaction (<0.02 µm). As a result, the oxidative damage caused by the singlet oxygen takes place in the immediate vicinity of where the photosensitive molecule is located. Thereby the photochemical process causes chemical destruction of the malignant cells.

As Fig. 6.8 shows, the wavelength range from 620 to 760 nm is a region of high penetration depth into tissue. Thus for treatment efficiency of the PDT process, it is advantageous to select drugs that are sensitive to light in this spectral range. An advantage of using laser diode sources is that the light can be transmitted directly from the laser to an internal treatment site (such as the esophagus, the stomach or intestines, the urinary bladder, and the lungs) by means of optical fibers [39]. The exit port of the fiber can be a flat or an angle-polished bare end or some type of diffuser can be attached on the fiber tips for homogeneous spot or cylindrical light distribution. The diffuser could be a microlens, a side-firing fiber, or a short length of side-emitting fiber, as shown in Fig. 6.22. A cylindrical side-emitting light diffuser typically is a plastic or a specially constructed glass fiber that produces a radial light pattern, which is homogeneous along its entire length. Standard cylindrical diffuser fiber sections range from 10 to 70 mm in length, have a nominal core diameter of 500 µm, a NA of 0.48, and a minimum bend radius of 10 mm.

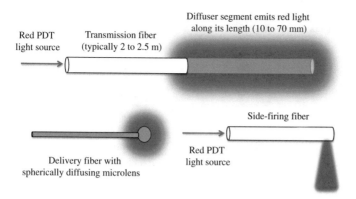

Fig. 6.22 Examples of light distribution schemes from optical fibers for PDT

Their main uses are for intraluminal PDT, such as in arteries or veins, in the esophagus, in gastrointestinal tracts, in urinary tracts, or in bronchi in the lungs.

In addition to using fibers to deliver the therapeutic light, optical fibers also are used to monitor the therapeutic light fluence, the concentration of the tumor sensitizer, and the oxygen levels produced by the PDT process [39]. A wide range of optical fiber types can be used in PDT depending on the medical application and the desired light distribution pattern. Bare-ended fibers typically are silica glass fibers with large NAs (normally less than 0.60) and core diameters ranging from 200 to 1000 μm. These fiber types allow the setting of a well-defined distance between the fiber end and the tissue being treated or examined, and thus can enable precise irradiations to be administered or accurate values of tissue properties to be collected.

Example 6.11 Consider the microlens diffuser device shown in Fig. 6.22. Assume the diffuser produces a spherical distribution of the light emerging from the fiber. If P is the optical power level emitted from the fiber, what is the irradiance at a distance R from the center of the microlens?

Solution: For a sphere of radius R, the surface area is $4\pi R^2$. Therefore at a distance R from the center of the microlens, the optical power per unit area (the irradiance) is $P/4\pi R^2$.

6.5.3 Thermal Interaction

Thermal interactions are characterized by an increase in tissue temperature in a local region [43, 44]. The temperature change is the main variable in these

Table 6.4 Effects on tissue at different temperature levels

Temperature (°C)	Biological effect
37	Normal body temperature
45	Hyperthermia
50	Enzyme activity reduction, cell immobility
60	Coagulation
80	Membranes become permeable
100	Vaporization and ablation
>100	Carbonization
>300	Melting

interactions and can be generated by either CW or pulsed laser light. The energy that produces thermal reactions is significantly larger than the optical energy associated with irradiance levels that produce photochemical reactions. The irradiance associated with thermal interactions ranges from approximately 10 W/cm^2 in application times of seconds to 10^6 W/cm^2 over time periods of microseconds. Depending on the duration and the peak value of the temperature increase of the tissue, the resulting effects in the tissue can include coagulation, vaporization, carbonization, and melting. Table 6.4 lists the effects on tissue at different temperature levels.

When a light beam impinges on a tissue area, most of the light that is absorbed is converted to heat. The rate of heat production per volume of tissue at a specific point r is the product of the total fluence rate (given in J/cm^2) at r and the absorption coefficient at r. For a short pulse of light of duration τ and irradiance E_0 impinging on a tissue with an absorption coefficient μ_a, the resultant maximum temperature rise ΔT is

$$\Delta T = \frac{\mu_a E_0 \tau}{\rho C} \tag{6.24}$$

Fig. 6.23 Thermal effect on tissues at different temperature levels include hyperthermia, coagulation, carbonization, and vaporization

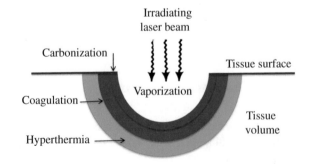

Here ρ is the tissue density (given in gm/cm^3) and C is the heat capacity (given in J/gm/°C). To a good approximation the value $\rho C \approx 4.2$ J/(cm^3/°C) holds for tissue.

The thermal effect on tissues at different temperature levels is illustrated in Fig. 6.23. As a tissue increases in temperature from its normal state of 37 °C, various stages of tissue denaturation are manifested. *Denaturation* refers to the disruption and possible destruction of both the secondary and tertiary structures of proteins. Cell activity and possibly cell death can result if proteins in a living cell are denatured. As shown in Table 6.4, at temperatures up to about 45 °C and beyond, a condition known as *hyperthermia* results from weakening or destruction of molecular bonds and alterations in membrane structures. If such a hyperthermia condition persists for several minutes, the tissue will undergo necrosis (death of living cells or tissue). Starting around 50 °C there is a noticeable reduction in enzyme activity. This effect results in cell immobility, reduced energy transfer in the cell, and a diminishing of certain cell repair mechanisms.

The desired therapeutic effect of coagulation commences around 55 °C and can be used up to 90 °C to heal cuts and wounds. The *coagulation* effect is due to the denaturation of proteins and collagen. Typically a Nd:YAG laser operating at 1064 nm is a good candidate for the irradiating process, because this wavelength offers a good penetration depth with sufficient tissue absorption. A common coagulation procedure is known as the *percutaneous technique*, and also is known as percutaneous laser ablation (PLA), laser-induced interstitial thermotherapy (LITT), or interstitial laser-induced thermotherapy (ILT). The localized heating is produced by means of optical fibers that carry the laser energy into the designated tissue area.

Increasing the temperature further to 100 °C initiates *vaporization* of the water molecules contained in most tissues. During the vaporization process, the high value of the latent heat of vaporization for water (2270 kJ/kg) helps in carrying away excess heat from the surrounding region, thereby preventing heating damage to the adjacent tissue. As the water molecules undergo a phase transition, they create large gas bubbles that induce mechanical rupture and expulsion of tissue fragments.

Once the water molecules have evaporated, the tissue heats up beyond 100 °C and the tissue starts to become carbonized. Because *carbonization* is not a desirable condition, the tissue can be cooled with water or gas following the vaporization treatment. When the tissue is heated beyond 300 °C, then melting can occur in certain materials for sufficient optical power densities and exposure times.

To calculate the desired level of tissue modification during a thermal heating process is a complex task, which involves knowing both precise tissue characteristics and laser operational parameters [2]. First the task requires accurate predictions of temperature changes with time. These temperature predictions necessitate knowledge of the rate of heat production, which depends on an accurate estimate of the fluence rate throughout the tissue.

A common application of thermal-based laser-tissue interactions is *laser resurfacing* or *photorejuvenation* for the treatment of certain skin conditions and for wrinkle removal [39]. The dermatological conditions include skin atrophy, skin

Fig. 6.24 Concept of the effects from a fractional laser treatment

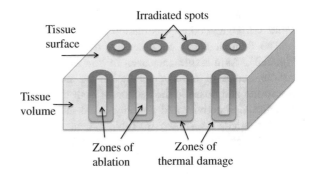

thickening, varicose veins, vascular lesions, unwanted hair, age spots, surgical and acne scars, and pigmented lesions. The photorejuvenation process uses intense pulses of light to ablate a skin area through thermal interactions by inducing controlled wounds on the skin, thereby stimulating the skin to heal itself through the creation of new cells. Two lasers that have been used historically in free-space or direct-beam setups are the Er:YAG laser (emitting at 2940 nm) and the CO_2 laser (emitting at 10.6 μm). Other sources such as a neodymium-doped yttrium ortho-vanadate laser also have been used.

To reduce the side effects associated with large-area irradiation, techniques using tools such as fractional lasers and fiber-coupled lasers have been introduced [39]. In the *fractional laser technique*, the skin is irradiated with a dense array of pinpoints of light, as shown in Fig. 6.24 [45]. This process leaves healthy skin between the ablated areas so that more rapid healing can take place.

The use of an optical fiber for transmitting the laser light enables a precise delivery of this light to a localized skin area. Various types of fiber-coupled lasers are available for research on the basic cell photorejuvenation mechanisms and for clinical use [46–52]. The following devices are some examples of such lasers [39]. The particular optical fibers that are coupled to these sources depend on the efficiency with which the fibers transmit light in the spectral emission region of the source.

- A Q-switched alexandrite laser emits pulses of 50 to 100 ns duration at a wavelength of 755 nm. The spot size emerging from the coupled optical fiber is 2–4 mm in diameter. This laser can be used for treatment of superficial pigmented lesions and is effective at removing black, blue, and most green tattoo inks.
- GaAs-based laser diodes emitting at 808 and 810 nm are effective for treating dentine hypersensitivity, venous insufficiency, varicose veins, and hair removal.
- Micropulsed Nd:YAG lasers emitting at 1444 nm are used for fat removal during facial and body contouring procedures.
- InP-based laser diodes emitting at 1460 nm are effective for skin treatments such as collagen regeneration, removal of surgical and acne scars, and wrinkle removal.

- 1550-nm erbium-doped optical fiber lasers have been used for treatment of facial and nonfacial cutaneous photodamage and resurfacing of facial scars.
- Fiber-coupled Er:YAG lasers emit at 2940 nm and are effective for procedures such as skin resurfacing and acne treatments.
- Fiber-coupled CO_2 lasers can emit more than 10 W at 10.6 μm for a wide range of surgical tissue removal procedures in disciplines such as cardiology, dermatology, gynecology, and orthopedics.

Example 6.12 Consider a CO_2 laser that emits at a wavelength of 10.6 μm. This laser may be used to cut tissue by vaporizing it. (a) What is the optical penetration depth into the tissue if the absorption coefficient $\mu_a = 10^3$ cm^{-1}? (b) Suppose the power from a single 1.5-ms pulse is 0.5 W and that it is delivered to the tissue through an optical fiber with a core radius of 300 μm. Assuming that the factor $\rho C \approx 4.2$ J/(cm^3/°C) holds for tissue, what is the maximum temperature rise in the tissue due to this pulse?

Solution: (a) From Eq. (6.6) the penetration depth into the tissue is $1/\mu_a \approx 10$ μm. Thus the light energy is deposited in a small volume of tissue, which rapidly heats up from 37 °C and will vaporize at 100 °C.
(b) The irradiance from the fiber is $E_0 = P_0/\pi a^2$, where $a = 300$ μm is the fiber radius. From Eq. (6.24) the resultant maximum temperature rise ΔT is

$$\Delta T = \frac{\mu_a P_0 \tau}{\rho C \pi a^2} = \frac{(10^3 \text{ cm}^{-1})(0.5 \text{ W})(1.5 \text{ ms})}{[4.2 \text{ J}/(\text{cm}^3/^\circ \text{C})]\pi (3 \times 10^{-2} \text{ cm})^2} = 63.2 \,^\circ\text{C}$$

Thus the tissue temperature becomes $T = 37 \,^\circ\text{C} + \Delta T = 100.2 \,^\circ\text{C}$.

6.5.4 Photoablation

Tissue photoablation processes are based on the absorption of short pulses of high-energy laser UV light to break molecular bonds in the tissue. This molecular breaking up condition is called *photodissociation* and causes the tissue to volatize [53–56]. The use of photoablation allows the removal of tissue in a very clean and precise fashion. In this laser-tissue interaction mode there is no damage to adjacent tissue, such as can arise from thermal effects in the ablation methods used in coagulation and vaporization processes. A major use of the photoablation process is in eye (corneal) surgery.

Fig. 6.25 Photoablation can lead to either fluorescence or molecular dissociation

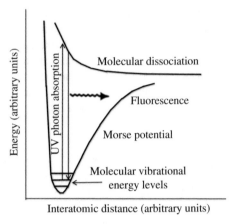

Fig. 6.26 Ablation depth as a function of incident irradiation in photoablation

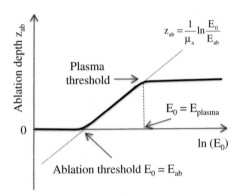

To have a phenomenological understanding of the tissue photoablation process, consider two atoms that are bound by a common electron in a molecule. If the energy of an incident photon is less than the atomic bonding energy, absorption of the photon excites the common electron to a higher molecular vibrational level. In the case when the photon energy is higher than the atomic bonding energy, then the electrons are raised to non-bonding orbitals, as is shown in Figs. 6.6 and 6.25. From this excited state condition either the molecule will fluoresce (that is, the electron drops back to a lower level and gives off a photon) or the two previously bonded atoms disassociate (separate) at the next immediate molecular vibration. This photodisassociation leads to a rapid expansion of the irradiated volume, which basically creates small explosions of vaporized tissue and ejection of the tissue from the surface.

For photodisassociation to take place, the energy of the incident photon must be greater than the bond energy. The photoablation irradiation values are in the 10^7–10^8 W/cm^2 range with pulse durations on the order of nanoseconds. The bond disassociation photoablation mode typically involves UV lasers, because high photon energies are needed to break molecular bonds, which then leads to ablation. A popular tool is a 193-nm ArF laser. For these lasers the photon energy of 6.4 eV is sufficient to break the C–C (3.6 eV), C–O (3.6 eV), and C–N (3.1 eV) bonds of polypeptide chains that are found in collagen. Although lasers with slightly lower photon energies (longer wavelengths) also could be used, such as 2.48-nm KrF lasers, the resulting photoablated surface is less smooth with a bit rougher edges than in the photoablations carried out with ArF lasers.

In order to initiate photoablation, a minimum threshold intensity must be applied to the tissue, as is shown in Fig. 6.26. At this threshold the rate of molecular bond disassociation must be greater than the rate of bond reformation. The rate of bond disassociation is directly related to the number of photons falling on the tissue per second, which is proportional to the irradiance.

Knowing the ablation threshold and assuming that absorption follows the exponential relationship described in Eq. (6.3) then allows an estimate of the photoablation depth in tissue for a given irradiance level E. Thus from Eq. (6.3) the light absorption in the tissue can be written as

$$E(z) = E_0 \exp(-\mu_a z) \tag{6.25}$$

where E_0 is the irradiance incident on the tissue surface, μ_a is the absorption coefficient of the material, and z is the depth into the tissue from the surface. If E_{ab} is the irradiance at the *photoablation threshold* (an irradiation level E_0 at which photoablation starts), then the depth z_{ab} to which material is ablated as a function of the incident irradiation E_0 is

$$z_{ab} = \frac{1}{\mu_a} \ln E_0 - \frac{1}{\mu_a} \ln E_{ab} = \frac{1}{\mu_a} \ln \frac{E_0}{E_{ab}} \tag{6.26}$$

A plot of the ablation depth z_{ab} as a function of the natural log of the incident irradiance E_0 is shown in Fig. 6.26. In addition to showing the irradiance level $E_0 = E_{ab}$ at the photoablation threshold, the figure also indicates that there is an irradiance level $E_0 = E_{plasma}$ at which a plasma threshold is reached. As described in Sect. 6.5.5, high irradiances create a plasma at which point molecules are torn apart. Further increases in irradiance levels beyond the plasma threshold do not lead to deeper ablation depths, because now the plasma absorbs all the incoming irradiation.

Example 6.13 Consider the case in which laser pulses of duration $\tau = 15$ ns from an ArF laser are used to irradiate a tissue sample for which the absorption coefficient is 16×10^3 cm^{-1}. Suppose that the ablation threshold occurs at a radiant exposure (energy density) of $H_{ab} = 110$ mJ/cm^2 and that the plasma threshold occurs at a radiant exposure of $H_{plasma} = 700$ mJ/cm^2. (a) What is the irradiance at the ablation threshold? (b) What is the irradiance

at the plasma threshold? (c) What is the ablation depth per pulse at the plasma
threshold?

Solution: (a) The irradiance is $E_{ab} = H_{ab}/\tau = 7.3 \times 10^6$ W/cm^2
(b) The irradiance is $E_{plasma} = H_{plasma}/\tau = 4.7 \times 10^7$ W/cm^2
(c) $z_{ab} = \frac{1}{\mu_a} \ln \frac{E_{plasma}}{E_{ab}} = \frac{1}{16 \times 10^3 \text{ cm}^{-1}} \ln \left(\frac{4.7 \times 10^7}{7.3 \times 10^6} \right) = 1.16 \, \mu\text{m}.$

6.5.5 Plasma-Induced Photoablation

In moving up the irradiance versus time curve shown in Fig. 6.19, a plasma consisting of ionized molecules and free electrons is produced through a phenomenon
called *optical breakdown*. This breakdown starts at the point where the irradiance
exceeds 10^{11} W/cm^2 and occurs when there are free electrons in an electric field.
When such high irradiances impinge on biological tissue in pulses of less than one
nanosecond, molecules are torn apart and the plasma is formed. In plasma-induced
photoablation a free electron is accelerated by an intense electric field that exists
around a tightly focused laser beam. As shown in Fig. 6.27, when this energetic
free electron gets accelerated and collides with a molecule, the electron transfers
some of its energy to the molecule. If the transferred energy is enough to free a
bound electron in this molecule, the second released free electron also gets accelerated along with the first free electron and together they can initiate additional
energetic collisions. Thereby a chain reaction of similar collisions is initiated and
the plasma is created.

As the plasma expands it ablates tissue material in a clean and precise fashion.
Applications include the treatment of cataracts, refractive corneal surgery, and
caries treatments. Typical parameter values are 100-fs to 500-ps pulses with irradiances varying from 10^{11} to 10^{13} W/cm^2. Example parameters in ophthalmology
are irradiances of 10^{11} W/cm^2 in 10-ps pulses.

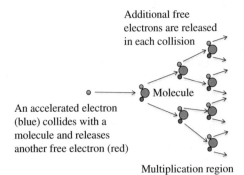

Fig. 6.27 Cascading of free
electrons in an intense electric
field

6.5.6 *Photodisruption*

Similar to plasma-induced photoablation, in *photodisruption* very high irradiance levels result in plasma formation. However, the amount of energy absorbed during photodisruption generally is at least two orders of magnitude higher than during plasma-induced photoablation. This condition produces a greater free electron density and also a higher plasma temperature than in the case of plasma-induced photoablation. When such a plasma has been newly formed, a condition called *plasma shielding* arises. In this effect the plasma absorbs and scatters further incident light, thereby protecting (shielding) the underlying tissue structure from the plasma effects. In ophthalmology, plasma shielding prevents potentially damaging light from reaching the retina when performing procedures such as lens capsulotomy. This is a treatment that involves breaking a secondary cataract, which is a membrane that developed at the back of a lens following cataract surgery.

The photodisruption laser-tissue interaction mode uses pulses of nanosecond or shorter duration and irradiances of 10^9–10^{14} W/cm^2. Because the molecules in the target tissue undergo rapid ionization from a high-intensity laser beam, the plasma-creation process in photodisruption creates localized mechanical effects such as shock waves, jetting of material, and bubble formation followed by cavitation. As a result of the high kinetic energy of free electrons, the temperature of the plasma rises rapidly and the plasma electrons diffuse into the surrounding tissue medium. The rapidly growing plasma creates a pressure wave or shock wave that travels outward and soon separates from the plasma boundary. About 1–5 % of the incident pulse energy is converted to shock wave energy, with short pulses on the order of picoseconds resulting in a weaker shock wave. High-energy pulses on the order of nanoseconds produce stronger shock waves. Such pulses are not desirable for ophthalmology applications, because they can cause damage at points that are remote from the focal region of the laser. However, pulses on the order of nanoseconds can be used in lithotripsy for breaking up kidney stones or gallstones.

Cavitation occurs when a vapor bubble that forms around the plasma grows to a critical size and then collapses violently, thereby sending out another shock wave. *Jetting* can occur when the cavitation bubble is formed close to a solid surface (for example, on a tooth surface). In this case a high-speed liquid jet is directed toward the wall of the surface as the bubble collapses.

In addition to its use in lithotripsy, photodisruption is widely used for minimally invasive surgery and in ophthalmology for drilling holes in the cornea or lenses and for cataract surgery.

Fig. 6.28 Example
illustration of a speckle
pattern

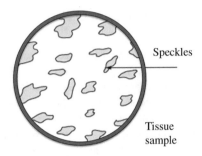

6.6   Formation of Speckles

When coherent light illuminates a rough or optically inhomogeneous material, such
as a biological tissue surface, the scattered light forms a random interference pattern
called *speckle*. A generic example is shown in Fig. 6.28. This temporally varying
speckle is caused by the interference of a large number of elementary electromagnetic waves with random phases when the coherent light is reflected from the
tissue surface or when the light passes through the tissue [57–60].

Speckles have a negative impact on biophotonics measurement techniques such
as optical coherence tomography in which speckle formations limit the interpretation of the observed images. Various speckle-reducing digital filters have been
examined to alleviate these distortions [57]. On the positive side, speckle formation
has been implemented successfully in biophotonics methodologies such as laser
speckle contrast imaging, which is an optical technique used to generate blood flow
maps with high spatial and temporal resolution. More details on laser speckle
imaging are given in Sect. 10.3.

Fig. 6.29 Example of a
typical Jablonski diagram

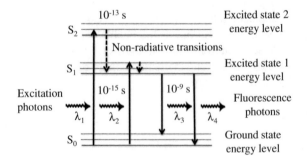

6.7 Fluorescence Basics

Fluorescence is the property of certain atoms and molecules to absorb light at a particular wavelength and, subsequently, to emit light of a longer wavelength after a short time [61–63]. The fluorescence process is characterized by three key events, all of which take place on time scales that are separated by several orders of magnitude. The first event generally involves using ultraviolet light ($\lambda \leq 300$ nm) to excite molecules to higher vibrational states. This transition occurs in femtoseconds (10^{-15} s). In the second event, the molecules then relax to slightly lower excited energy states in the order of 10^{-13} s. Subsequently, in the third event, the state of the molecule transitions to a lower ground-state level in the order of 10^{-9} s. This third time interval is called the *fluorescence lifetime*. A photon of lower energy (longer wavelength) than the excitation energy is emitted during the transition to the ground state. This fluorescence process can be used to analyze the characteristics of the molecule or to observe how the molecule interacts with other molecules.

The fluorescence process is illustrated in Fig. 6.29, which is called a *Jablonski diagram*. Here the label S_0 is the ground state energy with different vibrational energy levels shown by closely spaced dashed horizontal lines. The labels S_1 and S_2 are the first and second excited states, again with different vibrational energy levels shown by closely spaced dashed horizontal lines. Energy bands to which transitions are forbidden separate the ground state and the excited states. Transitions between the states occur only in terms of quantized energy units, that is, photons. The solid vertical lines represent either absorption of an excitation photon (rising line) or emission of a fluorescent photon (dropping line). The dashed vertical lines represent nonradiative internal relaxation process whereby some energy is converted to heat.

In the absorption process, an incoming photon (λ_1 or λ_2) will excite the molecule to some upper vibrational level of the first or second excited state. Subsequently, the molecule will rapidly relax through a nonradiative internal process to the lowest level of S_1. Then from this S_1 level the molecule will relax to the S_0 ground state, which results in a fluorescent photon being emitted (λ_3 or λ_4).

Fluorescent substances are known as *fluorophores*. These can be broadly classified as intrinsic fluorophores and extrinsic fluorophores. *Intrinsic fluorophores* are biological molecules that can fluoresce naturally through their interaction with incident excitation photons. This process is known as *autofluorescence* or *natural fluorescence*. Examples of intrinsic fluorophores are the aromatic amino acids (e.g.,

Table 6.5 Excitation and fluorescence wavelengths of some common intrinsic fluorophores

Molecule	Excitation (nm)	Fluorescence (nm)
Tryptophan	280	300–350
Tyrosine	270	305
Flavin	380–490	520–560
Collagen	270–370	305–450
Melanin	340–400	360–560
NADH	340	450

Table 6.6 Excitation and fluorescence wavelengths of some common extrinsic fluorophores

Molecule	Excitation (nm)	Fluorescence (nm)
Dansyl chloride	350	520
Fluorescein	480	510
Rhodamine	600	615
Prodan	360	440
BIODIPY-FL	503	512
Texas Red	589	615

tryptophan, tyrosine, phenylalanine), flavin, collagen, melanin, elastin, and NADH (a reduced form of nicotinamide adenine dinucleotide). Table 6.5 lists the excitation and fluorescence wavelengths of some common intrinsic fluorophores.

Many types of biological molecules are either nonfluorescent or the intrinsic fluorescence is not adequate for a particular desired molecular analysis. For example, substances such as DNA and lipids (e.g., fats, triglycerides, and vitamins A, D, E, and K) possess no intrinsic fluorescence characteristics. In these cases extrinsic fluorophores can be attached (generally through covalent bonding) to the molecules to provide fluorescence for experimental studies. An extrinsic fluorophore is known by various names such as a *label*, *marker*, *dye*, or *tag*. Thus the attachment process is known as *labeling*, *marking*, *dyeing*, or *tagging* of the molecule. Many varieties of extrinsic fluorophores are commercially available with individual peak wavelengths located throughout the visible spectrum. Table 6.6 lists the excitation and fluorescence wavelengths of some example extrinsic fluorophores.

The discovery of *green fluorescent protein* (GFP) started a new era in cell biology to monitor cellular processes in living cells and organisms using fluorescence microscopy and related methodologies [64–66]. Subsequent to the initial use of GFP, a number of genetic variants that exhibited fluorescence emission in different regions of the visible spectrum were developed from the original GFP nucleotide sequence. These include blue, cyan, and yellow fluorescent protein, which go by the acronyms BFP, CFP, and YFP. Longer wavelength fluorescent proteins emitting in the orange and red spectral regions also have been developed [67]. Consequently, a large number of fluorescent proteins are available with emission maxima ranging from 424 to 625 nm.

6.8   Summary

Light-tissue interaction is a complex process because the constituent tissue materials are multilayered, multicomponent, and optically inhomogeneous. The basic effects include reflection at a material interface, refraction when light enters a tissue structure that has a different refractive index, absorption of photon energy by the material, and multiple scattering of photons in the material. Because diverse intermingled biological tissue components have different optical properties, then along some path through a tissue volume, various physical parameters (such as the refractive index, absorption coefficients, and scattering coefficients) typically change at material boundaries.

Absorbed light can be converted into heat, be radiated in a fluorescent process, or be consumed in photochemical reactions. As shown in Figs. 6.8 and 6.18, the strength of the absorption coefficients for different tissue components determines how far light can penetrate into a specific tissue at a particular wavelength and also determines how much energy a specific tissue absorbs from a particular optical source.

Scattering of photons in tissue is another significant factor in the behavior of light-tissue interactions. Together, absorption and multiple scattering of photons cause light beams to broaden and decay as photons travel through tissue. Although light can penetrate several centimeters into a tissue, strong scattering of light can prevent observers from getting a clear image of tissue abnormalities beyond a few millimeters in depth. Scattering of photons can be either an elastic process or an inelastic process. Elastic scattering effects are used in many biophotonics applications such as optical coherence tomography, confocal microscopy, and elastic scattering spectroscopy. Raman scattering is a major inelastic scattering process used in biophotonics and is the basis for Raman vibrational spectroscopy. This technique is of importance for studying biological molecules and for diagnosing and monitoring the progress of diseases such as cataract formations, precancerous and cancerous lesions in human soft tissue, artherosclerotic lesions in coronary arteries, and bone and dental pathologies.

Light-tissue interactions can be classified into six generic categories that are commonly used to describe therapeutic and surgical applications. These interactions can be categorized as photobiomodulation, photochemical interactions, thermal interactions (e.g., coagulation and vaporization), photoablation, plasma-induced ablation, and photodisruption. The degree of light-tissue interaction depends on tissue characteristics (such as the coefficients of reflection, absorption, and scattering) and the parameters of the irradiating light.

The phenomenon of random interference patterns, or speckle fields, in relation to the scattering of laser light from weakly ordered media such as tissue can be both beneficial and a limitation in biophotonics imaging. The appearance of speckles arises from coherence effects in light-tissue interactions. Among the applications areas of this effect are the study of tissue structures and cell flow monitoring.

An important tool in biophotonics is the concept of fluorescence, which is the property of certain atoms and molecules to absorb light at a particular wavelength and, subsequently, to emit light of a longer wavelength after a short interaction time. This physical phenomenon is widely used in a variety of biophotonics sensing, spectroscopic, and imaging modalities.

6.9 Problems

6.1 Consider the interface between air (n_{air} = 1.00) and samples of the following three materials: hemoglobin (n_{Hb} = 1.37), dentin (n_d = 1.50), and tooth enamel (n_t = 1.73). Show that the Brewster angles are 53.9°, 56.3°, and 60.0°, respectively.

6.2 Verify the plots of the parallel and perpendicular reflection coefficients shown in Fig. 6.3 for light incident from air onto a material that has a refractive index of 1.33.

6.3 Consider a dermis skin tissue sample in which the absorption coefficient is 12 cm^{-1} at 308 nm and 2.7 cm^{-1} at 633 nm. Show that the absorption lengths for each of these wavelengths at which $I(x)/I_0$ = 1/e = 0.368 are 0.833 mm and 3.702 mm, respectively.

6.4 Consider an aorta tissue in which the absorption coefficient is 2.2 cm^{-1} at 1320 nm. Show that the penetration depth is 4.55 mm.

6.5 Because of its varying structure the absorption coefficient of skin changes with depth. For depths of 1 mm or less the absorption coefficient at 633 nm is 0.67 cm^{-1}, for depths between 1 and 2 mm the absorption coefficient at 633 nm is 0.026 cm^{-1}, and for depths > 2 mm the absorption coefficient at 633 nm is 0.96 cm^{-1}. Let a collimate light beam with a power level of 100 mW be incident on a thick skin tissue sample. If scattering effects are ignored, show that the optical powers at depths of 1, 2, and 3 mm are 93.5, 93.2, and 84.6 mW, respectively.

6.6. Consider a 1-mm thick optical filter that has a 10 cm^{-1} absorption coefficient at a HeNe laser wavelength of 633 nm. Suppose a collimated 5 mW light beam is incident perpendicular to the surface of the filter. Show that the power level of the attenuated beam that emerges from the filter is 1.84 mW if the attenuation is due only to absorption.

6.7. In the visible spectrum, the Rayleigh scattering effect is significantly smaller in the red region (around 680 nm) compared to the blue region (around 450 nm). Show that the ratio I_s/I_0 of Rayleigh scattering is 5.2 times smaller at 680 nm than at 450 nm.

6.8. Verify the values of the Rayleigh scattering cross section and the geometric cross section given in Example 6.7.

6.9. Consider a spherical particle that has a radius a = 10 nm, a refractive index n_s = 1.57, and that is in a background material of refractive index n_b = 1.33. Show that the scattering efficiency at a wavelength λ = 400 nm is Q_s = 6.83 × 10^{-5}.

6.10. Plot the albedo versus the scattering coefficient for attenuation coefficients of μ_a = 0.1, 1.0, and 10 cm^{-1}. Let the albedo on the vertical axis range from 0 to 1 and let the scattering coefficients on the horizontal axis be on a logarithmic scale ranging from 0.1 to 1000 cm^{-1}.

6.11. If the optical albedo of a biological tissue is a = 0.9, show that μ_s = 9μ_a.

6.12. Suppose that tooth dentin has an absorption coefficient of 6 cm^{-1} and a scattering coefficient of 1200 cm^{-1} for a HeNe laser emitting at a wavelength of 633 nm. Show that the optical albedo of this dental tissue is 0.995.

6.13. Consider the cylindrical diffuser device of length L shown in Fig. 6.22. Assume the diffuser produces a uniform radial distribution along the length of the fiber and that 80 % of the input power is radiated out of the diffuser. If P is the optical power level input to the diffuser, what is the irradiance at a distance R from the center of the cylinder?

6.14. Consider a Nd:YAG laser that emits at a wavelength of 1064 μm. (a) If the absorption coefficient of an irradiated abdominal tissue is $\mu_a = 18$ cm^{-1}, show that the optical penetration depth into the tissue is 0.56 mm. (b) Suppose the power from a single 16-ms Nd:YAG pulse is 2.6 W and that it is delivered to the tissue through an optical fiber with a core radius of 300 μm. Assuming that the factor $\rho C \approx 4.2$ J/(cm^3/°C) holds for tissue, show that the maximum temperature rise in the tissue due to this pulse is 63 °C.

6.15. Compare the radiant exposure in J/cm^2 for the following two optical pulses:

   (a) A 10-ps pulse with an irradiance of 8×10^{11} W/cm^2
   (b) A 100-ns pulse with an irradiance of 7.3×10^9 W/cm^2

   (Answer: 8 J/cm^2 and 730 J/cm^2.)

References

1. W.F. Cheong, S.A. Prahl, A.J. Welch, A review of the optical properties of biological tissues. IEEE J. Quantum Elec. **26**, 2166–2185 (1990)
2. R. Menzel, *Photonics: Linear and Nonlinear Interactions of Laser Light and Matter*, 2nd edn. (Springer, Berlin, 2007)
3. N.H. Niemz, *Laser-Tissue Interaction*, 3rd edn. (Springer, Berlin, 2007)
4. A.J. Welch, M.J.C. van Gemert (eds.), *Optical-Thermal Response of Laser-Irradiated Tissue*, 2nd edn. (Springer, Berlin, 2011)
5. M. Schmitt, T. Mayerhöfer, J. Popp, Light-matter interaction, Chap. 3, in *Handbook of Biophotonics: Vol. 1: Basics and Techniques*, ed. by J. Popp, V.V. Tuchin, A. Chiou, S.H. Heinemann (Wiley, London, 2011)
6. S.L. Jacques, Optical properties of biological tissues: a review. Phys. Med. Biol. **58**(11), R37–R61 (2013)
7. K. Kulikov, *Laser Interaction with Biological Material* (Springer, Berlin, 2014)
8. J. Mobley, T. Vo-Dinh, V.V. Tuchin, Optical properties of tissue, Chap. 2, in *Biomedical Photonics Handbook*, 2nd edn., ed. by T. Vo-Dinh (CRC Press, Boca Raton, FL, 2014), pp. 23–121
9. V.V. Tuchin, Light-tissue interactions, Chap. 3, in *Biomedical Photonics Handbook*, 2nd edn., ed. by T. Vo-Dinh (CRC Press, Boca Raton, Florida, 2014), pp. 123–167
10. M. Olivo, U.S. Dinish (eds.), *Frontiers in Biophotonics for Translational Medicine* (Springer, Singapore, 2016)
11. A.H.-P. Ho, D. Kim, M.G. Somekh (eds.), *Handbook of Photonics for Biomedical Engineering* (Springer, Berlin, 2016)

12. U. Fares, M.A. Al-Aqaba, A.M. Otri, H.S. Dua, A review of refractive surgery. Eur. Ophthalmol. Rev. **5**(1), 50–55 (2011)
13. P. Artal, Optics of the eye and its impact on vision: a tutorial. Adv. Opt. Photonics **6**, 340–367 (2014)
14. F. Guarnieri (ed.), *Corneal Biomechanics and Refractive Surgery* (Springer, New York, 2015)
15. S.L. Jacques, Monte Carlo modeling of light transport in tissue, Chap. 5, in *Optical-Thermal Response of Laser-Irradiated Tissue*, 2nd edn., ed. by A.J. Welch, M.J.C. van Gemert (Springer, Berlin, 2011)
16. L.V. Wang, H.I. Wu, *Biomedical Optics: Principles and Imaging* (Wiley, Hoboken, NJ, 2007)
17. W.M. Star, Diffusion theory of light transport, Chap. 6, in *Optical-Thermal Response of Laser-Irradiated Tissue*, 2nd edn., ed. by A.J. Welch, M.J.C. van Gemert (Springer, Berlin, 2011)
18. S.J. Norton, T. Vo-Dinh, Theoretical models and algorithms in optical diffusion tomography, Chap. 4, in *Biomedical Photonics Handbook; Vol. 1; Fundamentals, Devices, and Techniques*, 2nd edn., ed. by T. Vo-Dinh (CRC Press, Boca Raton, FL, 2014), pp. 253–279
19. B. Alberts, A. Johnson, J. Lewis, D. Morgan, M. Raff, K. Roberts, P. Walter, *Molecular Biology of the Cell*, 6th edn. (Garland Science, New York, 2015)
20. H. Lodish, A. Berk, C.A. Kaiser, M. Krieger, A. Bretscher, H. Ploegh, A. Amon, M.P. Scott, *Molecular Cell Biology*, 7th edn. (W.H. Freeman, San Francisco, CA, 2013)
21. T. Engel, P. Reid, *Physical Chemistry*, 3rd edn. (Prentice Hall, Englewood Cliffs, NJ, 2012)
22. D.B. Wetlaufer, Ultraviolet spectra of proteins and amino acids. Adv. Protein Chem. **17**, 303–390 (1963)
23. C.N. Pace, F. Vajdos, L. Fee, G. Grimsley, T. Gray, How to measure and predict the molar absorption coefficient of a protein. Protein Sci. **4**, 2411–2423 (1995)
24. A. Barth, The infrared absorption of amino acid side chains. Prog. Biophys. Mol. Biol. **74** (3–5), 141–173 (2000)
25. S.H. Tseng, P. Bargo, A. Durkin, N. Kollias, Chromophore concentrations, absorption and scattering properties of human skin in-vivo. Opt. Express **17**(17), 14599–14617 (2012)
26. S.A. Prahl, Tabulated molar extinction coefficient for hemoglobin in water. Available from http://omlc.ogi.edu/spectra/hemoglobin/summary.html. Accessed 25 July 2015
27. A. Wax, V. Backman, *Biological Applications of Light Scattering* (McGraw-Hill, New York, 2010)
28. R. Petry, M. Schmitt, J. Popp, Raman spectroscopy: a prospective tool in life sciences. Chemphyschem **4**, 14–30 (2003)
29. A. Downes, A. Elfick, Raman spectroscopy and related techniques in biomedicine. Sensors **10**, 1871–1889 (2010)
30. C. Krafft, B. Dietzek, M. Schmitt, J. Popp, Raman and coherent anti-Stokes Raman scattering microspectroscopy for biomedical applications. J. Biomed. Opt. **17**, 040801 (2012)
31. Q. Peng, A. Juzeniene, J. Chen, L.O. Svaasand, T. Warloe, K.-E. Giercksky, J. Moan, Lasers in medicine. Rpts. Prog. Phys. **71**, 056701 (2008)
32. E. Hahm, S. Kulhari, P.R. Arany, Targeting the pain, inflammation and immune (PII) axis: plausible rationale for LLLT. Photon. Laser Med. **1**(4), 241–254 (2012)
33. P.R. Arany, A. Cho, T.D. Hunt, G. Sidhu, K. Shin, E. Hahm, G.X. Huang, J. Weaver, A.C.-H. Chen, B.L. Padwa, M.R. Hamblin, M.H. Barcellos-Hoff, A.B. Kulkarni, D.J. Mooney, Photoactivation of endogenous latent transforming growth factor—b1 directs dental stem cell differentiation for regeneration. Sci. Transl. Med. **6**(238), 238ra69 (2014)
34. J.J. Anders, H. Moges, X. Wu, I.D. Erbele, S.L. Alberico, E.K. Saidu, J.T. Smith, B.A. Pryor, In vitro and in vivo optimization of infrared laser treatment for injured peripheral nerves. Lasers Surg. Med. **46**, 34–45 (2014)
35. J.D. Carroll, M.R. Milward, P.R. Cooper, M. Hadis, W.M. Palin, Developments in low level light therapy (LLLT) for dentistry. Dent. Mater. **30**, 465–475 (2014)
36. S. Wu, D. Xing, Intracellular signaling cascades following light irradiation. Laser Photonics Rev. **8**, 115–130 (2014)

37. M. Tschon, S. Incerti-Parenti, S. Cepollaro, L. Checchi, M. Fini, Photobiomodulation with low-level diode laser promotes osteoblast migration in an in vitro micro wound model. J. Biomed. Opt. **20**, 078002 (2015)
38. P. Cassano, S.R. Petrie, M.R. Hamblin, T.A. Henderson, D.V. Iosifescu, Review of transcranial photobiomodulation for major depressive disorder: targeting brain metabolism, inflammation, oxidative stress, and neurogenesis. Neurophotonics **3**(3), 031404 (2016)
39. G. Keiser, F. Xiong, Y. Cui, P.P. Shum, Review of diverse optical fibers used in biomedical research and clinical practice. J. Biomed. Opt. **19**, 080902 (2014)
40. R. Penjweini, B. Liu, M.M. Kim, T.C. Zhu, Explicit dosimetry for 2-(1-hexyloxyethyl)-2-devinyl pyropheophorbide-a-mediated photodynamic therapy: macroscopic singlet oxygen modeling. J. Biomed. Opt. **20**, 128003 (2015)
41. N.F. Gamaleia, I.O. Shton, Gold mining for PDT: great expectations from tiny nanoparticles. Photodiagn. Photodyn Ther **12**, 221–231 (2015)
42. I. Mfouo-Tynga, H. Abrahamse, Cell death pathways and phthalocyanine as an efficient agent for photodynamic cancer therapy. Intl. J. Molecular Sci. **16**, 10228–10241 (2015)
43. J.L. Boulnois, Photophysical processes in recent medical laser developments: a review. Lasers Med. Sci. **1**, 47–66 (1986)
44. H.Z. Alagha, M. Gülsoy, Photothermal ablation of liver tissue with 1940-nm thulium fiber laser: an *ex vivo* study on lamb liver. J. Biomed. Opt. **21**(1), 015007 (2016)
45. M.H. Gold, Update on fractional laser technology. J. Clin. Aesthet. Dermatol. **3**, 42–50 (2010)
46. G. Deka, K. Okano, F.-J. Kao, Dynamic photopatterning of cells *in situ* by Q-switched neodymium-doped yttrium ortho-vanadate laser. J. Biomed. Optics **19**, 011012 (2014)
47. Z. Al-Dujaiti, C.C. Dierickx, Laser treatment of pigmented lesions, Chap. 3, in *Laser Dermatology*, 2nd edn., ed. by D.J. Goldberg (Springer, Berlin, 2013), pp. 41–64
48. L. Corcos, S. Dini, D. De Anna, O. Marangoni, E. Ferlaino, T. Procacci, T. Spina, M. Dini, The immediate effects of endovenous diode 808-nm laser in the greater saphenous vein: Morphologic study and clinical implications. J. Vasc. Surg. **41**(6), 1018–1024 (2005)
49. Y.C. Jung, Preliminary experience in facial and body contouring with 1444 nm micropulsed Nd:YAG laser-assisted lipolysis: a review of 24 cases. Laser Ther. **20**(1), 39–46 (2011)
50. P.S. Tsai, P. Blinder, B.J. Migliori, J. Neev, Y. Jin, J.A. Squier, D. Kleinfeld, Plasma-mediated ablation: an optical tool for submicrometer surgery on neuronal and vascular systems. Curr. Opin. Biotechnol. **20**, 90–99 (2009)
51. A. Vogel, V. Venugopalan, Mechanisms of pulsed laser ablation of biological tissue. Chem. Rev. **103**, 577–644 (2003)
52. B.S. Biesman, M.P. O'Neil, C. Costner, Rapid, high-fluence multi-pass Q-switched laser treatment of tattoos with a transparent perfluorodecalin-infused patch: A pilot study. Lasers Surg. Med. **47**(8), 613–618 (2015)
53. C.V. Gabel, Femtosecond lasers in biology: nanoscale surgery with ultrafast optics. Contemp. Phys. **49**(6), 391–411 (2008)
54. H. Huang, L.-M. Yang, S. Bai, J. Liu, Smart surgical tool. J. Biomed. Opt. **20**, O28001 (2015)
55. B. Rao, J. Su, D. Chai, D. Chaudhary, Z. Chen, T. Juhasz, Imaging subsurface photodisruption in human sclera with FD-OCT. In: Proceedings SPIE 6429, Coherence Domain Optical Methods and Optical Coherence Tomography in Biomedicine XI, 642910, 12 Feb 2007
56. S.S. Harilal, J.R. Freeman, P.K. Diwakar, A. Hassanein, Femtosecond laser ablation: fundamentals and applications, in *Laser-Induced Breakdown Spectroscopy*, ed. by S. Musazzi, U. Perini (Springer, Berlin, 2014), pp. 143–166
57. A. Ozcan, A. Bilenca, A.F. Desjardins, B.E. Bouma, G.J. Tearney, Speckle reduction in optical coherence tomography images using digital filters. J. Opt. Soc. Am. A **24**, 1901–1910 (2007)
58. D.A. Boas, A.K. Dunn, Laser speckle contrast imaging in biomedical optics. J. Biomed. Opt. **15**(1), 011109 (2010)
59. L.M. Richards, S.M. Shams Kazmi, J.L. Davis, K.E. Olin, A.K. Dunn, Low-cost laser speckle contrast imaging of blood flow using a webcam. Biomed. Opt. Express **4**(10), 2269–2283 (2013)

60. I. Sigal, R. Gad, A.M. Caravaca-Aguirre, Y. Atchia, D.B. Conkey, R. Piestun, O. Levi, Laser speckle contrast imaging with extended depth of field for in-vivo tissue imaging. Biomed. Opt. Express **5**(1), 123–135 (2014)
61. J.R. Lakowicz, *Principles of Fluorescence Spectroscopy*, 3rd edn. (Springer, New York, 2006)
62. Y. Engelborghs, A.J.W.G. Visser (eds.), *Fluorescence Spectroscopy and Microscopy: Methods and Protocols* (Springer, New York, 2014)
63. J. Ge, C. Kuang, S.-S. Lee, F.-J. Kao, Fluorescence lifetime imaging with pulsed diode laser enabled stimulated emission. Opt. Express **20**(27), 28216–28221 (2014)
64. R.Y. Tsien, The green fluorescent protein. Annual Rev. Biochem. **67**, 509–544 (1998)
65. B. Seefeldt, R. Kasper, T. Seidel, P. Tinnefeld, K.F. Dietz, M. Heilemann, M. Sauer, Fluorescent proteins for single-molecule fluorescence applications. J. Biophotonics **1**(1), 74–82 (2008)
66. D.M. Chudakov, M.V. Matz, S. Lukyanov, K.A. Lukyanov, Fluorescent proteins and their applications in imaging living cells and tissues. Physiol. Rev. **90**, 1103–1163 (2010)
67. G.-J. Kremers, K.L. Hazelwood, C.S. Murphy, M.W. Davidson, D.W. Piston, Photoconversion in orange and red fluorescent proteins. Nat. Methods **6**, 355–358 (2009)

Chapter 7
Optical Probes and Biosensors

Abstract Optical probes and photonics-based biosensors are important tools in most biophotonics diagnostic, therapeutic, imaging, and health-status monitoring instrumentation setups. These devices can selectively detect or analyze specific biological elements, such as microorganisms, organelles, tissue samples, cells, enzymes, antibodies, and nucleic acids derived from human and animal tissue and body fluids, cell cultures, foods, or air, water, soil, and vegetation samples. Of particular interest for biosensing processes are optical fiber probes, nanoparticle-based sensors, optical fiber and waveguide substance sensors, photodetector arrays, fiber Bragg grating sensors, and surface plasmon resonance devices.

A biosensor is an essential component of most biophotonics diagnostic, therapeutic, imaging, and health-status monitoring instrumentation setups. Biosensors can be configured as electrical, optical, chemical, or mechanical devices with the capability to selectively detect specific biological elements. As indicated in Fig. 7.1, such elements can include microorganisms, organelles, tissue samples, cells, enzymes, antibodies, and nucleic acids derived from animal tissue and body fluids, human tissue and body fluids, cell cultures, foods, or air, water, soil, and vegetation samples.

Starting in the 1970s, the telecom industry initiated a major photonics technology evolution based on optical fibers and planar waveguides, laser diode sources, and high-performance photodetectors. The outcomes of these telecom activities resulted in a large collection of diverse optical fiber structures, lightwave couplers, photonic devices, wavelength-selective components, and the associated light sources and photodetectors that now are being applied to biomedicine. Numerous embodiments of optical fibers and optical waveguide structures using different materials and configurations are being utilized to form optical probes, molecular and disease analysis tools, health monitoring devices, and biosensors [1–8]. Of particular interest for biosensing processes are photonics-based devices and instruments such as optical fiber probes, nanoparticle-based sensors, optical fiber and waveguide substance sensors, photodetector arrays, fiber Bragg grating sensors, and surface plasmon resonance devices.

© Springer Science+Business Media Singapore 2016
G. Keiser, *Biophotonics*, Graduate Texts in Physics,
DOI 10.1007/978-981-10-0945-7_7

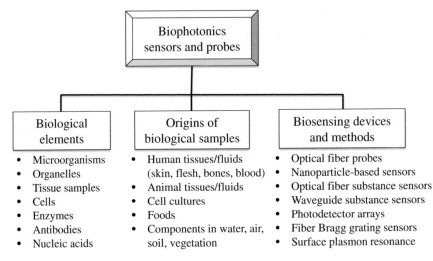

Fig. 7.1 Examples of biological elements, biological sample origins, and sensing devices and methods

7.1 Overview of Biosensors and Probes

This chapter addresses the basic designs of optical fiber probes and optical fiber-based and waveguide-based biosensor configurations that are necessary for the biomedical optics processes described in Chaps. 8–11. First Sect. 7.2 illustrates one-fiber, dual-fiber, and multiple-fiber probe configurations. Depending on the probe application, specific configurations could incorporate a number of peripheral optical elements such as optical filters, beam splitters, optical circulators, or alignment optics.

Next Sect. 7.3 describes optical fiber tip geometries used at the *distal end* of a fiber link (i.e., the farthest point from the optical source or from the final signal observation and analysis point). The tip configuration of a probe is an important design factor for efficient light delivery and collection in any biophotonics application. Implementations of these optical fiber probes in biophotonics disciplines are described in Chaps. 8–11.

In addition to the use of optical fibers for delivering light to and from a diagnostic or treatment area of a tissue, a number of optical fiber structures have been widely investigated and implemented for the diverse biosensing and biomedical measurement functions listed in Fig. 7.2. The sensing functions include detecting the presence of specific classes of molecules or measuring the characteristics of cells, proteins, DNA, and other biological species. These biophotonics applications are a result of optical fiber characteristics that include their small size and flexibility, excellent capability for integration into photonics components and instruments, and relatively high sensitivity of changes in the amplitude, phase, or polarization of propagating light in response to variations in physical factors such as stress, strain, temperature, the refractive index, and fiber movement or microbending. For example, as described in

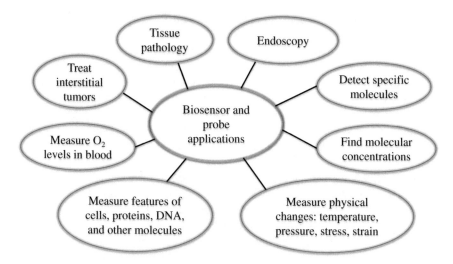

Fig. 7.2 Diverse biosensing and biomedical measurement applications

Sect. 7.4.1, an optical fiber-based device can sense perturbations in the evanescent field of the propagating modes in an optical fiber. These perturbations arise through the use of an optical fiber that is specially coated with an antibody layer. In a biosensor the antibody coated fiber attracts a specific class of molecules when it is immersed in a fluid containing these molecules, which are called *antigens*. The adherence of these antigen molecules to the fiber then causes a change in the effective index of the fiber cladding, which results in a variation of an optical signal in the fiber. This change in the signal characteristics then can be directly correlated to the concentration of the specific molecules being measured.

A variation on the antibody/antigen fiber sensor is described in Sec. 7.4.2. The detection method is based on using a sandwiched material structure consisting of an antibody layer, an antigen layer, and a detecting antibody layer. Following this discussion, next Sects. 7.4.3 and 7.4.4 illustrate how relative movements between two fibers or variations in the bend radius of a fiber, respectively, can measure changes in parameters such as pressure, stress, or temperature.

Other technology platforms that have been examined for waveguide biosensors include optical interferometers, photonic crystal fiber-based devices, surface plasmon resonance devices, an individual or an array of fiber Bragg gratings, and nanoparticle based configurations. For these technologies, Sect. 7.5 describes fiber-based Mach-Zehnder, Michelson, and Sagnac interferometers.

Some representative photonic crystal fiber (PCF) devices that are being used in different biosensor categories are described in Sect. 7.6. One reason for using a PCF in a biosensor system is that the air holes in a PCF serve as natural miniature chambers for a liquid sample to enter [3]. This feature greatly reduces the volume of a sample that is typically required in other categories of sensors. Another biophotonics fiber sensor configuration exploits intermodal interference between forward-propagating core and cladding modes in a PCF, which is one form of a compact inline Mach-Zehnder interferometer.

The fiber Bragg grating (FBG) that is described in Sect. 3.5.1 is a popular precise wavelength-selecting component that was enhanced by the telecom industry. For biophotonics applications, an external force (for example, from the weight of a person or through some variation in a muscular force) can slightly stretch the fiber in which the FBG is embedded. This stretching will change the FBG grating spacing and thus will change the value of the reflected Bragg wavelength. Applications of FBGs in various healthcare-monitoring disciplines are described in Sect. 7.7.

Next Sect. 7.8 addresses the concept and function of surface plasmon resonance, which is implemented in many biosensor applications and lab-on-a-chip sensors to provide a greatly enhanced sensitivity compared to other sensors. The surface plasmon resonance effect produces a collective oscillation of electrons when the surface of a solid or liquid is stimulated by polarized light. A resonance condition occurs when the frequency of the incident photons matches the natural frequency of surface electrons that are oscillating against the restoring force of positive nuclei in the solid or liquid material.

When a pattern of nanoparticles is deposited on the metalized tip of an optical fiber, surface plasmon waves can be excited by an illuminating optical wave. This technique is described in Sect. 7.9. If the fiber tip is inserted into a fluid that contains an *analyte* (a substance being identified and measured), the liquid will cover the nanoparticle pattern thereby changing the refractive index of the nanolayer-to-fluid interface. The result is a wavelength shift in the plasmon wave peak, which then can be used to sense the presence of a substance of interest.

7.2 Optical Fiber Probe Configurations

Optical fibers are used widely as biomedical probes for light delivery and collection functions [9–15]. The diameter of an optical fiber cable used as a probe is normally less than half a millimeter, which allows the cable to be inserted into most hollow medical needles and catheters. Two generic schematics of basic optical fiber-based probe systems that use the same optical path for both illumination and light collection functions are shown in Fig. 7.3. This optical path could consist of a fiber bundle or it could be selected from one of the fibers described in Chap. 3. In order to achieve bidirectional light flow in the same fiber, either a dichroic filter (see Sect. 5.6), a 3-dB coupler (see Sect. 5.7), or an optical circulator (see Sect. 5.7) is used to first direct the light from the source to the tissue and then to route the returning light to a photodetection device. The dichroic filters and their associated coupling optics for connecting to light sources, photodetectors, and optical fibers are available from a variety of vendors. Optical 3-dB couplers and optical circulators can be purchased as fiber-coupled devices for ease of attaching them to instrumentation optical fibers with optical connectors.

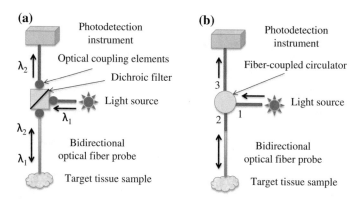

Fig. 7.3 Example single-fiber probe configurations using **a** a dichroic filter and **b** a fiber-coupled optical circulator

The excitation light can come from any one of the diverse optical sources that are described in Chap. 4, which include light-emitting diodes that emit optical power either in a broad white-light spectral band or in a narrow spectral band, an arc lamp, a laser diode, or a fiber laser. A dielectric optical bandpass filter or a monochromator can be used to select a very narrow spectral width for light delivery to a tissue sample. Candidate photodetection instruments that interpret the returned light include a viewing scope, a photomultiplier tube (PMT), a camera, or an optical spectrum analyzer. An *optical spectrum analyzer* (OSA), which also is called a *spectrometer*, is an instrument that measures some property of light (e.g., intensity per wavelength) contained within a selectable wavelength range.

For the example single-fiber setup shown in Fig. 7.3a, a dichroic optical filter reflects a specific spectral band from a light source into the fiber, which sends the light to the tissue target. As described in Sect. 5.6, a dichroic filter separates spectral bands by reflecting and transmitting light as a function of wavelength. The transmission and reflections are 50 % at the cutoff wavelength λ_c, that is, 50 % of the light below a specific wavelength is reflected from the filter and the other 50 % above the cutoff wavelength is transmitted through the filter. Thus, the configuration shown in Fig. 7.3a can be used in spectroscopy applications wherein the light is absorbed in a narrow spectral band (or at a specific wavelength, e.g., at λ_1) by a tissue sample and is remitted at a longer wavelength λ_2 (that is, at a lower energy). In this case, the filter reflects the source spectral band that is centered at λ_1, and the remitted light centered at λ_2 passes through the filter and falls on a photodetector.

In Fig. 7.3b the link uses an optical circulator in place of the optical filter. For this configuration, the light from an optical source enters the optical circulator at port 1 and exits into the transmission fiber at port 2. After being reflected or remitted from the tissue sample, the light travels back along the transmission fiber, reenters port 2, and emerges from port 3 toward a photodetector. In this setup, for spectroscopic applications a narrow-pass optical filter is needed after port 3 in order to suppress backscattered light from the source that can interfere with the desired optical signal from the tissue.

Example 7.1 As described in Chap. 9, in a fluorescent spectroscopy setup a laser emitting in a narrow spectral band is used to excite electrons in some substance to higher energy levels. The excited electrons lose some of their energy to molecular vibrations. Because the excited electrons now have a lower energy, during the transition back to the ground state they emit photons at a longer wavelength than that of the excitation light. Suppose the substance being excited is a *green fluorescent protein* (GFP) that absorbs light at a peak wavelength of 469 nm and emits light at a peak wavelength of 525 nm. In the setup in Fig. 7.1a, what characteristics should the dichroic filter have?

Solution: The following example characteristics can be found from commercial data sheets on dichroic edge filters, as Fig. 7.4 illustrates:

- Cutoff wavelength = 497 nm
- Reflection spectral band range: 450–490 nm
- Transmission spectral band range: 505–800 nm

The reflection and transmission bands could have an average ripple of 10 %.

Example 7.2 The use of a double-clad fiber (DCF) is another setup possibility for a single-fiber probe. Show a possible probe configuration using a combination of a 2 × 2 DCF coupler, a single-mode fiber (SMF), a multimode fiber (MMF), and a DCF single-fiber probe.

Solution: An example probe setup using a DCF coupler and three types of fibers is shown in Fig. 7.5. Here a single-mode fiber delivers laser light to the single-mode core in the port 1 DCF branch of the DCF coupler. Output port 2 of the DCF coupler is blocked and the fiber from port 3 is attached to a bidirectional DCF probe. The light is delivered to the tissue sample using the single-mode core of the DCF and the return path uses the large inner multimode cladding of the fiber at port 3 for increased light collection efficiency.

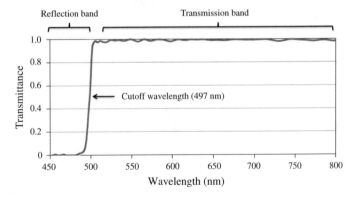

Fig. 7.4 Example of a dichroic edge filter with a 497-nm cutoff wavelength

Fig. 7.5 Single-fiber probe based on a DCF coupler and a DCF probe

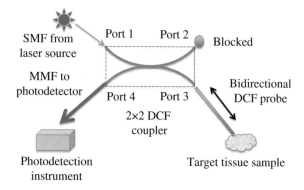

> In the return path, a MMF is attached to the DCF coupler at port 4 to collect the light from the multimode cladding and to deliver it to a photodetection instrument.

Instead of using a single-fiber probe, in fluorescence spectroscopy instruments often a more advantageous or necessary setup is to use separate optical fibers for the illumination and light collection functions, as is shown in Fig. 7.6. The excitation fiber typically would be a conventional single-mode fiber or some type of photonic crystal fiber. The collection and return fiber choice can be an optical fiber bundle, a SMF, a MMF, a polarization-preserving fiber, a PCF, or a DCF [3]. In the setup in Fig. 7.6, it is desirable to have a selectable wavelength for the incident radiation and to have the optical detector be capable of precise spectral manipulation and signal assessment. Therefore, a monochromator or an optical spectrum analyzer is provided for both the selection of the excitation light and the analysis of the optical emission from the sample. The function of a *monochromator* is to separate and transmit a selectable narrow spectral band of light from an optical signal that contains a wider range of wavelengths.

If the system can accommodate a larger diameter probe, then the efficiency of the illumination and collection functions can be increased by using more than two

Fig. 7.6 Schematic of a generic probe system consisting of one or more illumination and light-collection channels (*J. Biomed. Opt.* 19(8), 080902 (Aug 28, 2014). doi:10.1117/1.JBO.19.8. 080902)

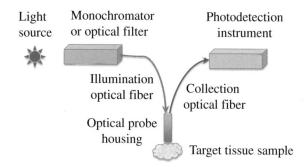

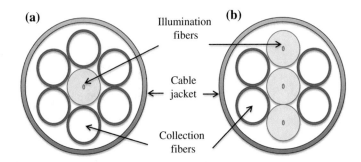

Fig. 7.7 End-face views of two different arrangements for hexagonal packing of multiple-fiber probes (*J. Biomed. Opt.* 19(8), 080902 (Aug 28, 2014). doi:10.1117/1.JBO.19.8.080902)

fibers. Figure 7.7 shows a classic example of hexagonal packing for a probe consisting of seven fibers [3]. The arrangement shown in Fig. 7.7a uses the central fiber as the light delivery channel for purposes such as illumination, scattering-based diagnosis, fluorescence excitation, or imaging. The surrounding six fibers are assigned as the collection channels. In Fig. 7.7b three fibers are assigned as the light delivery channels and the remaining fibers are used for light collection. Optionally, optical filters with different spectral pass bands can be deposited on the end faces of individual collection fibers for evaluating specific wavelength data.

Adding more rings of fibers around the central optical fiber can create larger bundles, as is shown in Fig. 7.8. The number of optical fibers N_{hex} that can be packaged hexagonally in a circular cross section is given by

$$N_{hex} = 1 + \sum_{n=0}^{m} 6n \qquad (7.1)$$

Fig. 7.8 Multiple hexagonally packaged rings of optical fibers

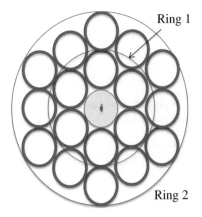

where the parameter m ≥ 0 designates the number of rings. Given that the diameter of an individual fiber is D_{fiber}, then the outer diameter of the fiber bundle D_{bundle} is given by

$$D_{bundle} = D_{fiber}(1 + 2m) \qquad (7.2)$$

The active end face area of the bundle depends on the core and cladding sizes. Larger core diameters and thinner claddings result in smaller dead spaces between fibers and thus give larger active areas.

Example 7.3 Consider an optical fiber that has a 100-μm core diameter and a 125-μm cladding diameter. (a) Find the number of fibers in bundles that have one and two rings. (b) What is the outer diameter of the bundle in each case?

Solution: (a) From Eq. (7.1) the number of optical fibers N_{hex} that can be packaged hexagonally in a single ring (m = 1) around a central fiber is

$$N_{hex} = 1 + 6 = 7$$

Similarly, for two rings of fibers

$$N_{hex} = 1 + \sum_{n=0}^{2} 6n = 1 + 6 + 12 = 19$$

(b) From Eq. (7.2) for one ring of fibers (m = 1) the outer diameter of the fiber bundle D_{bundle} is given by

$$D_{bundle} = D_{fiber}(1 + 2) = 3(125\,\mu m) = 375\,\mu m$$

Similarly, for two rings of fibers (m = 2)

$$D_{bundle} = D_{fiber}(1 + 2 \times 2) = 5(125\,\mu m) = 625\,\mu m$$

Example 7.4 Consider an optical fiber that has a 100-μm core diameter and a 125-μm cladding diameter. (a) Find the size of the active area in a bundle with one ring. (b) What is the ratio of the active area to the total cross sectional area of the bundle?

Solution: (a) Seven fibers can be packaged hexagonally in a single ring around a central fiber. Therefore the active area of this bundle is

$$7 \times \text{fiber-core area} = 7\pi(50\,\mu m)^2 = 5.50 \times 10^4\,\mu m^2 = 0.055\,mm^2$$

(b) From Example 7.3b, the radius of the bundle is 375/2 μm = 187.5 μm. Then the total cross sectional area of the bundle is

$$\text{Total area} = \pi(187.5\,\mu m)^2 = 1.10 \times 10^5\,\mu m^2 = 0.11\,mm^2$$

The ratio of the active area to the total cross-sectional area of the bundle then is

$$0.055/0.11 = 0.50 = 50\%$$

7.3 Optical Fiber Tip Geometries

An important design factor for optical probes is the tip geometry of the fiber at its distal end [3]. The tip shape controls the light distribution pattern on a tissue sample and also determines the light collection efficiency for viewing the scattered or fluorescing light emitted from the irradiated tissue sample.

When selecting the probe tip geometry, parameters that need to be considered include the sizes of the illumination and light-collection areas, the collection angle (related to the fiber NA), and the fiber diameter [3]. Another key point to remember is that *biological tissue has a multilayered characteristic from both compositional and functional viewpoints.* Because specific biological processes and diseases take place at different depths within this multilayered structure, it is important to ensure that the probing light penetrates the tissue down to the desired treatment or evaluation layer.

A multitude of fiber probe tip configurations have been analyzed, designed, and implemented [16–22]. The basic configurations when using a single optical fiber are illustrated in Fig. 7.9. The simplest end face is a flat surface that is orthogonal to the

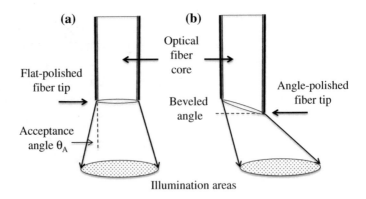

Fig. 7.9 Basic optical fiber probe tips: **a** flat-polished end face; **b** angle-polished end face

fiber axis, as shown in Fig. 7.9a. This end face will distribute the light emerging from the fiber in a circular pattern, which is determined by the core size and the numerical aperture. A handheld fiber-breaking tool that simply cleaves the fiber can be used to create the end face or the end face can be polished flat with a standard fiber-polishing machine. In a slightly more complex tip configuration shown in Fig. 7.9b, the exit surface can be polished at an angle relative to the fiber axis to deflect the light to a selectable area.

Example 7.5   Consider a conventional silica glass fiber that has a 50-μm core diameter and for which the NA = 0.20. If this fiber is used to illuminate a tissue sample, what is the diameter of the light spot on the tissue at a distance 1 mm from the end of the fiber?

Solution: From Eq. (3.3) the acceptance angle in air (n = 1.00) is

$$\theta_A = \sin^{-1}NA = \sin^{-1}0.20 = 11.5°$$

As shown in Fig. 7.10, at a distance d from the end of the fiber, the light will be projected onto a circular area of $\pi(a + x)^2$ where $x = d \tan \theta_A$ and a is the fiber radius. Therefore the diameter D_{spot} of the light spot on the tissue 1 mm (1000 μm) from the end of the fiber is $D_{spot} = 2[a + d \tan \theta_A]$ = $[50 + 2(1000) \tan 11.5°]$ μm = $[50 + 407]$ μm = 0.457 mm.

If the end face angle is beveled at the critical angle for total internal reflection as shown in Fig. 7.11, then the light will leave the fiber through its side [3]. This is the basis of what is called a *side-firing fiber*. For biomedical procedures the side-firing tip can be used with a needle (e.g., 0.5–0.9 mm diameter) for treating interstitial tumors. Other applications that involve directly using the side-emitted light emerging from the fiber include treating atrial fibrillation, prostate enlargements,

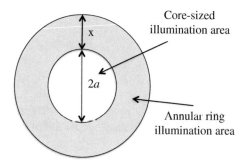

Fig. 7.10 Illuminated spot size of diameter 2(x + a) on a tissue sample

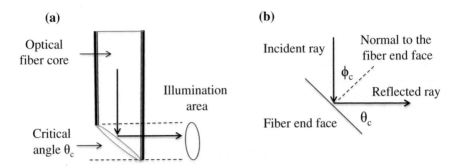

Fig. 7.11 Polishing the end face at the critical angle creates a side-firing fiber

dental diseases, and intracavity or superficial tumors in arteries or veins, in the esophagus, in gastrointestinal tracts, in urinary tracts, or in bronchi in the lungs.

As an alternative to using an angle-polished fiber tip, a flat-polished tip can be used in conjunction with a micro-optic 45° prism. An example of this configuration used within a needle probe is described in Chap. 11.

Example 7.6 Consider an optical fiber that has a core refractive index $n_1 = 1.450$ at a wavelength $\lambda = 680$ nm. At what angle should the end face be polished in order to have 680-nm light rays that are traveling parallel to the fiber axis be reflected sideways at the fiber end?

Solution: The end face must be oriented at angle that causes total reflection of the rays at a glass-air interface. If ϕ_c is the angle between the incident ray and the normal to the fiber end face as shown in Fig. 7.11b, then from Eq. (3.1) this angle is given by $\phi_c = \sin^{-1}(n_2/n_1)$. Here $n_2 = 1.000$ is the refractive index of air and $n_1 = 1.450$ is the refractive index of glass at 680 nm. Thus, from Eq. (3.2) the critical angle $\theta_c = \pi/2 - \phi_c$ at which the end face should be polished is

$$\theta_c = \pi/2 - \sin^{-1}(10.000/1.450) = 90° - 43.6° = 46.4°$$

Probe designs with two or more fibers can be used to increase the light illumination or collection efficiency. Figure 7.12 shows a probe design that is based on using two fibers that have flat-polished end faces. In this specific configuration, one fiber is used for illumination and the other fiber is used for light collection.

Fig. 7.12 Example of dual flat-polished fiber distal end tips

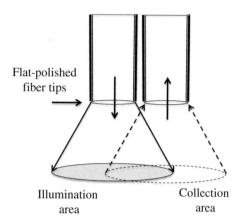

Flat-polished fiber tips →

Illumination area   Collection area

7.4   Optical Sensors

A biophotonics detection process can involve the sensing of perturbations in a transmitted optical signal caused by a change in some characteristic of an optical fiber or of a planar optical waveguide. These characteristics can include variations in the refractive index of the cladding material due to the adherence of molecules on the outside of the cladding, relative fiber movements induced by changes in an external parameter, or changes in the radius of curvature of the fiber axis. Here first Sects. 7.4.1 and 7.4.2 describe the sensing of externally induced evanescent field perturbations that cause a variation in the transmitted optical power level, which can be directly correlated to the concentration of specific molecules being measured. Next Sects. 7.4.3 and 7.4.4 illustrate how changes in an external physical parameter (e.g., pressure, stress, or temperature) can induce relative movements between two fibers or variations in the bend radius of a fiber, respectively. These changes also result in a fluctuation in the transmitted optical power level. The optical power level variation can be directly correlated to the value of the external physical parameter.

7.4.1   Biorecognition Optical Fiber Sensors

The attractions of optical fiber-based sensors are their small size, excellent integration capability within sensor packages, and relatively high sensitivity to detect diverse analytes. Figure 7.13 shows the basic operation of a class of biosensors that use absorbance measurements to detect any variations in the concentrations of substances that absorb a specific wavelength of light [3, 7, 23, 24]. First the fiber is coated with a *biorecognition material*, which can be an enzyme, an antibody, a DNA strand, a microorganism, or a cell. An *antibody* is a large Y-shaped molecule that typically is used by the immune system of a body to identify and neutralize

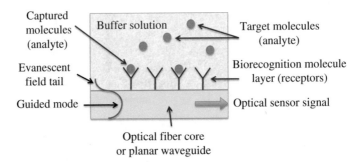

Fig. 7.13 Analyte sensing with an optical fiber or planar waveguide via an evanescent field perturbation (*J. Biomed. Opt.* 19(8), 080902 (Aug 28, 2014). doi:10.1117/1.JBO.19.8.080902)

pathogens such as bacteria and viruses. The biorecognition material is referred to as a *receptor*, which has the characteristic to absorb or capture a specific analyte. In immunology this analyte is called an *antigen*. An antigen is any substance that causes the immune system of the human body to produce antibodies in order to eliminate the antigen. Examples of antigens include chemicals that the immune system views as toxins and microorganisms such as bacteria, fungi, parasites, and viruses. The absorption process of the antigen by the receptor coating on the optical fiber or waveguide results in a physical-chemical alteration that can change the layer thickness, effective refractive index, degree of light absorption, or electrical charge at the coating surface. An optoelectronic sensor, such as a spectrum analyzer or a photodetector, then can measure variations in the optical signal parameters that result from these changes in the physical characteristics.

For example, as noted in Fig. 7.13 (also see Fig. 3.4), part of the optical power of a propagating mode is contained in an evanescent field that travels in the cladding or coating [3]. Thus a change in the index of the fiber coating will induce a slight perturbation in the mode near the fiber-to-coating interface. This modal perturbation results a change in the optical power level in the fiber core. The light power variation seen by the photodetector at that specific wavelength then can be related to the concentration of the absorbed analyte. Applications of this methodology have been used for sensing glucose levels, pH levels, oxygen levels, and the presence of antibodies.

7.4.2 ELISA

A technique called *enzyme-linked immunosorbent assay* (ELISA) is a variation on the class of antibody/antigen fiber sensor described above [23, 25, 26]. This procedure is a common serological test to check for the presence of certain antibodies or antigens. The term *serology* refers to the scientific and clinical study of plasma serum and other body fluids. The ELISA concept, which is illustrated in Fig. 7.14,

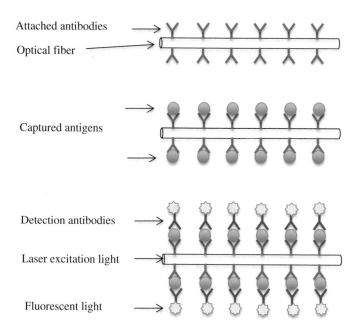

Fig. 7.14 Fiber-based biosensor that uses a fluorescing detection antibody attached to the antigens (*J. Biomed. Opt.* 19(8), 080902 (Aug 28, 2014). doi:10.1117/1.JBO.19.8.080902)

is based on observing the fluorescence emitted from a detection antibody that binds to an absorbed antigen [3]. The key operational characteristic of this sensing method is that the detection antibody will fluoresce at a specific excitation wavelength, which thereby indicates the presence and concentration of an analyte. To construct the sensor, first a fiber is coated with an antibody and then is immersed in a medium containing antigens, which are captured by the antibody layer. Next a layer of the detection antibodies is attached to the captured antigens. This procedure forms a sandwich structure consisting of the adhered capturing antibody layer, the captured antigen layer, and the detection antibody layer. Once the detection antibodies bind to the antigen, light sent through the fiber induces the detection antibodies to fluoresce. The antigen concentration then can be determined by measuring the degree of fluorescence produced.

7.4.3   Sensors Based on Optical Fiber Movements

Some common optical fiber-based sensors are based on measuring variations in back-reflected light or changes in light levels coupled between two fibers [27]. The optical power variations then can be directly correlated to changes in an external physical parameter (e.g., pressure, stress, or temperature). One method among many is shown in Fig. 7.15. In this case, light leaving a fixed fiber is coupled into a

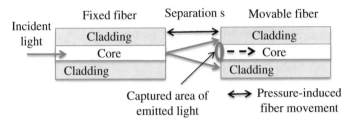

Fig. 7.15 Simple optical fiber-based pressure sensor based on using a moving acceptor fiber (*J. Biomed. Opt.* 19(8), 080902 (Aug 28, 2014). doi:10.1117/1.JBO.19.8.080902)

second movable fiber that is located an axial distance s from the fixed fiber [3]. The second fiber can move axially due to a variation in an external parameter, thus changing the distance s between the two fiber ends. If there is an increase in the distance s, then less light will enter the second fiber and vice versa. The optical power variation at the photodetector thus can be used to measure the change in an external parameter.

When light is coupled from one optical fiber to another, not all of the higher-mode optical power emitted in the ring of width x shown in Fig. 7.10 will be intercepted by the receiving fiber. The fraction of optical power coupled into the receiving fiber is given by the ratio of the cross-sectional area of the receiving fiber (πr^2) to the area $\pi(a + x)^2$ over which the emitted power is distributed at a distance s. From Figs. 7.10 and 7.15 it follows that x = s tan θ_A, where θ_A is the acceptance angle of the fibers, as defined in Eq. (2.2). From this ratio the loss in dB for an offset joint between two identical step-index fibers is found to be [28]

$$L_{\text{gap}} = -10\log\left(\frac{a}{a+x}\right)^2 = -10\log\left(\frac{a}{a+s\tan\theta_A}\right)^2$$
$$= -10\log\left[1 + \frac{s}{a}\sin^{-1}\left(\frac{NA}{n}\right)\right]^{-2} \tag{7.3}$$

where a is the fiber radius, NA is the numerical aperture of the fiber, and n is the refractive index of the material between the fiber ends (usually either air or an index matching gel).

Example 7.7 Suppose two identical step-index fibers each have a 25-μm core radius and an acceptance angle of 14°. Assume the two fibers are perfectly aligned axially and angularly. What is the variation in the insertion loss when the longitudinal separation changes from 0.020 to 0.025 mm?

Solution: The insertion loss due to a gap between fibers can be found by using Eq. (7.3).

For a 0.020-mm = 20-μm gap

$$L_{gap} = -10\log\left(\frac{25}{25 + 20\tan 14°}\right)^2 = 1.580\,\text{dB}$$

For a 0.025-mm = 25-μm gap

$$L_{gap} = -10\log\left(\frac{25}{25 + 25\tan 14°}\right)^2 = 1.934\,\text{dB}$$

Thus the loss variation is 0.354 dB.

Instead of examining transmitted light levels, a sensor can be configured to measure intensity changes in reflected light, as is shown in Fig. 7.16. In this case, light emerges from a sensing optical fiber and a mirror or a diaphragm located a distance d from the end of the fiber reflects the light [3]. Then a percentage of the reflected light is captured by the sensing fiber core and is transmitted to a photodetector that senses the returning optical power level. For example, if the pressure on the mirror or diaphragm increases or decreases, the distance d will decrease or increase correspondingly. Consequently the measured level of the captured reflected light will increase or decrease, respectively. Again, the optical power variation at the photodetector is a measure of the change in an environmental parameter.

The same analysis as is given in Eq. (7.3) is applicable to the case shown in Fig. 7.16. However, now the parameter s is replaced by 2d, because the light makes a round trip of 2d through the gap.

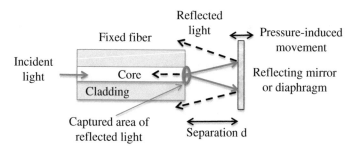

Fig. 7.16 Simple optical fiber-based pressure sensor based on using a reflecting mirror or diaphragm (*J. Biomed. Opt.* 19(8), 080902 (Aug 28, 2014). doi:10.1117/1.JBO.19.8.080902)

Example 7.8 Consider a step-index fibers with a 25-μm core radius and an acceptance angle of 14° in the reflective sensor shown in Fig. 7.16. What is the variation in the insertion loss when the longitudinal separation between the fiber end and the reflector changes from 0.020 mm to 0.025 mm?

Solution: The insertion loss due to a gap between fiber and the reflector can be found by using Eq. (7.3) with the parameter 2d substituted for the parameter s.
For a 0.020-mm = 20-μm gap

$$L_{gap} = -10 \log \left(\frac{25}{25 + 40 \tan 14°} \right)^2 = 2.916 \, dB$$

For a 0.025-mm = 25-μm gap

$$L_{gap} = -10 \log \left(\frac{25}{25 + 50 \tan 14°} \right)^2 = 3.514 \, dB$$

Thus the loss variation is 0.598 dB.

7.4.4   Microbending Fiber Sensors

As a multimode fiber is progressively bent into a tighter radius, more of the optical power from the higher-order modes gets radiated out of the fiber. This effect can be used to build a sensor based on measuring optical power level variations due to fiber bending [29–31]. An embodiment of one such device is illustrated in Fig. 7.17. Such a microbending sensor was one of the earliest fiber optic sensors. Here an optical fiber is run between two interleaved corrugated plates. As the

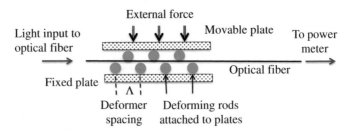

Fig. 7.17 Concept for sensing via fiber microbending

external force on one or both of the plates varies due to changes in factors such as pressure, temperature, or stress, the bend radius of the optical fiber changes. The result is that the optical power level transmitted through the optical fiber fluctuates.

For measuring changes ΔP in pressure, the operation of the sensor can be expressed in terms of the change in the transmission coefficient Δt of the light propagating through the bent fiber as

$$\Delta t = \left(\frac{\Delta t}{\Delta X}\right) A_p \left(k_f + \frac{A_s Y_s}{L_s}\right)^{-1} \Delta P \approx \left(\frac{\Delta t}{\Delta X}\right) A_p k_f^{-1} \Delta P \qquad (7.4)$$

Here ΔX is the displacement of the deformer plates, $(\Delta t/\Delta X)$ is the device sensitivity, A_p is the plate area, and k_f is the force constant or effective spring constant of the bent fiber. The parameters A_s, Y_s, and L_s are the cross-sectional area, Young's modulus, and the thickness, respectively, of the deformer spacers. The approximation on the right-hand side of Eq. (7.4) is valid for the design condition $A_s Y_s / L_s \ll k_f$. The effective spring constant k_f can be expressed as

$$k_f^{-1} = \frac{\Lambda^3}{3\pi Y d^4 \eta} \qquad (7.5)$$

where Λ is the spacing of the deformers, Y is the effective Young's modulus, d is the diameter of the fiber, and η is the number of deformation intervals.

Similarly, for changes ΔT in temperature, the operation of the sensor can be expressed as

$$\Delta t = \left(\frac{\Delta t}{\Delta X}\right) A_s \alpha_s Y_s \left(k_f + \frac{A_s Y_s}{L_s}\right)^{-1} \Delta T \approx \left(\frac{\Delta t}{\Delta X}\right) \alpha_s L_s \Delta T \qquad (7.6)$$

Here α_s is the thermal expansion coefficient of the spacers and the approximation on the right-hand side of Eq. (7.5) is valid for the design condition $k_f L_s \ll A_s Y_s$.

Example 7.9 Consider a microbending pressure sensor that has the following characteristics: $A_p = 1$ cm^2 and $k_f^{-1} = 33\times10^{-8}$ cm/dyn. If the minimum measurable displacement is $\Delta X_{min} = 10^{-10}$ cm, what is the minimum detectable pressure ΔP_{min}?

Solution: From Eq. (7.4), the minimum detectable pressure ΔP_{min} is

$$\Delta P_{min} = \frac{\Delta X_{min}}{A_p k_f^{-1}} = \frac{10^{-10} \text{cm}}{(1 \text{ cm}^2)(33 \times 10^{-8} \text{ cm/dyn})} = 3 \times 10^{-4} \text{dyn/cm}^2$$
$$= 3 \times 10^{-5} \text{Pa (pascals)}$$

[Note: 1 dyne = 10^{-5} N; 1 Pa (Pa) = 1 N/m^2 = 10^5 dyne/10^4 cm^2 = 10 dyne/cm^2].

For comparison purposes, atmospheric pressure at sea level is 101.325 kPa.

7.5   Interferometric Sensors

Interferometric sensors are based on measuring the phase difference between two superimposed light beams that have the same frequency. As is shown by the generic interferometer diagram in Fig. 7.18, typically an incident light beam is split into at least two parts by a 3-dB coupler or beamsplitter in the interferometer. These parts follow different paths through the interferometer system and then another coupler recombines the parts to create an interference pattern. If the optical path lengths differ by an integer number of wavelengths then a constructive interference pattern will be displayed. If the optical path lengths differ by an odd number of half wavelengths then a destructive interference pattern will appear.

The optical paths could be in the same optical fiber if two or more distinguishable optical fiber modes are used. Here each mode defines one optical path. An example is the Sagnac interferometer where the different optical paths are defined by clockwise and counter clockwise modes. Another option is to have the optical paths pass through separate optical fibers. For example, this is the case in the commonly deployed Mach-Zehnder optical fiber interferometer. This section describes three interferometer architectures commonly used in biophotonics. These are the Mach-Zehnder, Michelson, and Sagnac interferometers [32].

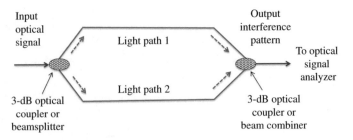

Fig. 7.18 Operational concept of an interferometer sensor

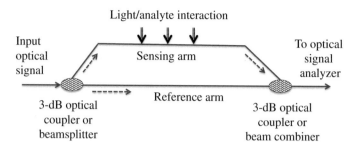

Fig. 7.19 Operation of a basic Mach-Zehnder interferometer

7.5.1 Mach-Zehnder Interferometer

Owing to their versatile setup configurations, Mach-Zehnder interferometers are popular for diverse biosensing applications, for example, reflectometric interference spectroscopy. In a basic *Mach-Zehnder interferometer* (MZI) light from a laser source is divided into two paths by means of a beamsplitter or a 3-dB optical coupler, as shown in Fig. 7.19. These paths are called the *arms* of the interferometer. One path of the MZI is a *sensing arm* and the other path is a *reference arm*. In the sensing arm an interaction between light and a biosensor, for example, a receptor-analyte complex as described in Sect. 7.4, takes place by means of evanescent field waves. The reference arm is kept isolated from the external parameter being analyzed. After the light passes through the two arms, a beam combiner or a 3-dB optical coupler recombines the two waves. The constructive and destructive interference pattern resulting from the recombination of the two lightwaves then is directed to an optical signal analyzer, which makes an assessment of the parameter being evaluated. For example, the optical signal variations can directly measure the concentration of the analyte interacting with the sensing arm.

In order to create a more compact and robust interferometer design, biosensors based on an all-fiber inline MZI scheme have been proposed and implemented. The designs include a tapered fiber configuration [33, 34], a microcavity imbedded in a fiber [35], the use of fiber Bragg gratings [36], cascaded segments of different fibers [37–39], and photonic crystal fibers that are selectively filled with a liquid.

As one example using an FBG, Fig. 7.20 shows a sensor structure with two long-period gratings embedded in a fiber core. A *long-period grating* (LPG) is a periodic structure in a fiber core that couples co-propagating modes in optical fibers [40–42]. The coupling is realized through phase matching between the fundamental LP_{01} guided core mode and higher-order cladding modes. When light traveling in the fiber core encounters the first grating, part of the light is coupled into the fiber cladding and the other part continues propagating in the fiber core. The light in the cladding then interacts with an external parameter through evanescent wave coupling. Subsequently, the perturbed cladding waveform reenters the fiber core by means of the second grating and is combined with the light traveling in the core reference path. The

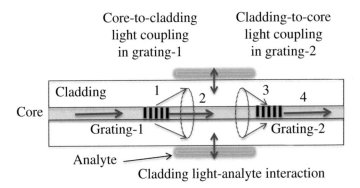

Fig. 7.20 Sensor structure with two long-period gratings embedded in a fiber core

resulting observed constructive and destructive interference patterns at the sensor output then allow an assessment of the external parameter being evaluated.

Similarly, as a second example, cascaded segments of fibers with different core diameters can be used to couple light from the core in one fiber into the cladding of an adjoined fiber. This is illustrated in Fig. 7.21 where fiber type 1 could be a multimode fiber and fiber type 2 a single-mode fiber. The cladding modes that now travel in the inserted fiber segment (type 2) then interact with an external parameter through evanescent wave coupling. At the second fiber junction, the cladding modes are coupled back into the core of fiber type 1 to create an interference pattern at the fiber sensor output.

A third example is the use of photonic crystal fibers. The flexibility of photonic crystal fiber structures enables them to be used in a wide variety of biosensors. In this case, fiber type 2 can be a PCF. More details on this structure and some applications are discussed in Sect. 7.6.

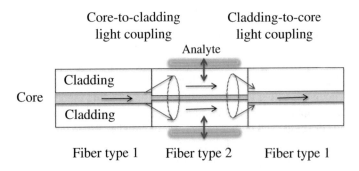

Fig. 7.21 Sensor using cascaded segments of different fibers

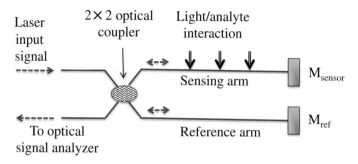

Fig. 7.22   Operation of a basic Michelson interferometer

7.5.2   *Michelson Interferometer*

A Michelson Interferometer (MI) is similar to an MZI and is widely used for optical coherence tomography (see Chap. 10) and radial keratectomy procedures (removal or shaving of the cornea with a laser to correct vision). Similar to the MZI, the basic MI concept is the implementation of interference between light beams traveling in two arms of the interferometer. However, the difference from the MZI is that in the MI each beam is reflected by a mirror at the end of each arm, as is shown in Fig. 7.22. After the reference and sensor beams are reflected by mirrors M_{ref} and M_{sensor}, respectively, they are recombined with an optical coupler. As with the MZI, the resulting observed constructive and destructive interference patterns then allow an assessment of the external parameter being examined.

When the two interfering beams are of equal amplitude, the relationship describing the interference pattern is given by

$$I = 4I_0\cos^2(\delta/2) \tag{7.7}$$

Here I_0 is the intensity of the input light and the phase difference δ between the sensing and reference beams is defined by

$$\delta = 2\pi\Delta/\lambda \tag{7.8}$$

with λ being the wavelength of the input light and Δ is the difference in the optical path lengths between the two beams, which is defined by

$$\Delta = 2d\cos\theta + \lambda/2 = (m + 0.5)\lambda \tag{7.9}$$

Here 2d is the difference in the path lengths from the optical splitter, m is the number of interference fringes, and θ is the angle of incidence ($\theta = 0°$ for a normal or on-axis beam). Consider the case when a thin slice of tissue that has a uniform refractive index n_s is inserted in one of the beam paths. Assume that there is minimal absorption and scattering in this tissue slice. Then

$$d = (n_s - n_{air})L \tag{7.10}$$

where L is the thickness of the tissue and n_{air} is the refractive index of air.

Example 7.10 Consider a thin slice of tissue that is inserted normally to the light beam in one path of a Michelson interferometer. Assume that the tissue has a uniform refractive index $n_s = 1.33$. Using a test wavelength of 620 nm, the fringe pattern shifts by 50 fringes. What is the thickness of the tissue slice?

Solution: With $\theta = 0°$ for a normal beam, then Eq. (7.9) becomes $d = m\lambda/2$. Thus the optical path length is $d = 50(0.620 \ \mu m)/2 = 15.50 \ \mu m$. Then from Eq. (7.10), the thickness of the sample is

$$L = d/(n_s - n_{air}) = (15.50 \ \mu m)/(1.33 - 1.00) = 46.97 \ \mu m$$

7.5.3   Sagnac Interferometer

A Sagnac interferometer traditionally is used as a sensor for measuring rotation, stress, and temperature [43, 44]. The interferometry technique also has found biomedical applications in spectral imaging for disease detection, optical polarimetry procedures in pharmaceutical drug testing, quality control for food products, and for noninvasive glucose sensing in diabetic patients.

The basic configuration of a Sagnac interferometer consists of a single fiber loop and a 3-dB optical fiber coupler, as shown in Fig. 7.23. The 3-dB coupler divides the input light into two counter-propagating directions and also recombines the two counter-rotating beams. The recombined beams then are sent to an optical signal analyzer. The operational concept is based on measuring the difference in the polarization-dependent modal propagating speeds between the two beams. This can be achieved by using either a highly birefringent fiber or a polarization-maintaining fiber in the sensing section. The polarization in the interferometer can be adjusted by means of a polarization controller situated at the entrance to the fiber loop.

The measured signal at the output of the 3-dB coupler is determined by the interference between the beams that are polarized along the slow axis and the fast axis. The phase δ_{SI} of the interference in the Sagnac interferometer is given by

$$\delta_{SI} = \frac{2\pi}{\lambda} B_f L = \frac{2\pi}{\lambda} |n_f - n_s| L \tag{7.11}$$

where B_f is the birefringence coefficient of the sensing fiber of length L. The parameters n_f and n_s are the effective refractive indices of the fast and slow modes, respectively.

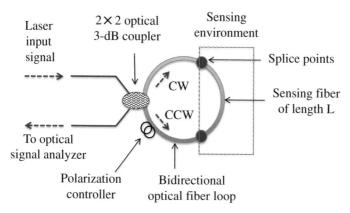

Fig. 7.23 Operation of a basic Sagnac interferometer

Measuring the difference in transmissivity through the two directions of optical signal rotation is an alternative use of a Sagnac interferometer. In either case, the Sagnac interferometer has a linear relationship between the output changes and the parameter being measured. Thus, a Sagnac interferometer can be calibrated by plotting the phase difference or transmissivity variations as a function of these measured parameters.

7.6   Photonic Crystal Fiber Biosensors

Photonic crystal fiber-based (PCF) devices are being applied increasingly in various types of biosensors [45–51]. One attractive feature is that the air holes in a PCF can be used as natural miniature chambers for holding a liquid sample [3]. This feature greatly reduces the required sample volume that generally is much larger in other biosensors. The mechanism for evaluating the analyte solution inside the PCF can be based on the use of a FBG, interferometry, mode coupling, evanescent field monitoring, or observation of a bandgap shift.

7.6.1   Interferometry Sensing Methods

One fiber sensor type that uses a PCF as the sensing element is based on intermodal interference between forward-propagating core and cladding modes in a PCF [38, 45]. The basic configuration of a two-beam interference method uses the core mode as the reference beam and the cladding modes as the sensing beam [3]. An example of such a PCF-based interferometer is shown in Fig. 7.24, which is similar to the

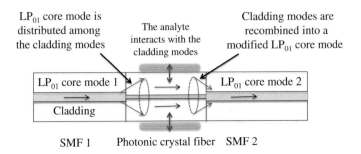

Fig. 7.24 Concept of an inline core-cladding intermodal interferometer based on the mode-field mismatch method

setup illustrated in Fig. 7.20. Here a short length of PCF is sandwiched between input and output SMFs (such as the conventional industry standard G.652 fibers) to form an inline core-cladding intermodal interferometer. Different higher-order cladding modes can be excited in the PCF when the fundamental mode (the LP_{01} mode) in the single-mode fiber encounters abrupt fiber tapers formed at the interfaces between the PCF segment and the single-mode input and output fibers.

When light from a broadband source is sent through the device, the interference between the core and cladding modes in the PCF sensing section will result in a transmission-versus-wavelength interference pattern [3]. Since the higher-order mode is a cladding mode of a glass-air waveguide, any change of the outer refractive index caused by immersion of the PCF sensing segment in a liquid will change the propagation constant of the higher-order mode and consequently will cause a shift in the interference fringes.

7.6.2 Liquid Infiltration Sensor

The infiltration of a fluid sample into the PCF can be achieved by simply immersing the fiber end in the liquid. The liquid then enters the fiber holes through capillary action. However, afterwards if the liquid needs to be removed from the PCF, which could be required in multi-step biochemical processes, then either a vacuum or a pressure needs to be applied to the fiber end.

Because the air holes in a PCF run along the entire length of the fiber, this feature is being used to enable long interaction lengths between the excitation light and small volumes of an analyte. One such embodiment involves fabricating a PCF that has a side-channel opening running along the fiber, which can be used for containing a liquid analyte that can be analyzed by means of surface enhanced Raman scattering (SERS) sensing [47–49].

7.7 Fiber Bragg Grating Sensors

A highly successful precise wavelength selection component is a fiber Bragg grating (FBG). As Sect. 4.2.1 describes, a FBG is a narrowband reflection-grating filter that is constructed within an optical fiber core [3]. For biophotonics applications an external force, such as a strain induced on the device by the weight of a person, will slightly stretch the fiber thereby changing the length of the FBG and thus changing the reflected wavelength. To measure the induced strain, the FBG sensor typically is glued to or embedded in a specimen that responds to the external strain. The one precaution that needs to be taken when using such a sensor is to realize that the FBG is temperature sensitive. Thus, either the substrate on which the FBG is glued has to be temperature insensitive or some type of temperature compensating method has to be deployed along with the strain sensor.

A wide selection of biosensor applications to diagnose or monitor the health status in a person has been realized using FBGs [3]. Among these are the following:

- Measuring the polymerization contraction of common dental composite resins [52]
- Use of a non-invasive FBG-based optical fiber probe for carotid pulse (a heartbeat felt through the wall of a carotid artery) waveform assessment [53]
- The development of a smart-bed healthcare system for monitoring patient movements [54–56]
- The use of FBG sensors in biomechanics and rehabilitation applications, for example, (a) monitoring contact stress, contact area, and joint alignment during knee joint replacement and (b) simultaneous measurements of contact force or stress and fluid pressure in articular joints [57–59]
- FBG-embedded sensor pads for human-machine interface surfaces for rehabilitation and biomechanics applications, such as medical beds, wheelchairs, and walkers [60]
- In vivo use of a high-resolution FBG-based optical fiber manometer for diagnosing gastrointestinal motility disorders [61, 62]
- Measuring the shock absorption ability of laminate mouth guards [63]
- Smart textiles for respiratory monitoring [64]
- Simultaneous pressure and temperature measurements [65]

More details for two of these examples are given below. Both applications used wavelength division multiplexing techniques for simultaneous querying of an array of FBG sensors along a single fiber line with different wavelengths.

7.7.1 Smart-Bed FBG System

One healthcare implementation of an array of FBG sensors is a smart-bed or smart-chair system for monitoring movements of patients [54–56]. In this

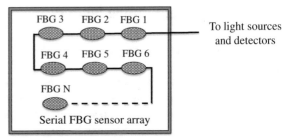

Fig. 7.25 FBG pressure-pad sensor array for patient condition monitoring

To light sources and detectors

Sensor pad with N FBG elements

application, a mattress pad or a chair cushion can have a series of embedded FBG sensors with different Bragg wavelengths cascaded along a single fiber, as is shown in Fig. 7.25. In one setup, a line of twelve FBG sensors was mounted on the surface of a bed to form a 3 by 4 matrix array, which then was covered by a standard mattress [3, 55]. Each FBG sensor was specially packaged into an arc-shaped elastic bending beam using a carbon fiber reinforced plastic material, which ensures excellent sensitivity and good linear translation from a lateral force exerted to the apex of the sensor into axial strain of the FBG when a subject is on the bed. Thereby patient movements and respiratory rate cause different pressures on individual FBG sensors, so that changes in the Bragg wavelength of specific FBG sensors allows the monitoring of both healthy and abnormal conditions of a patient.

7.7.2 Distributed FBG-Based Catheter Sensor

A second example of an FBG-based biosensor is a catheter that does distributed in vivo sensing of pressure abnormalities due to motility disorders in the gastrointestinal tract [3, 61]. The catheter was formed from a serial array of twelve FBG sensors that were fabricated into a continuous length of single mode fiber. As shown in Fig. 7.26, each FBG was attached to a localized pressure-sensitive structure consisting of a rigid metallic substrate and a flexible diaphragm. Each FBG element was 3 mm long and had a full-width half-maximum spectral response of 0.6 nm for Bragg wavelengths spaced 1.3 nm apart in the 815–850 nm sensing range. The FBG sensor elements were spaced 10 mm apart, thereby resulting in a catheter with a 12-cm sensing length. The device was designed to measure pressure changes between −50 and +300 mmHg, which adequately covers the range of pressures normally encountered in gastrointestinal tracts. As shown in Fig. 7.26, a circulator was used to insert light from a broadband optical source into the sensor array. Optical signal variations returning from the sensor array reentered the circulator and were sent to an optical detector from the third port of the circulator.

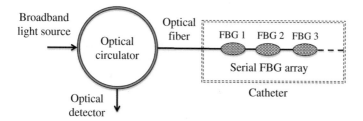

Fig. 7.26 Data acquisition equipment to record pressure changes along a catheter

7.8 Surface Plasmon Resonance Biosensors

Surface plasmons are electromagnetic waves that propagate along the surface of a metallic-dielectric or a metallic-air interface and are evanescent in all other directions [66–71]. Because these waves travel along the surface of a thin metallic layer (on the order of several ten nanometers thickness for visible light), they are very sensitive to any changes in the boundary characteristic at the surface. For example, changes in the boundary characteristic can occur when molecules suspended in a liquid sample are adsorbed on the metal surface. A *surface plasmon resonance* (SPR) is the collective oscillation of the surface electrons when they are stimulated by incident light. The resonance condition occurs when the frequency of the photons matches the natural frequency of surface electrons that are oscillating against the restoring force of positive nuclei in the surface material. By making use of the SPR effect, biosensors with a high sensitivity can be created for applications such as imaging, medical diagnostics, drug discovery, food safety analysis, and environmental monitoring.

The basic concept of a commonly used SPR configuration is illustrated in Fig. 7.27. In this setup, first a thin metallic film, such as gold, is deposited on a dielectric substrate. This film is then functionalized with a layer of specific selective biosensing receptors, for example antibodies. By using a prism, polarized light that

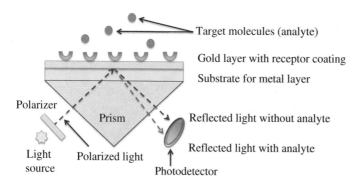

Fig. 7.27 Concept of a biosensor using a surface plasmon resonance effect

is incident on the metal surface at a specific angle (called the *plasmon resonance angle*) is coupled to the surface plasmon modes in the metal film and is totally internally reflected toward a photodetector. The resonance angle is very sensitive to small changes in the refractive index at the interface between the metal and the receptor layer. For example, these changes occur when the chemically activated surface captures biological samples. These events will modify the surface plasmon mode thereby causing a shift in the resonance peak of the reflected light. This shift can be viewed by a photodetector, for example, a charge-coupled device (CCD). This shifting process enables the construction of a biosensor that correlates the shifts in the resonance angle to quantitative molecular binding data at the sensor surface.

7.9 Optical Fiber Nanoprobes

The recent joining of SPR, nanotechnology, and optical fiber technology has led to the *lab-on-fiber* concept [72–78]. In this currently evolving technology, major efforts include attaching patterned layers of nanoparticles either on the tip of a fiber or as a nanocoating over a Bragg grating written inside a microstructured optical fiber [3]. When a patterned nanostructure is deposited on the tip of a standard optical fiber, *localized surface plasmon resonances* (LSPRs) can be excited by an illuminating optical wave, because of a phase matching condition between the scattered waves and the modes supported by the nanostructure. Figure 7.28 shows an example in blue of the surface plasmon resonance wave associated with a nanostructure pattern for the condition when no analyte is covering the pattern on the end of a fiber. These LSPRs are very sensitive to the surrounding refractive index (SRI) at the fiber tip. Thus, inserting the fiber tip into a fluid will cause the analyte liquid to cover the nanoparticle pattern. This action changes the refractive index (RI) of the nanolayer-fluid interface, thereby causing a wavelength shift in the LSPR peak due to a change in the phase matching condition, as shown by the red

Fig. 7.28 Example of the shift in the surface plasmon resonant peak when there is a relative index change at the tip of a fiber covered with a nanoarray pattern (*J. Biomed. Opt.* 19(8), 080902 (Aug 28, 2014). doi:10.1117/1.JBO.19.8.080902)

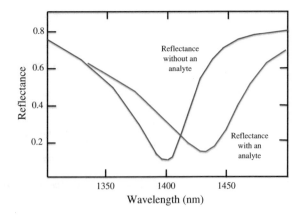

curve in Fig. 7.28. Thereby highly sensitive physical, chemical, and biological sensing functions can be enabled with a compact device.

7.10   Summary

Numerous embodiments of optical fibers and optical waveguide structures using different materials and configurations have been investigated to form optical probes, molecular and disease analysis tools, health monitoring devices, and biosensors. A wide variety of biosensors are available to selectively detect specific biological elements, such as microorganisms, organelles, tissue samples, cells, enzymes, antibodies, and nucleic acids derived from animal tissue and body fluids, human tissue and body fluids, cell cultures, foods, or air, water, soil, and vegetation samples. In addition to analyzing biological elements, diverse types of photonics-based biosensors are being used in the healthcare field for assessments of dental conditions and materials, in equipment for biomechanics and rehabilitation applications, in the diagnosis of gastrointestinal disorders, and in smart textiles for respiratory monitoring. Supporting optical and photonics technologies include fiber Bragg gratings, optical filters, interferometry methodologies, nanoparticle arrays, and surface plasmon resonance techniques.

7.11   Problems

7.1   Consider the probe configuration shown in Fig. 7.3a. Suppose that the dichroic filter has the following characteristics: a loss of 1 dB at each coupling junction, reflection loss of 1.1 dB at a wavelength λ_1, and a transmission loss of 1.1 dB at a wavelength λ_2. (a) If the optical source launches 10 mW of power into the first coupling junction of the dichroic filter, show that the optical power level at the tissue surface is 6.9 dBm (4.9 mW). (b) If the optical fiber probe at the tissue surface collects a power level of 10 μW at a wavelength λ_2, show that the power level at the photodetector is −23.1 dBm (4.9 μW).

7.2   Consider the probe configuration shown in Fig. 7.3b. Suppose that the optical circulator has the following characteristics: a loss of 1 dB at each coupling junction and an insertion loss of 1.7 dB at both wavelength λ_1 and wavelength λ_2. (a) If the optical source launches 10 mW of power into the first coupling junction of the circulator, show that the optical power level at the tissue surface is 6.3 dBm (4.3 mW). (b) If the optical fiber probe at the tissue surface collects a power level of 10 μW at a wavelength λ_2, show that the power level at the photodetector is −23.7 dBm (4.3 μW).

7.3   Consider the experimental setup shown in Fig. 7.5. Suppose the internal intrinsic loss plus the power splitting loss of the coupler add up to 3.7 dB and assume there is a power loss of 0.8 dB at each port of the coupler. If a power level of 10 mW enters the coupler from the SMF fiber attached to the light source, show that the power level at the tissue surface is 4.7 dBm (2.95 mW).

7.4 Suppose an optical fiber has a 365-μm core diameter and a 400-μm cladding diameter. (a) Show that the outer diameter is 1.2 mm for a hexagonally packed bundle that has one ring of fibers. (b) Show that the outer diameter is 2.0 mm for a bundle that has two rings.

7.5 Consider an optical fiber that has a 200-μm core diameter and a 240-μm cladding diameter. (a) Show that the size of the active area in a bundle with one ring is 0.22 mm^2. (b) Show that the ratio of the active area to the total cross sectional area of the bundle is 54 %.

7.6 Consider an optical fiber that has a 200-μm core diameter and a 240-μm cladding diameter. (a) Show that the size of the active area in a bundle with two rings is 1.13 mm^2. (b) Show that the ratio of the active area to the total cross sectional area of the bundle is 53 %.

7.7 Consider a conventional silica glass fiber that has a 100-μm core diameter and for which the NA = 0.26. If this fiber is used to illuminate a tissue sample, show that the diameter of the light spot on the tissue at a distance of 1 mm from the end of the fiber is 0.64 mm.

7.8 Two identical step-index fibers each have a 50-μm core radius and an acceptance angle of 15°. Assume the two fibers are perfectly aligned axially and angularly. Show that the variation in the insertion loss when the longitudinal separation changes from 0.020 to 0.025 mm is 0.21 dB.

7.9 Suppose that a step-index fiber with a 100-μm core radius and an acceptance angle of 14° is used in the reflective sensor shown in Fig. 7.16. Show that the variation in the insertion loss when the longitudinal separation between the fiber end and the reflector changes from 0.040 mm to 0.050 mm is 0.18 dB.

7.10 Consider a microbending pressure sensor that has the following characteristics: $A_p = 1$ cm^2 and $k_f^{-1} = 33 \times 10^{-8}$ cm/dyn. If the minimum measurable displacement is $\Delta X_{min} = 10^{-8}$ cm, show that the minimum detectable pressure ΔP_{min} is 3×10^{-3} Pa (pascals).

7.11 Consider a thin slice of tissue that is inserted normally to the light beam in one path of a Michelson interferometer. Assume that the tissue has a uniform refractive index $n_s = 1.36$. Using a test wavelength of 760 nm, the fringe pattern shifts by 45 fringes. Show that the thickness of the tissue slice is 47.5 μm.

References

1. D. Hoa, A.G. Kirk, M. Tabrizian, Towards integrated and sensitive surface plasmon resonance biosensors: a review of recent progress. Biosens. Bioelectron. **23**(2), 151–160 (2007). (Review paper)
2. X. Fan, I.M. White, S.I. Shopova, H. Zhu, J.D. Suter, Y. Sun, Sensitive optical biosensors for unlabeled targets: A review. Anal. Chim. Acta **620**, 8–26 (2008). (Review paper)
3. G. Keiser, F. Xiong, Y. Cui, P.P. Shum, Review of diverse optical fibers used in biomedical research and clinical practice. J. Biomed. Optics **19**, art. 080902 (2014) (Review paper)
4. L.C.L. Chin, W.M. Whelan, I.A. Vitkin, Optical fiber sensors for biomedical applications (Chap. 17), in *Optical-Thermal Response of Laser-Irradiated Tissue*, 2nd edn., ed. by A. J. Welch, M.J.C. van Gemert (Springer, New York, 2011)
5. X.D. Fan, I.M. White, Optofluidic microsystems for chemical and biological analysis. Nat. Photonics **5**(10), 591–597 (2011). (Biosensors review paper)
6. F. Taffoni, D. Formica, P. Saccomandi, G. Di Pino, E. Schena, Optical fiber-based MR-compatible sensors for medical applications: an overview. Sensors **13**, 14105–14120 (2013). (Review paper)
7. X.D. Wang, O.F. Wolfbeis, Review: fiber-optic chemical sensors and biosensors (2008–2012). Anal. Chem. **85**(2), 487–508 (2013). (Review paper)
8. O. Tokel, F. Inci, U. Demirci, Advances in plasmonic technologies for point of care applications. Chem. Rev. **114**, 5728–5752 (2014). (Biosensors review paper)
9. T.J. Pfefer, K.T. Schomacker, M.N. Ediger, N.S. Nishioka, Multiple-fiber probe design for fluorescence spectroscopy in tissue. Appl. Opt. **41**(22), 4712–4721 (2002)
10. P.R. Bargo, S.A. Prahl, S.L. Jacques, Optical properties effects upon the collection efficiency of optical fibers in different probe configurations. IEEE J. Sel. Topics Quantum Electron. **9**(2), 314–321 (2003)
11. U. Utzinger, R.R. Richards-Kortum, Fiber optic probes for biomedical optical spectroscopy. J. Biomed. Opt. **8**, 121–147 (2003). (Review paper)
12. L. Wang, H.Y. Choi, Y. Jung, B.H. Lee, K.T. Kim, Optical probe based on double-clad optical fiber for fluorescence spectroscopy. Opt. Express **15**(26), 17681–17689 (2007)
13. G.K. Bhowmick, N. Gautam, L.M. Gantayet, Design optimization of fiber optic probes for remote fluorescence spectroscopy. Opt. Commun. **282**(14), 2676–2684 (2009)
14. R.A. McLaughlin, D.D. Sampson, Clinical applications of fiber-optic probes in optical coherence tomography. Opt. Fiber Technol. **16**(6), 467–475 (2010). (Review paper)
15. R. Pashaie, Single optical fiber probe for fluorescence detection and optogenetic stimulation. IEEE Trans. Biomed. Eng. **60**, 268–280 (2013)
16. P. Svenmarker, C.T. Xu, S. Andersson-Engels, J. Krohn, Effects of probe geometry on transscleral diffuse optical spectroscopy. Biomed. Opt. Express **2**, 3058–3071 (2011)
17. P. Gregorčič, M. Jezeršek, J. Možina, Optodynamic energy-conversion efficiency during an Er:YAG-laser-pulse delivery into a liquid through different fiber-tip geometries. J. Biomed. Opt. **17**, article 075006 (2012)
18. D. Lorenser, B.C. Quirk, M. Auger, W.J. Madore, R.W. Kirk, N. Godbout, D.D. Sampson, C. Boudoux, R.A. McLaughlin, Dual-modality needle probe for combined fluorescence imaging and three-dimensional optical coherence tomography. Opt. Lett. **38**(3), 266–268 (2013)
19. I. Latka, S. Dochow, C. Krafft, B. Dietzek, J. Popp, Fiber optic probes for linear and nonlinear Raman applications—current trends and future development. Laser Photon. Rev. **7**(5), 698–731 (2013). (Review paper)
20. A.J. Gomes, V. Backman, Algorithm for automated selection of application-specific fiber-optic reflectance probes. J. Biomed. Opt. **18**, article 027012 (2013)
21. C.R. Wilson, L.A. Hardy, J.D. Kennedy, P.B. Irby, M.N. Fried, Miniature ball-tip optical fibers for use in thulium fiber laser ablation of kidney stones. J. Biomed. Opt. **21**(1), article 018003 (2016)

22. U. Utzinger, Fiber optic probe design (Chap. 7), in *Biomedical Photonics Handbook; Vol 1; Fundamentals, Devices, and Techniques*, 2nd edn., ed. by T. Vo-Dinh (CRC Press, Boca Raton, 2014), pp. 253–279

23. D.V. Lim, Detection of microorganisms and toxins with evanescent wave fiber-optic biosensors. Proc. IEEE **91**(6), 902–907 (2003)

24. A. Leung, P.M. Shankar, R. Mutharasan, A review of fiber-optic biosensors. Sensors and Actuators B **125**, 688–703 (2007). (Review paper)

25. R.M. Lequin, Enzyme immunoassay (EIA)/enzyme-linked immunosorbent assay (ELISA). J. Clin. Chem. **51**(12), 2415–2418 (2005)

26. S.D. Gan, K.R. Patel, Enzyme immunoassay and enzyme-linked immunosorbent assay. J. Invest. Dermatol. **133**(9), article 287 (2013)

27. P. Roriz, O. Frazão, A.B. Lobo-Ribeiro, J.L. Santos, J.A. Simões, Review of fiber-optic pressure sensors for biomedical and biomechanical applications. J. Biomed. Opt. **18**(5), article 050903 (2013) (Review paper)

28. G. Keiser, *Optical Fiber Communications* (Chap. 5) (McGraw-Hill, New York), 4th US edn., 2011; 5th international edn. 2015

29. N. Lagakos, J.H. Cole, J.A. Bucaro, Microbend fiber-optic sensor. Appl. Opt. **26**(11), 2171–2180 (1987)

30. Z. Chen, D. Lau, J.T. Teo, S.H. Ng, X. Yang, P.L. Kei, Simultaneous measurement of breathing rate and heart rate using a microbend multimode fiber optic sensor. J. Biomed. Opt. **19**, article 057001 (2014)

31. M. Karimi, T. Sun, K.T.V. Grattan, Design evaluation of a high birefringence single mode optical fiber-based sensor for lateral pressure monitoring applications. IEEE Sens. J. **13**(11), 4459–4464 (2013)

32. B.H. Lee, Y.H. Kim, K.S. Park, J.B. Eom, M.J. Kim, B.S. Rho, H.Y. Choi, Interferometric fiber optic sensors. Sensors **12**(3), 2467–2486 (2012)

33. R. Yang, Y.S. Yu, X. Yang, C. Chen, Q.D. Chen, H.B. Sun, Single S-tapered fiber Mach-Zehnder interferometers. Opt. Lett. **36**(33), 4482–4484 (2011)

34. Z.B. Tian, S.S.-H. Yam, J. Barnes, W. Bock, P. Greig, J.M. Fraser, H.P. Loock, R.D. Oleschuk, Refractive index sensing with Mach-Zehnder interferometer based on concatenating two single-mode fiber tapers. IEEE Photon. Technol. Lett. **20**(8), 626–628 (2008)

35. L. Jiang, L. Zhao, S. Wang, J. Yang, H. Xiao, Femtosecond laser fabricated all-optical fiber sensors with ultrahigh refractive index sensitivity: modeling and experiment. Opt. Express **19**, 17591–17598 (2011)

36. C.R. Liao, Y. Wang, D.N. Wang, M.W. Yang, Fiber in-line Mach-Zehnder interferometer embedded in FBG for simultaneous refractive index and temperature measurement. IEEE Photon. Technol. Lett. **22**(22), 1686–1688 (2010)

37. Y. Jung, S. Lee, B.H. Lee, K. Oh, Ultracompact in-line broadband Mach-Zehnder interferometer using a composite leaky hollow-optical-fiber waveguide. Opt. Lett. **33**(24), 2934–2936 (2008)

38. W.J. Bock, T.A. Eftimov, P. Mikulic, J. Chen, An inline core-cladding intermodal interferometer using a photonic crystal fiber. J. Lightw. Technol. **27**(17), 3933–3939 (2009)

39. L.C. Li, L. Xia, Z.H. Xie, D.M. Liu, All-fiber Mach-Zehnder interferometers for sensing applications. Opt. Express **20**(10), 11109–11120 (2012)

40. K.S. Chiang, F.Y.M. Chan, M.N. Ng, Analysis of two parallel long-period fiber gratings. J. Lightw. Technol. **22**(5), 1358–1366 (2004)

41. D.J.J. Hu, J.L. Lim, M. Jiang, Y. Wang, F. Luan, P.P. Shum, H. Wei, W. Tong, Long period grating cascaded to photonic crystal fiber modal interferometer for simultaneous measurement of temperature and refractive index. Opt. Lett. **37**(12), 2283–2285 (2012)

42. L. Marques, F.U. Hernandez, S.W. James, S.P. Morgan, M. Clark, R.P. Tatam, S. Korposh, Highly sensitive optical fibre long period grating biosensor anchored with silica core gold shell nanoparticles. Biosens. Bioelectron. **75**, 222–231 (2016)

43. X.Y. Dong, H.Y. Tam, P. Shum, Temperature-insensitive strain sensor with polarization-maintaining photonic crystal fiber based Sagnac interferometer. Appl. Phys. Lett. **90**(15), article 151113 (2007)
44. B. Dong, J. Hao, C.Y. Liaw, Z. Xu, Cladding-mode-resonance in polarization maintaining photonics crystal fiber based Sagnac interferometer and its application for fiber sensor. J. Lightw. Technol. **29**(12), 1759–1762 (2011)
45. D.J.J. Hu, J.L. Lim, M.K. Park, L.T.-H. Kao, Y. Wang, H. Wei, W. Tong, Photonic crystal fiber-based interferometric biosensor for streptavidin and biotin detection. IEEE J. Sel. Top. Quantum Electron. **18**(4), 1293–1297 (2012)
46. U.S. Dinish, G. Balasundaram, Y.T. Chang, M. Olivo, Sensitive multiplex detection of serological liver cancer biomarkers using SERS-active photonic crystal fiber probe. J. Biophotonics **7**(11–12), 956–965 (2014)
47. T. Gong, Y. Cui, D. Goh, K.K. Voon, P.P. Shum, G. Humbert, J.-L. Auguste, X.-Q. Dinh, K.-T. Yong, M. Olivo, Highly sensitive SERS detection and quantification of sialic acid on single cell using photonic-crystal fiber with gold nanoparticles. Biosens. Bioelectron. **64**, 227–233 (2015)
48. N. Zhang, G. Humbert, T. Gong, P.P. Shum, K. Li, J.-L. Auguste, Z. Wu, D.J.J. Hu, F. Luan, Q.X. Dinh, M. Olivo, L. Wei, Side-channel photonic crystal fiber for surface enhanced Raman scattering sensing. Sens. Actuators B Chem. **223**, 195–201 (2016)
49. Z. Xu, J. Lim, D.J.J. Hu, Q. Sun, R.Y.-N. Wong, M. Jiang, P.P. Shum, Investigation of temperature sensing characteristics in selectively infiltrated photonic crystal fiber. Opt. Express **24**(2), 1699–1707 (2016)
50. G.J. Triggs, M. Fischer, D. Stellinga, M.G. Scullion, G.J.O. Evans, T.F. Krauss, Spatial resolution and refractive index contrast of resonant photonic crystal surfaces for biosensing. IEEE Photonics J. **7**(3), article 6801810 (2015)
51. A. Candiani, A. Bertucci, S. Giannetti, M. Konstantaki, A. Manicardi, S. Pissadakis, A. Cucinotta, R. Corradini, S. Selleri, Label-free DNA biosensor based on a peptide nucleic acid-functionalized microstructured optical fiber-Bragg grating. J. Biomed. Opt. **18**, article 057004 (2013)
52. H. Ottevaere, M. Tabak, K. Chah, P. Mégret, H. Thienpont, Dental composite resins: measuring the polymerization shrinkage using optical fiber Bragg grating sensors, in *Proceedings of SPIE 8439, Optical Sensing and Detection II*, paper 843903, May 2012
53. C. Leitão, L. Bilro, N. Alberto, P. Antunes, H. Lima, P.S. André, R. Nogueira, J.L. Pinto, Development of a FBG probe for non-invasive carotid pulse waveform assessment, in *Proceedings of SPIE on Biophotonics, Proceedings of SPIE 8439*, paper 84270J, May 2012
54. G.T. Kanellos, G. Papaioannou, D. Tsiokos, C. Mitrogiannis, G. Nianios, N. Pleros, Two dimensional polymer-embedded quasi-distributed FBG pressure sensor for biomedical applications. Opt. Express **18**(1), 179–186 (2010)
55. J. Hao, M. Jayachandran, N. Ni, J. Phua, H.M. Liew, P.W. Aung Aung, J. Biswas, S.F. Foo, J. A. Low, P.L.K. Yap, An intelligent elderly healthcare monitoring system using fiber-based sensors. J. Chin. Inst. Eng. **33**(5), 653–660 (2010)
56. M. Nishyama, M. Miyamoto, K. Watanabe, Respiration and body movement analysis during sleep in bed using hetero-core fiber optic pressure sensors without constraint to human activity. J. Biomed. Opt. **16**, 017002 (2011)
57. L. Mohanty, S.C. Tjin, D.T.T. Lie, S.E.C. Panganiban, P.K.H. Chow, Fiber grating sensor for pressure mapping during total knee arthroplasty. Sens. Actuators A **135**(2), 323–328 (2007)
58. E.A. Al-Fakih, N.A. Abu Osman, F.R.M. Adikan, The use of fiber Bragg grating sensors in biomechanics and rehabilitation applications: the state-of-the-art and ongoing research topics. Sensors **12**, 12890–12926 (2012) (Review paper)
59. C.R. Dennison, P.M. Wild, D.R. Wilson, M.K. Gilbart, An in-fiber Bragg grating sensor for contact force and stress measurements in articular joints. Meas. Sci. Technol. **21**(11), 115803 (2010)

60. G.T. Kanellos, G. Papaioannou, D. Tsiokos, C. Mitrogiannis, G. Nianios, N. Pleros, Two dimensional polymer-embedded quasi-distributed FBG pressure sensor for biomedical applications. Opt. Express **18**(1), 179–186 (2010)
61. J.W. Arkwright, N.G. Blenman, I.D. Underhill, S.A. Maunder, M.M. Szczesniak, P.G. Dinning, I.J. Cook, In-vivo demonstration of a high resolution optical fiber manometry catheter for diagnosis of in gastrointestinal motility disorder. Opt. Express **17**(6), 4500–4508 (2009)
62. P.G. Dinning, L. Wiklendt, L. Maslen, V. Patton, H. Lewis, J.W. Arkwright, D.A. Wattchow, D.Z. Lubowski, M. Costa, P.A. Bampton, Colonic motor abnormalities in slow transit constipation defined by high resolution, fibre-optic manometry. Neurogastroenterol. Motil. **27** (3), 379–388 (2015)
63. A. Bhalla, N. Grewal, U. Tiwari, V. Mishra, N.S. Mehla, S. Raviprakash, P. Kapur, Shock absorption ability of laminate mouth guards in two different malocclusions using fiber Bragg grating (FBG) sensor. Dent. Traumatol. **29**(3), 218–225 (2013)
64. M. Ciocchetti, C. Massaroni, P. Saccomandi, M.A. Caponero, A. Polimadei, D. Formica, E. Schena, Smart textile based on fiber Bragg grating sensors for respiratory monitoring: design and preliminary trials. Biosensors **5**, 602–615 (2015)
65. S. Poeggel, D. Duraibabu, K. Kalli, G. Leen, G. Dooly, E. Lewis, J. Kelly, M. Munroe, Recent improvement of medical optical fibre pressure and temperature sensors. Biosensors **5**, 432–449 (2015)
66. J. Homola, Surface plasmon resonance sensors for detection of chemical and biological species. Chem. Rev. **108**(2), 462–493 (2008)
67. M. Bauch, K. Toma, M. Toma, Q. Zhang, J. Dostalek, Plasmon-enhanced fluorescence biosensors: a review. Plasmonics **9**(4), 781–799 (2014)
68. C.L. Wong, M. Olivo, Surface plasmon resonance imaging sensors: a review. Plasmonics **9**(4), 809–824 (2014)
69. O. Tokel, F. Inci, U. Demirci, Advances in plasmonic technologies for point of care applications. Chem. Rev. **114**(11), 5728–5752 (2014)
70. E. Seymour, G.G. Daaboul, X. Zhang, S.M. Scherr, N.L. Ünlü, J.H. Connor, M.S. Ünlü, DNA directed antibody immobilization for enhanced detection of single viral pathogens. Anal. Chem. **87**(20), 10505–10512 (2015)
71. Y.T. Long, C. Jing, *Localized Surface Plasmon Resonance Based Nanobiosensors* (Springer, Berlin, 2014)
72. M. Consales, M. Pisco, A. Cusano, Review: lab-on-fiber technology: a new avenue for optical nanosensors. Photonic Sens. **2**(4), 289–314 (2012)
73. A. Ricciardi, M. Consales, G. Quero, A. Crescitelli, E. Esposito, A. Cusano, Lab-on-fiber devices as an all around platform for sensing. Opt. Fiber Technol. **19**(6), 772–784 (2013)
74. J. Cao, T. Sun, K.T.V. Grattan, Gold nanorod-based localized surface plasmon resonance biosensors: a review. Sens. Actuators B Chem. **195**, 332–351 (2014)
75. C.-K. Chu, Y.-C. Tu, Y.-W. Chang, C.-K. Chu, S.-Y. Chen, T.-T. Chi, Y.-W. Kiang, C.C. Yang, Cancer cell uptake behavior of Au nanoring and its localized surface plasmon resonance induced cell inactivation. Nanotechnology **26**(7), article 075102 (2015)
76. J. Albert, A lab on fiber. IEEE Spectr. **51**, 49–53 (2014)
77. K.Z. Kamili, A. Pandikumar, G. Sivaraman, H.N. Lim, S.P. Wren, T. Sun, N.M. Huang, Silver@graphene oxide nanocomposite-based optical sensor platform for biomolecules. RSC Adv. **5**(23), 17809–17816 (2015)
78. N. Lebedev, I. Griva, W.J. Dressick, J. Phelps, J.E. Johnson, Y. Meshcheriakova, G. P. Lomonossoff, C.M. Soto, A virus-based nanoplasmonic structure as a surface-enhanced Raman biosensor. Biosens. Bioelectron. **77**, 306–314 (2016)

Chapter 8
Microscopy

Abstract Many technical developments have appeared in recent years to enhance imaging performance and instrument versatility in optical microscopy for biophotonics and biomedical applications to visualize objects ranging in size from millimeters to nanometers. These developments include increasing the penetration depth in scattering media, improving image resolution beyond the diffraction limit, increasing image acquisition speed, enhancing instrument sensitivity, and developing better contrast mechanisms. This chapter first describes the basic concepts and principles of optical microscopy and then discusses the limitations of distinguishing two closely spaced points. Next, the functions of confocal microscopes, fluorescence microscopy, multiphoton microscopy, Raman microscopy, light sheet microscopy, and super-resolution fluorescence microscopy are given.

Microscopy applications can be found in all branches of science and engineering. The diverse instruments used in these fields include acoustic microscopes, atomic force microscopes, scanning electron microscopes, x-ray microscopes, and optical microscopes. This chapter addresses optical microscopes and their applications to biophotonics. Many technical developments have appeared in recent years to enhance imaging performance and instrument versatility in optical microscopy for biophotonics and biomedical applications. These developments include increasing the penetration depth in scattering media, improving image resolution beyond the diffraction limit, increasing image acquisition speed, enhancing instrument sensitivity, and developing better contrast mechanisms. Different embodiments of these optical microscopic techniques can visualize objects ranging in size from millimeters to nanometers. Microscopes can be categorized by their physical structure, functional principle, illumination method, and image processing procedure. Applications of microscopy instruments and techniques are found in all branches of biomedicine and biophotonics, for example, analyses of biological samples, brain research, cancer research, cytology, drug discovery, food analyses, gynecology, healthcare, hematology, and pathology.

First Sect. 8.1 describes the basic concepts and principles of optical microscopy. This is followed in Sect. 8.2 by discussions of the limitations of distinguishing two

© Springer Science+Business Media Singapore 2016
G. Keiser, *Biophotonics*, Graduate Texts in Physics,
DOI 10.1007/978-981-10-0945-7_8

closely spaced points. Next, the functions of confocal microscopes, fluorescence microscopy, multiphoton microscopy, Raman microscopy, light sheet microscopy, and super-resolution fluorescence microscopy are given in Sect. 8.3 through Sect. 8.8, respectively. The following two chapters then describe spectroscopic and imaging techniques that make use of microscopic instruments and methodologies.

8.1   Concepts and Principles of Microscopy

This section addresses the operational concepts of basic optical microscopes, which are used widely to obtain both topographical and dynamic information from biological samples. The functions and parameters of interest include viewing and illumination techniques, observation methods, numerical aperture, field of view, and depth of field [1–8]. In basic microscopes the entire sample is illuminated simultaneously with a light source such as a pure white mercury lamp. Thus these instruments are known as *wide-field microscopes*. The light coming from the illuminated object is viewed with a device such as a CCD camera. A limitation of wide-field microscopes is that light emitted by the specimen from above and below the focal plane of interest interferes with the resolution of the image being viewed. Consequently, images of specimens that are thicker than about 2 μm appear blurry. Finer resolutions can be obtained with other types of instruments, such as the confocal microscopes described in Sect. 8.3.

8.1.1   Viewing and Illumination Techniques

Modern conventional microscopes show a magnified two-dimensional image that can be focused axially in successive focal planes. This capability enables the examination of the fine structural details of a specimen in both two and three dimensions. The basic setup for a microscope is shown in Fig. 8.1. The key components are the following:

1. A movable stage that includes clips for holding a specimen and a manual or automated translation mechanism to accurately position and focus the specimen
2. A set of selectable objective lenses with different magnifications mounted on a rotating *turret* (also called the *nosepiece*) for magnifying the specimen
3. One or more lamps, lasers, or LED sources that can include optical elements such as collector lenses, field diaphragms, heat filters, and various light filters
4. One or more mirrors for directing light onto the specimen
5. A viewing device such as an eyepiece, a photodetector array, a photomultiplier tube, or a video camera
6. Various optical focusing elements that are strategically placed to achieve a desired illumination intensity and contrast in the specimen image

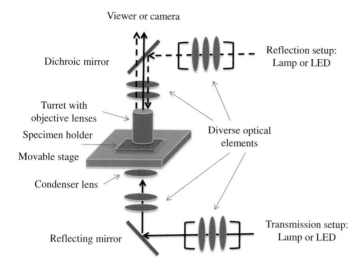

Fig. 8.1 Basic transmission and reflection setups for a microscope

7. Special measurement devices such as grids, scales, and pointers that are intended to be in focus and registered on the specimen are placed at the locations of a set of field conjugate planes, which are simultaneously in focus and are superimposed on one another when observing specimens through the microscope

The illumination methods illustrated in Fig. 8.1 include both transmission and reflection techniques, which are described in the following paragraphs.

The two fundamental viewing and illumination techniques used in microscopes for examining different specimens are the following:

(1) *Transmitted light microscopy* is the simplest illumination and viewing technique. The general concept is shown in Fig. 8.1. First a thin slice of a sample is encapsulated in a specimen holder such as a thin glass or plastic microscope slide. Standard glass slides have smoothed edges and typically are 2.5 cm by 7.5 cm (1 × 3 in.) in size with 1 to 1.2 mm thicknesses. After a prepared slide is placed on the movable stage, it can be illuminated from below with white light and observed from above, as shown by the bottom lighting configuration in Fig. 8.1. The image is formed through the absorption of some of the transmitted (also called *diascopic*) illuminating light by dense areas of the sample. This method creates a contrast pattern consisting of a dark sample on a bright background. The advantage of transmitted light microscopy is that it is a simple setup that requires only basic equipment. This simplicity results in some limitations, which include a very low *contrast* (i.e., a low ability to distinguish differences in color or intensity) of most biological samples, a low optical resolution due to a blur caused by some areas being out of focus, and the fact that the sample often needs to be stained (colored) to enhance

visualization of certain cells or cellular components under a microscope. Because of this staining requirement, live cells usually cannot be viewed with this technique.

(2) *Reflected light microscopy* is a popular illumination method used to examine opaque specimens that are highly reflective and thus do not readily absorb or transmit light. As the top lighting configuration in Fig. 8.1 shows, in this technique the incident light originates from above the specimen and is often referred to as *episcopic illumination, epi-illumination*, or *vertical illumination*. After the incident light has been specularly or diffusively reflected, it can be viewed with the eye or a light-imaging system. An advantage of reflected light microscopy compared to the transmitted light method is that the image created through the reflected light can be viewed as a three-dimensional representation, provided that the microscope optics can distinguish between regions of different heights in the specimen.

Many advanced modern optical microscopes have both illumination methods built into the same instrument. This feature allows the viewer to examine a specimen either by alternating between the two illumination modes independently or by using both techniques simultaneously. In either situation, the same specimen-holding platform and viewing optics are used. Recent instrument enhancements include the use of energy efficient high-intensity light-emitting diodes (LEDs) in place of the standard illumination lamps.

Three common microscope structures are the upright microscope, the inverted microscope, and the stereomicroscope. In the *upright microscope* shown in Fig. 8.2 the viewer looks down at the stage and the specimen through a pair of eyepiece lenses. The stage is a vertically and horizontally movable platform for mounting specimen holders such as microscope slides. The translation mechanisms that control the stage movements allow the viewer to accurately position, orient, and focus the specimen to optimize visualization and recording of images. Above the stage is a rotatable *nosepiece* or *turret* that contains three or four *objective lenses* of different magnifying strengths. Typically the lens magnifications are 4X, 10X, 40X, and 100X, which are coupled with an eyepiece magnification of 10X to yield total magnifications of 40X (4X times 10X), 100X, 400X, and 1000X. Light from the optical source located at the bottom of the microscope travels horizontally toward a mirror located below the stage and then is transmitted upward after being reflected by the mirror. The function of the *condenser lens* shown in Fig. 8.2 is to focus light onto the specimen. The image then can be viewed directly by the eye or by means of an image-capturing device, for example, a camera or an image-processing CCD array. A variety of spectral filtering, light intensity control, polarizing, and light focusing optical elements (such as diaphragms, mirrors, prisms, beam splitters, polarizers, and lenses) are located within the optical path from the light source to the viewer. Many modern microscopes have motorized lens changeover, motor-driven stages for precision positioning, and automatic adjustment of illumination intensity capabilities.

Fig. 8.2 Example of an upright microscope

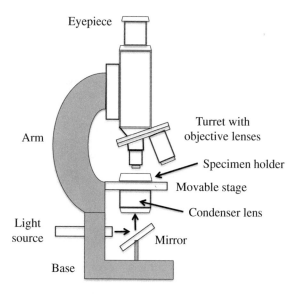

Eyepiece

Arm

Turret with objective lenses

Specimen holder

Movable stage

Condenser lens

Light source

Mirror

Base

In the *inverted microscope* shown in Fig. 8.3, the stage and the specimen are viewed from the bottom of the instrument and the transmitted illumination of the specimen comes from the top. When a specimen is placed on the stage, the surface of interest that is to be examined faces downward. The advantage of this specimen orientation is that large samples, such as a petri dish, can be positioned and viewed easier than with an upright microscope where the size of the sample area is more restricted. The objective lenses are mounted on a nosepiece under the stage and thus face upward towards the specimen. To focus the microscope on the specimen, either the nosepiece or the entire stage is moved up and down. Similar to the upright microscope, the illuminating light source can be either a high-intensity LED or some other source, such as a halogen, mercury, or xenon lamp.

The *stereomicroscope* is illustrated in Fig. 8.4. This instrument is useful for examining large objects, such as plants, zebra fish, drosophila (a genus of small flies, for example, the common fruit fly), or small animals (such as mice). A key feature of the stereomicroscope is that the viewer sees a three-dimensional erect image. This feature is useful for performing interactive manipulations of the specimen being examined, such as dissecting biological specimens, microinjection of biological components into live specimens, or microscopic bonding of tissue.

8.1.2 Observation Methods

The use of microscopes can be classified according to the observation method for various types of image enhancements. Five different microscopy observation methods are described below and summarized in Table 8.1.

Fig. 8.3 Example of an inverted microscope

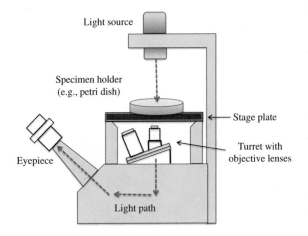

Fig. 8.4 Example of a stereomicroscope

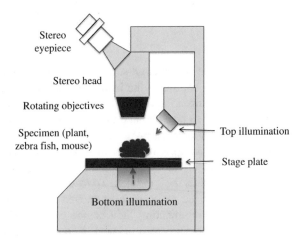

- *Bright field microscopy* is commonly used to observe color and light intensity (brightness) information from a stained specimen. The entire specimen section being examined is illuminated and appears bright to the viewer. Most specimens are fairly transparent under the microscope and consequently are difficult to analyze visually against a plain white background. Thus, *cell staining* can be used to add color to the picture, thereby making objects such as cells and their components much easier to see. Through the use of different stains, certain cell components (e.g., a nucleus, a cell wall, or the entire cell) can be selectively or preferentially stained to examine the component structure. Staining may also be used to highlight metabolic processes or to differentiate between live and dead cells in a sample. However, it is important to note that because the specimen is stained prior to being viewed under a microscope, the observed colors are not necessarily those of the actual specimen.

Table 8.1 Summary of five different microscopy observation methods

Microscopy method	Features	General applications
Bright field microscopy	• Common observation method • Light illuminates the entire field of view • Cell staining adds color	Used to observe color and light intensity information from a stained specimen
Dark field microscopy	• Oblique illumination enhances contrast • Transparent specimens seen as bright objects on a black background	Used for viewing and imaging living bacteria, cells, and tissues
Phase contrast microscopy	Uses tiny refractive index differences between cells and their surrounding aqueous solutions and within cells	Suitable for viewing very thin colorless and transparent unstained specimens (e.g., culture cells on glass) and for live cells
Differential interference contrast microscopy	Uses plane-polarized light and light-shearing prisms to exaggerate minute differences in thickness gradients and in refractive indices of a specimen	Suitable for viewing very thin colorless and transparent unstained specimens and for live cells
Polarized light microscopy	Uses crossed polarizing elements to dramatically improve the quality of an image obtained from birefringent materials	Used for specimens such as plant cell walls, starch granules, and protein structures formed during cell division

- *Dark field microscopy* employs oblique illumination to enhance contrast in specimens that are not imaged well when using bright field illumination. In this observation method an opaque ring-shaped diaphragm with a small central opening is used together with the condenser lens to direct a hollow inverted cone of light onto the specimen at high azimuthal angles. Thus, first-order wave fronts do not directly enter the objective lens, which creates an artificial dark background in the microscope. The result is that the specimen features are seen as bright objects on a black background. Dark field microscopy is a popular tool for biophotonics and medical investigations, such as viewing and imaging living bacteria, cells, and tissues.
- *Phase contrast microscopy* makes use of tiny refractive index differences between cells and their surrounding aqueous solutions and within the cells between the cytoplasm and the cell nucleus. This observation method is suitable for viewing very thin colorless and transparent unstained specimens (e.g., culture cells on glass) and for live cells. Typically these specimens are approximately 5–10 μm thick in the central region, but less than 1 μm thick at the periphery. Such specimens absorb extremely little light in the visible portion of the spectrum and they cannot be viewed with the human eye when using bright field and dark field illumination methods. The phase contrast method translates

the index differences into intensity variations that can be visually observed and recorded.

- *Differential interference contrast* (DIC) *microscopy* uses plane-polarized light and light-shearing prisms, known as Nomarski prisms, to exaggerate minute differences in thickness gradients and in refractive index variations of a specimen. When light passes through a specimen, DIC microscopy utilizes the phase differences generated in the regions where there is a thickness gradient. This process adds bright and dark contrast to images of transparent specimens. DIC allows a fuller use of the numerical aperture of the system than in other observation methods. Thereby the microscope achieves an excellent resolution and the user can focus on a thin plane section of a thick specimen without getting interference from images that lie above or below the plane being examined. One limitation is that because DIC utilizes polarized light, plastic Petri dishes cannot be used.

- *Polarized light microscopy* is an image contrast-enhancing technique that, compared to other observation methods, dramatically improves the quality of an image obtained from birefringent materials. In polarized light microscopy the specimen is viewed between crossed polarizing elements inserted into the optical path before the condenser lens and after the objective lens. Muscle tissue and structures within the cell that have birefringent properties (e.g., plant cell walls, starch granules, and protein structures formed during cell division) rotate the plane of light polarization and thus appear bright on a dark background.

8.1.3   Numerical Aperture

Analogous to the discussion in Chap. 3 about the light capturing capability of an optical fiber, the *numerical aperture* (NA) of an objective lens of a microscope measures the ability of the lens to gather light and to resolve fine detail at a fixed distance from a specimen. Given that α is the half-angle of the light cone captured by an objective lens of diameter D and focal length f as shown in Fig. 8.5, and letting n be the index of refraction of the medium between a specimen and the lens (referred to as the *immersion medium*), then the NA is defined by

$$NA = n \sin \alpha \approx D/2f \qquad (8.1)$$

The immersion medium commonly is either air (n = 1.00) or a transparent medium such as water (n = 1.33), glycerin (n = 1.47), or *immersion oil* (n = 1.51) that is used to increase the NA. Values of NA range from 0.025 for very low magnification objectives up to 1.6 for high-performance objectives utilizing specialized transparent immersion oils. For an illustration, consider a series of objectives (shown in blue in Fig. 8.5) that have the same lens diameter D but varying values of the focal length. As the focal length decreases, the angle α increases and the light cone shown in yellow becomes wider. Thus, the lenses with a shorter focal

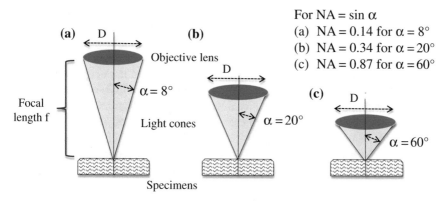

Fig. 8.5 The NA increases for short objective focal lenses

length have a greater NA (i.e., a greater light-gathering ability) and yield a finer resolution of the sample.

Example 8.1 Consider a series of objective lenses in which the half-angle of the captured light cone varies from 7° to 60°. What is the range of numerical apertures if the immersion medium is air?

Solution: From Eq. (8.1) with n = 1.00, as the angle α increases from 7° to 60° (as the light cones grow larger) the numerical aperture increases from 0.12 to 0.87.

Example 8.2 Consider a series of objective lenses in which the half-angle of the captured light cone varies from 7° to 60°. What is the range of numerical apertures if the immersion medium is oil with n = 1.51?

Solution: From Eq. (8.1) with n = 1.51, as the angle α increases from 7° to 60° (as the light cones grow larger) the numerical aperture increases from 0.18 to 1.31.

Note: Objective lenses with a magnification range between 60x and 100x typically are designed for use with immersion oil. In practice, most oil immersion objectives have a maximum numerical aperture of 1.4, with the most common numerical apertures ranging from 1.00 to 1.35.

8.1.4 Field of View

When looking into a microscope, it is useful to know the diameter of the viewed field in millimeters measured at the intermediate image plane. Knowing this number, which is called the *field of view* (FOV) or the *field diameter*, allows the viewer to estimate the size of the object being examined. The field of view can be calculated from a number that is listed on the microscope eyepiece and from the magnification of the objective. A typical example of two numbers on the eyepiece might be 10X/22. In this case, 10X is the magnification of the eyepiece and 22 is called the *field-of-view number*, or simply the *field number* (FN), which is given in mm. Typically the field number increases as the magnification of the eyepiece decreases. For example, a 5X eyepiece might have a FN of 26 mm, for a 10X eyepiece FN may be 22 mm, and for a 30X eyepiece the FN might decrease to 7 mm. To calculate the diameter of the field of view (measured in mm), one divides the FN on the eyepiece by the magnification M_{obj} of the objective lens, as given by

$$FOV = FN/M_{obj} \qquad (8.2)$$

By knowing the diameter of the FOV and counting how many times the object fits across this FOV, then the following equation can be used to calculate the object size:

$$\text{Object size} = \frac{\text{Diameter of FOV}}{\text{Number of times the object fits across the FOV}} \qquad (8.3)$$

Example 8.3 Consider an objective lens that has a 10X magnification and which is used with an eyepiece of field number 22.

(a) What is the field of view?
(b) Suppose an object that is viewed by this eyepiece and objective combination fits across the FOV 2.75 times. What is the size of the object?

Solution:

(a) From Eq. (8.2) the FOV in mm is: FOV = FN/M = (22 mm)/ 10 = 2.2 mm
(b) From Eq. (8.3) the object size is FOV/2.75 = (2.2 mm)/2.75 = 0.80 mm

8.1.5 Depth of Field

The parameter *depth of field* (DOF) refers to the thickness of the plane of focus. That is, the depth of field indicates the range of depth over which the nearest and farthest object planes of a viewed specimen are in acceptable focus simultaneously.

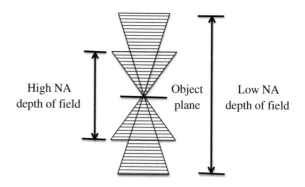

Fig. 8.6 Depth of field ranges for a layered specimen for low and high NA values

For example, consider a specimen consisting of many thin layers of material, as shown in Fig. 8.6. With a large depth of field (i.e., a thick plane of focus resulting from a low NA), many layers of the specimen will be in focus at the same time. Fewer layers can be focused simultaneously for a shorter depth of field (i.e., a thinner plane of focus resulting from a high NA). Every other layer of the specimen will be out of focus. In order to examine the other lower and higher layers of a thick specimen, a microscope with a short DOF must be focused downward or upward continuously. In microscopy the DOF typically is measured in units of micrometers.

As is shown in Fig. 8.6, the DOF becomes larger as the NA decreases. Several different formulas have been proposed to calculate the DOF of a microscope. A commonly used equation is

$$DOF = \frac{n\lambda}{NA^2} \tag{8.4}$$

where n is the refractive index of the immersion material, λ is the wavelength, and NA is the numerical aperture of the objective.

Example 8.4 Consider two objective lenses that have NAs of 0.20 and 0.55, respectively. What is the depth of field at a wavelength of 780 nm if the immersion material is oil with an index n = 1.51?

Solution: From Eq. (8.4) the DOF for NA = 0.20 is found to be

$$DOF = \frac{n\lambda}{NA^2} = \frac{1.51(0.780\,\mu m)}{(0.20)^2} - 29.4\,\mu m$$

Similarly, from Eq. (8.4) the DOF for NA = 0.55 is 3.89 μm.

8.2   Resolution and Diffraction Limit

When viewing an object in a microscope, the light coming from the specimen can be represented as consisting of multiple point sources of light. As the light from these point sources passes through the various optical elements and aperture diaphragms of a microscope, the optical waveforms start to spread out due to scattering and diffraction effects [9–11]. This spreading results in a small, broadened diffraction pattern that appears in the image of the point source. Because of the image broadening, there is a lower limit below which the instrument cannot resolve the separation of the images of two adjacent point sources to show the real structural features and details of a specimen. This condition occurs when two source points in the specimen are separated by *a lateral distance that is less than approximately half the wavelength of the imaging light*. This restriction is referred to as the *Abbe criterion* or the *diffraction limit*. Thus, according to the diffraction limit, the minimal distances $(\delta x_{min}, \delta y_{min})$ that can be resolved in the lateral xy-plane are approximated by

$$(\delta x_{min}, \delta y_{min}) \approx \lambda/2 \tag{8.5}$$

Another consequence of the diffraction limit is that it is not possible to focus a laser beam to a spot with a dimension smaller than about $\lambda/2$. However, note that Sect. 8.8 describes super-resolution fluorescence microscopic methods that enable the capture of images with a higher resolution than the diffraction limit.

The broadened image of a point source is called an *Airy pattern*. The three-dimensional distribution of light intensity in the Airy pattern is called the *point-spread function* (PSF) of the microscope lens. An Airy pattern and the corresponding xz distribution of the PSF are illustrated in Fig. 8.7. For an ideal, aberration-free lens, the size of the light distribution is determined only by the wavelength of the light, the NA of the lens, and diffraction. The point-spread function shows an oscillatory behavior with a large main central peak and low-intensity side lobes surrounding the main peak. The central maximum, which is the region enclosed by the first minimum (the first side lobe) of the Airy pattern, is called the *zeroth-order maximum Airy disk*. This disk contains 84 percent of the luminous energy and is surrounded by concentric rings of sequentially decreasing brightness that make up the total intensity distribution. The intensity of the first side lobe is 3 % of the intensity of the main peak. The first side lobe is located at a distance $x_0 = 2n\lambda/(NA)^2$ from the center of the pattern.

The term *resolution* or *resolving power* is used to describe the ability of an imaging system to distinguish the Airy disks of two point sources that are close together. The value of the resolution of a microscope is defined in terms of the numerical aperture. The value of the resolution is not a precise number because the total NA of different microscope setups depends on the NA values of both the objective lens and additional lenses in the optical path to the viewer. However, *a higher total NA yields a better resolution*. Note that *smaller values of the resolution*

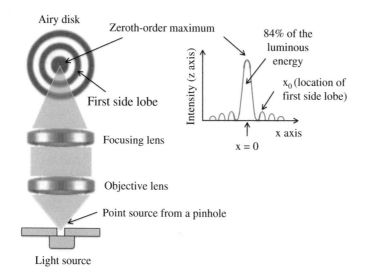

Fig. 8.7 Example of an Airy disk diffraction pattern and the point-spread function

indicate a better ability to distinguish two closely spaced points. Resolution also depends on the wavelength of the illumination used to examine a specimen. *Shorter wavelengths can resolve details better than longer wavelengths.* Two commonly used expressions for the *wide-field resolution* R are

$$R = 0.61\lambda/NA_{obj} \tag{8.6}$$

and

$$R = \frac{1.22\lambda}{NA_{obj} + NA_{cond}} \tag{8.7}$$

where λ is the wavelength of the illuminating light and NA_{obj} and NA_{cond} are the numerical apertures of the objective lens and the condenser lens, respectively. In order to get good imaging results, the NA of the condenser lens should match the NA of the objective lens.

Example 8.5 What are some ways to increase the resolution of a microscope?

Solution: Smaller values of resolution indicate a better ability to distinguish two closely spaced points. Thus Eqs. (8.6) and (8.7) indicate that a better resolution can be obtained by using a shorter wavelength or by increasing the numerical aperture.

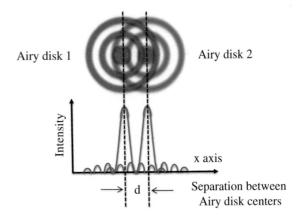

Fig. 8.8 The separation of two closely spaced Airy disks determines the diffraction limit

The *limit of resolution* refers to the smallest separation between two closely spaced Airy disks for which a microscope objective can distinguish the individual Airy patterns. For example, Fig. 8.8 shows two Airy disks and their intensity distributions that are separated by a short distance. If the separation d between the two disks is greater than their radii, then the two corresponding image points are resolvable. The limiting separation for which the two Airy disks can be resolved into separate entities is often called the *Rayleigh criterion*. That is, two Airy disks usually cannot be resolved when the center-to-center distance between the zeroth-order maxima is less than the Airy resolution given in Eqs. (8.6) or (8.7).

If several objective lenses have the same focal length but different NAs, then the Airy disks in the images become smaller as the NA becomes larger. This is illustrated in Fig. 8.9 where the NA increases from left to right. As the Airy disk projected in the image becomes smaller, then more details within the specimen can be seen. That is, smaller Airy disks allow the viewer to better distinguish two closely spaced points in the specimen.

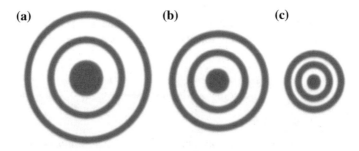

Fig. 8.9 As the NA increases, the Airy disks get smaller for objectives of the same focal length. **a** Small NA large Airy disk. **b** Intermediate NA medium Airy disk. **c** Large NA small Airy disk

Example 8.6 Consider two microscope setups being used at a 700 nm wavelength. Suppose for one setup $NA_{obj1} = 0.80$ and for the other setup $NA_{obj2} = 1.25$. Which setup yields a finer resolution?

Solution: From Eq. (8.6), $R_1 = 0.61\lambda/NA_{obj} = 0.61(700 \text{ nm})/0.80 = 533$ nm. For the second setup, $R_2 = 0.61(700 \text{ nm})/1.25 = 341$ nm. Therefore the second setup has a finer resolution.

8.3 Confocal Microscopy

When using the conventional microscopes described in Sect. 8.1, the entire specimen being examined, or a large portion of it, is illuminated and viewed simultaneously. In this method, the viewer concentrates on the segment of the specimen that is located at the focal point of the objective lens of the microscope. The simplicity of this setup imposes a limitation on the quality of the image, because unfocused light from other portions of the specimen, plus other unwanted scattered light, also will show up in the image. The result is that the image can be blurred or some details can be obscured.

These limitations are mitigated through the use of confocal microscopy [12–14]. A confocal microscope creates sharp images of a specimen that normally appear blurred when viewed with a conventional microscope. This image-enhancing characteristic is achieved by an optical method that excludes most of the light that does not come from the focal plane of the specimen.

The key feature of a confocal microscope is the use of two spatial filters with pinhole apertures, as Fig. 8.10 shows. One aperture is located near the light source

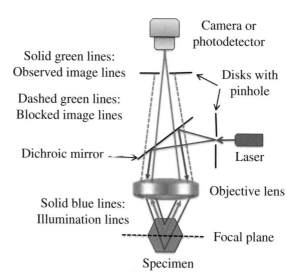

Fig. 8.10 Basic concept of confocal microscopy

Camera or photodetector

Solid green lines: Observed image lines

Dashed green lines: Blocked image lines

Disks with pinhole

Dichroic mirror

Laser

Solid blue lines: Illumination lines

Objective lens

Focal plane

Specimen

and other in front of the photodetector. With this setup, the focal plane of the specimen and the image plane at the retina of the eye or the surface of a camera detector belong to a set of image-forming conjugate planes. By definition, an object that is in focus at one conjugate plane also is in focus at the other conjugate planes of that light path. That is, *conjugate planes* are *paired sets of in-focus planes*. This definition gives rise to the term *confocal microscope*.

The feature of the conjugate planes allows only the light that is emitted from the desired focal spot to pass through the microscope to the viewer. The screen at the photodetector blocks all other diffracted or scattered light that is outside of the desired focal plane. In Fig. 8.10 the blue lines represent the light cone that comes from the pinhole at the laser, strikes the dichroic mirror, and illuminates a volume within the sample. The solid green lines represent rays that come from spots on the focal plane in the specimen, pass through the lens and dichroic mirror system, and then pass through the pinhole aperture at the photodetector. The dashed green lines originate from out-of-focus points in the sample and are blocked by the pinhole aperture.

The confocal technique allows the specimen to be imaged through the use of a rapid two-dimensional point-by-point serial scanning method in the xy-plane. To obtain a three-dimensional image, the objective lens is focused to another depth and the two-dimensional scan is repeated. Thereby a layered stack of virtual, confocal image planes can be stored in a computer and used later to make three-dimensional tomographic images of the specimen.

Slit scanning is an alternative method to the pinhole-based point-by-point scanning for high-speed confocal imaging. This slit scanning method uses a 1-pixel wide scan bar of excitation light to rapidly sweep across the sample. This procedure allows video-rate image acquisition (for example, 120 images/s) by repeatedly sweeping the bar at a high rate across the sample and then detecting the fluorescence with high-speed detectors through a slit aperture.

Because of its optical sectioning capability, confocal microscopy is a well-established and widely used methodology in biophotonics. This feature enables the analysis of morphologic changes in thick biologic tissue specimens with sub-cellular resolution. Traditionally, confocal microscopy was limited to in vitro studies of biopsy specimens and to in vivo analyses of easily accessible sites such as the cornea, the skin, and lip and tongue areas, because it required large microscope objectives and relatively long image acquisition times. These limitations can be overcome through the use of optical fiber-based techniques such as laser scanning through a coherent-imaging fiber optic bundle or by means of a needle-based imaging probe [15–18].

A confocal setup using a laser scanning mechanism with an optical fiber bundle is shown in Fig. 8.11. First the light from a confocal pinhole aperture is collimated and sent through a beam splitting mirror to a xy laser scanning mechanism. This mechanism focuses the light sequentially onto each individual fiber in the coherent bundle, so that the light gets transmitted to the sample. The light coming from the

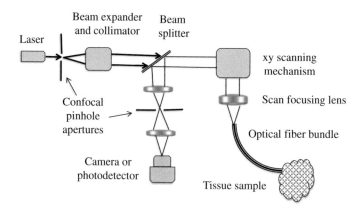

Fig. 8.11 Setup for an optical fiber-based confocal microscope system

sample returns through the fiber bundle, passes through the scanning and dichroic mirror system, and travels to the photodetector through the confocal pinhole aperture.

8.4 Fluorescence Microscopy

Fluorescence microscopy has become an important tool in biophotonics for clarifying many questions in the life sciences and in medicine. As Sect. 6.7 describes and as Fig. 8.12 shows, certain atoms and molecules have the ability to absorb light at a particular wavelength, for example, λ_1 or λ_2. This absorption occurs on the order of 10^{-15} s (femtoseconds) and elevates the molecule to a higher energy level as indicated by the solid upward arrows. Immediately following the absorption the molecules drop to a more stable intermediate energy level through a non-radiative transition shown by the dashed downward arrows. This occurs on the order of 10^{-14}–10^{-11} s. Subsequently, as indicated by the solid downward arrows, the molecule drops back to a lower energy level after a time interval in the order of 10^{-9}–10^{-7} s, thereby emitting light of a longer wavelength (lower energy), for example, λ_3 or λ_4. This spectroscopic property is called *fluorescence* and is widely used as an imaging mode in biological laser-scanning confocal microscopy. The advantage of this technique is the ability to target specific in vitro and in vivo structural components and dynamic chemical and biological processes [19–22]. Further details on its application in areas such as fluorescence spectroscopy, fluorescence lifetime imaging, and fluorescence correlation spectroscopy are given in Chap. 9.

Recall from Sect. 6.7 that the fluorescing light can originate either from internal fluorophores or selective external fluorophores. These external fluorophores basically are probes that are known as dyes, labels, markers, stains, or tags. Thus,

Fig. 8.12 Example of a molecular fluorescence processes

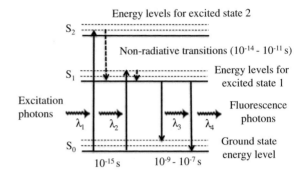

fluorescence microscopy is a technique for viewing specimens that are *dyed* (also referred to as *stained*), *labeled*, *marked*, or *tagged* with one or more fluorophores.

Both internal and external fluorescent probes can be utilized in a variety of ways. *Internal fluorophores* include aromatic amino acids, collagen, elastin, flavins, and NADH. Many *extrinsic fluorophores* (for example, dansyl chloride and the fluorescent dyes fluorescein, rhodamine, prodan, and BIODIPY) are available for fluorescent labeling of macromolecules such as proteins, amino acids, peptides, and amines to assess molecular characteristics and dynamics. For example, dansyl chloride can be excited by 350-nm light (where proteins do not absorb light) and typically emit near 520 nm. The dyes fluorescein and rhodamine have absorption maxima near 480 and 600 nm and emission peaks ranging from 510 to 615 nm, respectively. On a smaller scale, these fluorophores can be localized within a cell to examine specific cell structures such as the cytoskeleton, endoplasmic reticulum, Golgi apparatus, and mitochondria.

Other applications of fluorophores in fluorescent spectroscopy include the following:

- Antibody/antigen fiber sensors for an enzyme-linked immunosorbent assay (ELISA) (see Sect. 7.3.2)
- Monitors of dynamic molecular processes
- Monitors of cellular integrity, endocytosis and exocytosis (the movement of material into and out of a cell, respectively), membrane fluidity, and enzymatic activity
- Genetic mapping

It should be noted that many fluorophores are susceptible to *photobleaching* effects. This is the result of a photochemical alteration of a dye or fluorophore molecule that causes it to no longer fluoresce. This photobleaching effect can occur anytime from after a few cycles to after many thousands of fluorescent cycles.

The basic setup, components, and light paths of an upright fluorescence microscope are illustrated in Fig. 8.13. The four main elements are the following:

- An excitation light source, for example, a xenon arc lamp, a mercury vapor lamp, or a selected laser operating in a specific spectral band

Fig. 8.13 Basic setup and components a fluorescence microscope

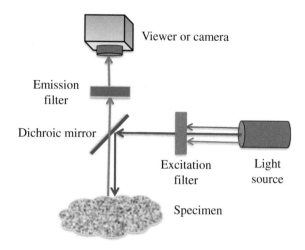

Viewer or camera

Emission filter

Dichroic mirror

Excitation filter

Light source

Specimen

- An excitation optical filter, which passes only the selected spectral band (≈40 nm wide) needed for absorption by the fluorophore being used
- A dichroic mirror (or beamsplitter), which deflects short-wavelength light from the excitation source to the specimen and passes longer-wavelength fluorescent light from the fluorophore to the photodetector
- An emission filter, which blocks light from the excitation wavelengths and passes only the spectral region in which the fluorescent emission occurs

As described in Chap. 9, *fluorescence lifetime imaging microscopy* (FLIM) is an advanced version of the fluorescence microscopy technique. By generating fluorescence images based on the differences in the exponential decay times of fluorescence from excited molecules, FLIM can detect fluorophore-fluorophore interactions and environmental effects on fluorophores.

8.5 Multiphoton Microscopy

In conventional fluorescence microscopy, a single photon is used to excite a fluorophore from a ground state level to a higher energy state. Typically this requires using photons from the ultraviolet or blue-green spectral range. Because such photons have high energies that could damage certain biological materials, the concept of *multiphoton microscopy* was devised [23–25]. For example, consider the case of two-photon as shown in Fig. 8.14. This excitation process yields the same results as single-photon excitation, but is generated by the simultaneous absorption of two less energetic photons (typically in the infrared spectral range) if there is sufficiently intense laser illumination. For multiphoton excitation the sum of the energies of the two photons needs to be greater than the energy gap between the ground state and the excited states of the molecule under investigation. Three-photon

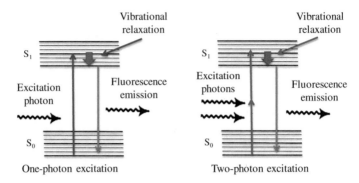

Fig. 8.14 Examples of one-photon excitation and two-photon-excitation

and higher photon excitations are also possible if the density of the excitation photons is high enough to ensure a sufficient level of fluorophore excitation.

In two-photon microscopy the photon concentration must be approximately a million times that required for an equivalent number of single-photon absorptions. This is accomplished with high-power mode-locked pulsed lasers, which typically are used to achieve the high photon concentrations. Such lasers generate a significant amount of power during pulse peaks, but have an average power that is low enough not to damage the specimen. Example lasers used for multiphoton microscopy configurations emit short pulses of around 100 fs (10^{-13} s) with a repetition rate of 80–100 MHz.

Example 8.7 Because the energy of a photon is inversely proportional to its wavelength, the wavelengths of the two photons should be about twice that required for single-photon excitation. What two-photon wavelengths are needed to excite an ultraviolet-absorbing fluorophore in the 320-nm region?

Solution: The two photons need to have a wavelength of 640 nm (red light). This red light will result in secondary fluorescence emission of longer (blue or green) wavelengths in the blue or green spectral region.

Example 8.8 What are some optical sources that can be used for multiphoton microscopy?

Solution: Some common lasers and their selected emission wavelengths that are applicable to multiphoton microscopy include the following:

- Ti-sapphire: 100 fs pulses of 1.5 W at 1050 nm
- Ytterbium-doped fiber lasers: 100 fs pulses of 2 W at 1050 nm
- Nd:YAG (neodymium: yttrium aluminum garnet) laser: 50 ns pulses 1064 nm

A key application of multiphoton microscopy has been for noninvasive imaging deep within scattering media, such as biological tissue. Because multiphoton microscopy falls within the broader field of *nonlinear optics* or *nonlinear optical microscopy*, it also provides several contrast mechanisms [24]. These mechanisms include two-photon excitation fluorescence (TPEF), second-harmonic generation (SHG), third-harmonic generation (THG), sum-frequency generation (SFG), stimulated Raman scattering (SRS), and coherent anti-Stokes Raman spectroscopy (CARS). More details on CARS and SRS are given in Chap. 9. These contrast modalities enable the extraction of information about the structure and function of the specimen under consideration, which is not available in other optical imaging techniques.

8.6 Raman Microscopy

As described in Chap. 6, Raman scattering occurs when photons undergo an inelastic scattering process with a molecule [26–28]. *Raman microscopy* is the combination of Raman scattering with optical microscopy. This is a widely used analysis technique for biological specificity (which describes the selective attachment or influence of one substance on another, for example, the interaction between an antibody and its specific antigen) and provides lateral resolution down to the subcellular level. The detection and analysis of this inelastically scattered light is the key function in Raman spectroscopy and is used to obtain information about molecular compositions, structures, and interactions in biological tissue samples.

When monochromatic laser light impinges on a specimen sample that is being examined, in a Raman inelastic scattering event a small amount of the incident light (a fraction of about 10^{-6}) interacts with molecular vibrations in the sample and is scattered at a slightly different wavelength. That is, there is an energy shift between the excitation light and the Raman-scattered photons. In this process, either a small amount of energy is transferred from the photon to the vibrational modes of the molecule (called *Stokes scattering*), or the molecular vibrations can transfer some energy to the photon (called *anti-Stokes scattering*). Thus, as shown in Fig. 8.15, for Stokes scattering the deflected photon has a lower energy (longer wavelength) than the incident photon, whereas for anti-Stokes scattering the deflected photon has a higher energy (shorter wavelength) than the incident photon. This effect is the basis of inelastic light scattering. Because the energy shift is a function of the mass of the involved atoms and the molecular binding strength, every molecular compound has a unique Raman spectrum. Thus, a plot of the intensity of the inelastically scattered light versus its frequency can be used to identify the sample. More details on advanced setups and applications are given in Chap. 9 for Raman spectroscopy, surface enhanced Raman scattering (SERS) spectroscopy, coherent anti-Stokes Raman scattering (CARS) spectroscopy, and stimulated Raman scattering (SRS) spectroscopy.

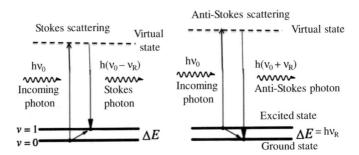

Fig. 8.15 Concepts of Stokes and anti-Stokes scattering

8.7   Light Sheet Fluorescence Microscopy

As described in the sections above, in a conventional microscope a sample is illuminated and observed along the same optical axis. This illumination and viewing method provides a high lateral resolution but a limited axial resolution, plus it produces significant optical background noise. In contrast to this conventional method, *light sheet fluorescence microscopy* (LSFM) uses a plane of light perpendicular to the viewing axis to create observation slices of a sample in an optical manner, as Fig. 8.16 illustrates [29–33]. In this setup the light sheet is fixed and the sample is moved up and down through the light sheet to capture different image slices in order to form 3D images. Thus fluorescence only takes place in a thin layer through the sample. Note that in this process the focus of the observing microscope objective has to synchronize with the area scanned by the light sheet.

Compared to other techniques, the LSFM method greatly reduces out-of-focus light and improves the signal-to-noise ratio of the images. Because LSFM scans a

Fig. 8.16 Equipment setup for light sheet fluorescence microscopy

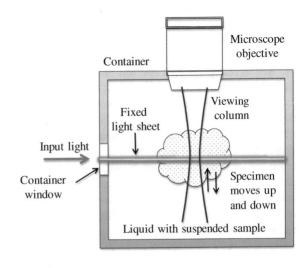

sample by using a plane of light instead of a point as in confocal microscopy, LSFM can acquire images at speeds ranging from 100 to 1000 times faster than those obtainable by point-scanning methods.

8.8   Super-Resolution Fluorescence Microscopy

Super-resolution fluorescence microscopy techniques allow the capture of images with a resolution of just tens of nanometers, which is better by an order of magnitude than is possible with diffraction-limited microscopic methods. A number of different methods have been proposed and implemented [34–41]. Two general groups that categorize these methods are the following:

1. *Deterministic super-resolution microscopy* is based on the nonlinear dependence of the emission rate of fluorophores on the intensity of the excitation laser. These techniques typically require the simultaneous application of several high-intensity pulsed lasers with specialized modulation filters to control the excitation beam geometry. These methods are referred to as *ensemble focused light imaging techniques*. For example, by using two lasers, one laser provides the energy for excitation of an ensemble of fluorophores to their fluorescent state. The other laser is used for de-excitation of the fluorophores by means of stimulated emission. The methods include the following:

   a. *Stimulated emission depletion microscopy* (STED)
   b. *Saturated structured illumination microscopy* (SSIM)

2. *Stochastic super-resolution microscopy* is a single-molecule approach in comparison to the ensemble methods of deterministic super-resolution microscopy. It functions by making several localized fluorophores emit light at separate times and thereby allowing these fluorophores to become resolvable in time. The methods include the following:

   a. *Photo activated localization microscopy* (PALM)
   b. *Stochastic optical reconstruction microscopy* (STORM)

8.9   Summary

Many developments in microscopy have appeared in recent years to enhance imaging performance and instrument versatility for biophotonics and biomedical applications. These developments include increasing the penetration depth in scattering media, improving image resolution beyond the diffraction limit, increasing image acquisition speed, enhancing instrument sensitivity, and developing better contrast mechanisms. Different embodiments of these optical microscopic techniques can visualize objects ranging in size from millimeters to nanometers.

This chapter addressed the basic functions of confocal microscopes, fluorescence microscopy, multiphoton microscopy, Raman microscopy, light sheet microscopy, and super-resolution fluorescence microscopy. Applications of microscopy instruments and techniques are found in all branches of biomedicine and biophotonics, for example, analyses of biological samples, brain research, cancer research, cytology, detection and assessment of diseases, drug discovery, food analyses, gynecology, healthcare, hematology, and pathology. As such, microscopic techniques are fundamental tools for the fields of spectroscopy and for imaging modalities, which are the topics of the next two chapters, respectively.

8.10   Problems

8.1   Write a simple Java applet to show how the size and angular aperture of objective light cones change with numerical aperture.

8.2   Suppose that the half-angle of the light cone captured by an objective lens is $20°$. Show that the NA is 0.45, 0.50, and 0.52 for an immersion medium of water, glycerin, and immersion oil, which have refractive indices of 1.33, 1.47, and 1.51, respectively.

8.3   Consider a microscope objective that has NA = 0.60. Show that the half-angle of the light cone captured by an objective lens with an immersion medium of water is $26.8°$.

8.4   Consider an objective lens that has a 30X magnification and which is used with an eyepiece of field number 7. (a) Show that the field of view is 0.23 mm. (b) Suppose that only half of an object that is viewed by this eyepiece and objective combination fits across the FOV (i.e., only half of the object is seen in the FOV). Show that the size of the object is 0.46 mm.

8.5   Consider two objective lenses that have NAs of 0.20 and 0.55, respectively. Show that the depths of field at a wavelength of 650 nm are 24.5 and 3.24 μm, respectively, if the immersion material is oil with an index n = 1.51.

8.6   Consider two microscope setups being used at a 480 nm wavelength. Suppose for one setup $NA_{obj1} = 0.80$ and for the other setup $NA_{obj2} = 1.25$. (a) Show that the resolution is 366 and 234 nm, respectively. (b) How do these resolution values compare with a test setup at 700 nm?

8.7   The key feature of a confocal microscope is the use of image-forming conjugate planes, which are paired sets of in-focus planes. Using reference texts or Web resources, use about one page to describe the operating principle of conjugate planes and show in a diagram where these sets of planes are located in a confocal microscope.

8.8   Consider the optical fiber-based confocal microscope system shown in Fig. 8.5. Here the light from a confocal pinhole aperture is collimated and sent through a beam splitting mirror to a xy laser scanning mechanism. This mechanism focuses the light sequentially onto each individual fiber in the coherent bundle, so that the light gets transmitted to the sample. Using vendor literature, reference texts, or Web resources, describe the operating principle of commercially available xy laser scanning mechanism for such an application. Include factors such as scanning speed, incremental stepping from one fiber to another, and the lens requirements to focus onto individual fibers in a fiber bundle.

8.9 Consider the optical fiber-based confocal microscope system shown in Fig. 8.5. Describe various options for selecting an optical fiber bundle for this application.

References

1. W.F.P. Török, F.J. Kao (eds.), *Optical Imaging and Microscopy: Techniques and Advanced Systems*, 2nd edn. (Springer, Berlin, 2007)
2. J. Mertz, *Introduction to Optical Microscopy* (Roberts and Company Publishers, Greenwood Village, 2009)
3. T.S. Tkaczyk, *Field Guide to Microscopy*. (SPIE Press, 2010)
4. D.B. Murphy, M.W. Davidson, *Fundamentals of Light Microscopy and Electronic Imaging*, 2nd edn. (Wiley-Blackwell, Hoboken, 2013)
5. P. Xi, *Optical Nanoscopy and Novel Microscopy Techniques*. (CRC Press, 2015)
6. http://zeiss-campus.magnet.fsu.edu/index.html. Accessed 12 Dec 2014
7. http://microscopy.fsu.edu/primer/anatomy/numaperture.html. Accessed 08 Dec 2014
8. www.microscopyu.com/articles/formulas/formulasna.html. Accessed 06 Dec 2014
9. F. Jenkins, H. White, *Fundamentals of Optics*, 4th ed., chap. 15. (McGraw-Hill, New York, 2002)
10. C.A. Diarzio, *Optics for Engineers* (CRC Press, Boca Raton, 2012)
11. E. Hecht, *Optics*, 5th edn., chap. 8. (Addison-Wesley, 2016)
12. J.W. Pawley (ed.), *Handbook of Biological Confocal Microscopy*, 3rd edn. (Springer, Berlin, 2006)
13. C.J.R. Sheppard, S. Rehman, "Confocal microscopy", chap. 6, in *Biomedical Optical Imaging Technologies: Design and Applications*, ed. by R. Liang (Springer, Berlin, 2013)
14. T. Wilson, Confocal microscopy, chap. 11 in *Biomedical Photonics Handbook* vol. I, 2nd edn. ed. by T. Vo-Dinh, (CRC Press, Boca Raton, FL, 2015)
15. H.J. Shin, M.C. Pierce, D. Lee, H. Ra, O. Solgaard, R.R. Richards-Kortum, Fiber-optic confocal microscope using a MEMS scanner and miniature objective lens. Opt. Express **15** (15), 9113–9122 (2007)
16. R.S. Pillai, D. Lorenser, D.D. Sampson, Fiber-optic confocal microscopy using a miniaturized needle-compatible imaging probe, in *Proceedings of SPIE 7753*, 21st International Conference Optical Fiber Sensors, paper 77534 M, 17 May 2011
17. L.W. Zhang, N.A. Monteiro-Riviere, Use of confocal microscopy for nanoparticle drug delivery through skin, J. Biomed. Opt. **18**(6), art. 061214 (2013)
18. B.N. Ozbay, J.T. Losacco, R. Cormack, R. Weir, V.M. Bright, J.T. Gopinath, D. Restrepo, E. A. Gibson, Miniaturized fiber-coupled confocal fluorescence microscope with an electrowetting variable focus lens using no moving parts. Opt. Express **40**(11), 2553–2556 (2015)
19. M. Mueller, *Introduction to Confocal Fluorescence Microscopy*, SPIE Press, 2nd ed. (2006)

20. U. Kubitscheck, *Fluorescence Microscopy: From Principles to Biological Applications* (Wiley-Blackman, Weinheim, 2013)
21. Y. Engelborghs, A.J.W.G. Visser (eds.), *Fluorescence Spectroscopy and Microscopy: Methods and Protocols* (Springer, New York, 2014)
22. P.P. Mondal, A. Diaspro, *Fundamentals of Fluorescence Microscopy* (Springer, Dordrecht, 2014)
23. K. König, Multiphoton microscopy in life sciences: review article. J. Microsc **200**, 83–104 (2002)
24. M.D. Young, J.J. Field, K.E. Sheetz, R.A. Bartes, J. Squier, A pragmatic guide to multiphoton microscope design. Adv. Opt. Photonics **7**, 276–378 (2015)
25. P.T.C. So, Two-photon excitation fluorescence microscopy, chap. 12, in *Biomedical Photonics Handbook*, ed. by T. Vo-Dinh, vol. I, 2nd edn. (CRC Press, Boca Raton, 2015)
26. T. Dieing, O. Hollricher, J. Toporski (eds.), *Confocal Raman Microscopy* (Springer, Berlin, 2010)
27. C. Krafft, B. Dietzek, M. Schmitt, J. Popp, Raman and coherent anti-Stokes Raman scattering microspectroscopy for biomedical applications. J. Biomed. Opt. **17**(4), article 040801, (2012)
28. K.A. Antonio, Z.D. Schultz, Advances in biomedical Raman microscopy. Anal. Chem. **86**(1), 30–46 (2014)
29. P.A. Santi, Light sheet fluorescence microscopy: a review. J. Histochem. Cytochem. **59**(2), 129–138 (2011)
30. L. Gutiérrez-Heredia, P.M. Flood, G.R. Emmanuel, Light sheet fluorescence microscopy: beyond the flatlands, pp. 838–847 in ed. by *Current microscopy contributions to advances in science and technology* A. Méndez-Vilas, ed., vol. 2 (Formatex Research Center, Badajoz, Spain, 2012)
31. M.B. Ahrens, M.B. Orger, D.N. Robson, J.M. Li, P.J. Keler, Whole-brain functional imaging at cellular resolution using light-sheet microscopy. Nat. Methods **10**, 410–413 (2013)
32. O.E. Olarte, J. Andilla, D. Artigas, P. Loza-Alvaez, Decoupled illumination detection in light sheet microscopy for fast volumetric imaging. Optica **2**, 702–705 (2015)
33. A.K. Glaser, Y. Wang, J.T.C. Liu, Assessing the imaging performance of light sheet microscopies in highly scattering tissues. Biomed. Opt. Express **7**(2), 454–466 (2016)
34. B.O. Leung, K.C. Chou, Review of super-resolution fluorescence microscopy for biology. Appl. Spectrosc. **65**(9), 967–980 (2011)
35. S.W. Hell, J. Wichmann, Breaking the diffraction resolution limit by stimulated emission: stimulated emission depletion microscopy. Opt. Lett. **19**(11), 780–782 (1994)
36. S.W. Hell, Toward fluorescence nanoscopy. Nat. Biotechnol. **21**, 1347–1355 (2003)
37. B. Harke, J. Keller, C.K. Ullal, V. Westphal, A. Schönle, S.W. Hell, Resolution scaling in STED microscopy. Opt. Express, **16**(6), 4154–4162 (2008)
38. E. Betzig, G.H. Patterson, R. Sougrat, O.W. Lindwasser, S. Olenych, J.S. Bonifacino, M.W. Davidson, J. Lippincott-Schwartz, H.F. Hess, Imaging intracellular fluorescent proteins at nanometer resolution. Science **313**, 1642–1645 (2006)
39. R. Henriques, M.M. Mhlanga, Review: PALM and STORM: what hides beyond the Rayleigh limit? J. Biotechnol. J. **4**, 846–847 (2009)
40. P. Annibale, S. Vani, M. Scarselli, U. Rothlisberger, A. Radenovic, Quantitative photo activated localization microscopy: unraveling the effects of photoblinking. PLoS ONE **6**, e22678 (2011)
41. A. Shivanandan, H. Deschout, M. Scarselli, A. Radenovic, Review: challenges in quantitative single molecule localization microscopy. FEBS Lett. **588**(19), 3595–3602 (2014)

Chapter 9
Spectroscopic Methodologies

Abstract Numerous viable optical spectroscopic methodologies are being implemented in biophotonics. Each spectroscopic discipline is progressively adopting more sophisticated photonics and optical fiber-based systems for delivering probing light to a tissue analysis site, for collecting light emitted from a specimen, and for returning this light to photodetection, recording, and analysis instruments. A key technological advance of spectroscopic methodologies is for rapid, accurate, and noninvasive in vivo detection and diagnosis of various health conditions. Examples of spectroscopic techniques used in biophotonics include fluorescence spectroscopy, fluorescent correlation spectroscopy, elastic scattering spectroscopy, diffuse correlation spectroscopy, Raman spectroscopy, surface-enhanced Raman scattering spectroscopy, coherent anti-Stokes Raman scattering spectroscopy, stimulated Raman scattering spectroscopy, photon correlation spectroscopy, Fourier transform infrared spectroscopy, and Brillouin scattering spectroscopy.

A number of optical spectroscopic methodologies that make use of advanced photonics and optical fiber technology are being used worldwide in research laboratories and medical clinics. In addition to in vitro applications, a key technological advance of these methodologies is for rapid, accurate, and noninvasive in vivo detection and diagnosis of various health conditions [1–4]. These procedures include

- Observing either short-term or long-term optical fluorescence or reflectance variations to discover premalignant and malignant changes in tissue for cancer diagnoses
- Diagnosing various diseases in a wide range of organs, such as the brain, skin, colon, breast, esophagus, pancreas, and oral cavity
- Quantifying microvascular blood flow in highly scattering biological tissues to assess their function and health
- Sensing glucose concentrations, oxygen levels, hemoglobin concentration, and other constituents in blood for assessments of conditions such as diabetes and anemia

- Rapid diagnosis of bacterial infections in blood samples to determine the correct selection and administration of antibiotics
- Monitoring the progress of wound healing and skin disease treatments
- Healthcare diagnosis of infectious diseases caused by microorganisms such as bacteria, viruses, fungi, and parasites

Selections of applications from among numerous viable optical spectroscopic methodologies are listed in Table 9.1. Each spectroscopic discipline is progressively adopting more sophisticated photonics and optical fiber-based systems for delivering probing light to a tissue analysis site, for collecting light emitted from a specimen, and for returning this light to photodetection, recording, and analysis instruments.

One point to note about notation is that in spectroscopy it is common to use *wavenumber* units instead of wavelength. In spectroscopy the wavenumber υ

Table 9.1 Examples of spectroscopic techniques used in biophotonics

Spectroscopic technique	Description and function
Fluorescence spectroscopy	Based on examining the fluorescence spectra of molecules to determine their basic molecular behavior characteristics, to identify infectious diseases, and to perform noninvasive biopsies
Fluorescent correlation spectroscopy (FCS)	Examines spontaneous fluorescent intensity fluctuations to determine concentrations and diffusion coefficients of molecules and large molecular complexes
Elastic scattering spectroscopy (ESS)	Also called *diffuse reflectance spectroscopy* and *light scattering spectroscopy*; based on analyzing the relative intensity of elastic backscattered light to distinguish diseased from healthy tissue
Diffuse correlation spectroscopy (DCS)	A noninvasive technique that probes deep into tissue to measure blood flow by using the time-averaged intensity autocorrelation function of the fluctuating diffuse reflectance signal
Raman spectroscopy	A non-invasive, label-free biomedical optics tool for evaluating the chemical composition of biological tissue samples (variations: CARS; time-resolved; wavelength-modulated)
Surface-enhanced Raman scattering (SERS) spectroscopy	Combines Raman scattering effects with surface plasmon resonance to identify a molecular species and to quantify different targets in a mixture of different types of molecules
Coherent anti-Stokes Raman scattering (CARS) spectroscopy	A nonlinear optical four-wave-mixing process for label-free imaging of a wide range of molecular assemblies based on the resonant vibrational spectra of their constituents
Stimulated Raman scattering (SRS) spectroscopy	Uses two laser beams to coherently excite a sample for straightforward chemical analyses
Photon correlation spectroscopy (PCS)	Uses dynamic light scattering to measure density or concentration fluctuations of small particles in a highly diluted suspending fluid to examine sizes and movements of scattering particles
Fourier transform infrared (FTIR) spectroscopy	Precisely measures light absorption per wavelength over a broad spectra range to identify materials, determine their constituent elements, and check their quality
Brillouin scattering spectroscopy	Optical technique for noninvasively determining the elastic moduli or stiffness of materials

(Greek letter nu) of electromagnetic radiation is defined by $\upsilon = 1/\lambda$, with typical units being cm^{-1}. Thus, for example, the wavenumber $\upsilon = 1500$ cm^{-1} is equivalent to a wavelength of $\lambda = 6.67$ μm.

Example 9.1 What are the wavelength equivalents in nm to the wavenumbers $\upsilon = 6500$ cm^{-1}, 15,000 cm^{-1}, and 25,000 cm^{-1}?

Solution: Using the relationship $\lambda(nm) = 10^7/\upsilon(cm^{-1})$ yields 1538 nm, 667 nm, and 400 nm, respectively.

9.1 Fluorescence Spectroscopy

Fluorescence spectroscopy involves the observation and analysis of the fluorescence spectrum either from a naturally fluorescing molecule or from an extrinsic fluorophore that is attached to the molecule. The interpretations of these spectra include studying the characteristics of molecules, identifying infectious diseases, and performing noninvasive biopsies. In a standard biopsy procedure for *soft tissue pathology* (the diagnosis and characterization of diseases in soft tissue), typically one or more tissue samples are physically removed for later laboratory evaluation. In contrast, optical spectroscopic methods reduce the need for surgical removal of tissue. Instead, by using an optical probe placed on or near the surface of the tissue to be evaluated, an imaging system records some in vivo form of spectral analysis of the tissue [2]. This feature of an optical biopsy enables an immediate diagnosis of the tissue characteristics instead of having to wait hours or days for a standard laboratory evaluation of a tissue sample. Depending on the spectroscopic method used, the diagnostic information obtained from the tissues can be at the biochemical, cellular, molecular structural, or physiological levels [5–9].

An advantage of fluorescence spectroscopy for soft tissue pathology is that the emitted spectra are sensitive to the biochemical composition of the tissue being examined. This feature is helpful in assessing whether the tissue is in a normal or diseased state. For example, Fig. 9.1 shows generic plots of fluorescence intensity as a function of wavelength for spectroscopic measurements of healthy and malignant samples of a specific tissue. Distinct fluorescence intensity variations are clearly seen. The wavelength range and the exact fluorescence spectral intensity response will depend on the tissue type being examined and the nature of the disease. The curves in Fig. 9.1 show that for this particular tissue type the fluorescence intensity increases with the progression of the tissue disease. In other cases the fluorescence intensity might decrease with the onset and progression of a specific disease.

Fig. 9.1 Generic examples of the differences in fluorescence spectra of normal and diseased tissues

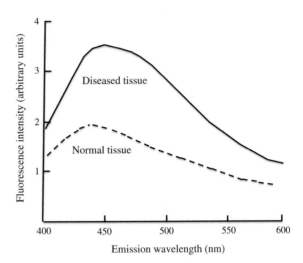

As Sects. 6.7 and 8.4 describe, certain types of fluorophores within molecules have the ability to absorb external excitation light at a particular wavelength, thereby boosting the molecule to a higher-energy excited state. In general, this excited state is energetically unstable and the molecule rapidly relaxes from the excited state back to the ground state. This relaxation can take place through processes such as energy dissipation into heat, internal transitions between different excitation states of the molecule, or by means of a fluorescence process. During the fluorescence process, the fluorophore will emit light of a longer wavelength (lower energy). The terms singlet state (designated by S) and triplet state (designated by T) are used in describing molecular energy states in fluorescence. In quantum mechanics, a *singlet state* has only one allowed value of the spin component, which is 0. A *triplet state* can have three allowed values of the spin component, which are -1, 0 and $+1$.

A generic fluorescence process is illustrated in Fig. 9.2. First a molecule can absorb incoming light, which occurs on the order of 10^{-15} s (femtoseconds). As indicated by the solid upward arrows, this action elevates the molecule from the S_0 ground state to a higher vibrational singlet energy level such as S_2, according to quantum mechanical transition rules. Immediately following the absorption, the molecule drops to the lowest more stable vibrational energy level (the $v = 0$ level) within the S_2 state by means of a nonradiative transition shown by the dashed downward arrows. This transition occurs on the order of 10^{-12} s and is called a *vibrational relaxation.*

Subsequently, because the $v = 0$ level within the S_2 state is very close in energy to an excited vibrational level of the S_1 state, a rapid energy transfer occurs between the S_2 and S_1 states. The wavy horizontal line shows this process. This process is referred to as *internal conversion* and takes place on the order of 10^{-13} s. Next another vibrational relaxation process takes place from the excited vibrational level to the lowest more stable $v = 0$ vibrational energy level of the S_1 state. This process occurs on the order of 10^{-12} s.

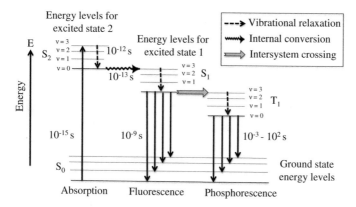

Fig. 9.2 Energy transition diagram of a generic fluorescence process

Several relaxation transition possibilities then can take place to return the molecule back to the S_0 state from the S_1 state. The two main ones are a radiative transition wherein a fluorescence photon is emitted or there can be an *intersystem crossing* to an excited level in the first triplet state T_1. The fluorescence emission takes place from the lowest vibrational level of the singlet S_1 state, as indicated by the solid downward arrows going to the S_0 state. This fluorescence occurs on the order of 10^{-9} s. The transition from the excited level to the lowest level in the first triplet T_1 state also takes place by means of vibrational relaxation. The eventual transition from the lowest level of the T_1 state to excited vibrational levels of the S_0 state are called *phosphorescence* and take place in times of 10^{-3}–10^2 s.

9.2 FRET/FLIM

The phenomenon of raising molecules to an excited state and then observing their fluorescent characteristics are important methods for investigating the biological and chemical properties of biological tissues and systems. This section describes two widely used techniques in this discipline. These are Förster resonance energy transfer and fluorescence lifetime imaging microscopy, which are popularly known as FRET and FLIM, respectively.

9.2.1 Förster Resonance Energy Transfer

Förster resonance energy transfer (FRET) describes a process of energy transfer between two light-sensitive molecules (fluorophores) [10–13]. This process also is known as *fluorescence resonance energy transfer* (again designated by FRET),

Fig. 9.3 Molecular
orientations and basic FRET
process

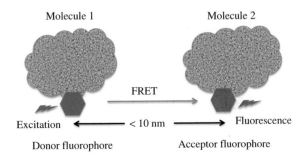

resonance energy transfer (RET), or electronic energy transfer (EET). The FRET
technique is used in fluorescence microscopy and molecular biology disciplines to
quantify molecular dynamics such as protein-protein and protein—DNA interactions, to examine protein conformational changes, and for monitoring the complex
formation between two molecules.

The FRET theory can be explained by treating an excited fluorophore as an
energetically oscillating dipole. This dipole can undergo an energy exchange with a
second nearby dipole that has a similar resonance frequency. To illustrate the FRET
process, consider the interactions between two molecules that are to be studied by
means of FRET. The molecular orientations and the FRET process are illustrated in
Fig. 9.3. Here the molecule on the left contains a donor fluorophore shown in blue,
which is capable of releasing (donating) its excited state energy to an acceptor
fluorophore shown in red in the molecule on the right. As is described below, the
degree of FRET interaction between the two fluorophores depends strongly on their
separation distance, which typically is less than 10 nm.

The state transitions in a FRET process are illustrated in Fig. 9.4. First a donor
fluorophore undergoes photoexcitation from the ground state S_0 to its first excited
singlet state S_1, as denoted by the solid blue upward arrows. The dashed downward
green arrows denote the possible fluorescence transitions of the donor. If an
acceptor molecule is within 1 to 10 nm of the donor, the donor fluorophore may
transfer the excited-state energy to the acceptor fluorophore through nonradiative
dipole–dipole coupling (see Sect. 2.7). That is, the energy that is transferred
between the two molecules does not require the spontaneous emission of a photon
by the donor. The energy transfer can only take place if there is an exact energy
matching of the donor fluorescence transitions with the excitation transitions in the
acceptor shown by the dashed orange upward arrows. Following the energy
transfer, the acceptor may drop to the ground state thereby emitting fluorescent
photons, as indicated by the solid red downward arrows. The degree of fluorescence
from the acceptor then can be used to measure molecular activity.

In addition to an exact energy matching of the acceptor and donor fluorescence
transitions, the absorption spectrum of the acceptor must overlap the fluorescence
emission spectrum of the donor. The absorption and emission spectra of the donor
and acceptor fluorophores are shown conceptually in Fig. 9.5. The blue triangular

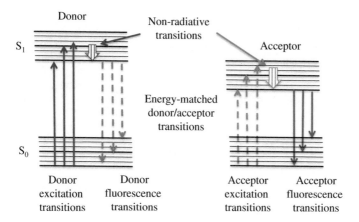

Fig. 9.4 State transitions in a FRET process

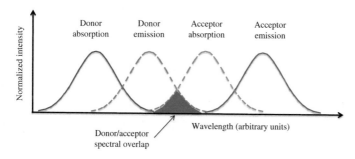

Fig. 9.5 Absorption and emission spectra of the donor and acceptor fluorophores

area indicates the overlap between the donor emission and the acceptor excitation spectra.

The efficiency E of the energy transfer is given by the Förster relation

$$E = \frac{R_0^6}{R_0^6 + R^6} \tag{9.1}$$

where the Förster distance R_0 is the fluorophore separation at which 50 % of the energy is transferred and R is the actual distance between the fluorophores. Thus, because the energy transfer efficiency is inversely proportional to the sixth power of the distance between the donor and the acceptor, the FRET process is extremely sensitive to small changes in the separations of the donor and acceptor fluorophores (see Example 9.2 and Prob. 9.3). In FRET applications the experimental separation distances typically are on the order of the Förster distance.

Example 9.2 Suppose the distance between the donor and acceptor is increased by a factor of three from $R = R_1 = R_0$ to $R = R_2 = 3R_0$. What is the decrease in energy transfer efficiency?

Solution: Letting the parameter $x = R/R_0$, then Eq. (9.1) can be written as $E = 1/(1 + x^6)$. Thus, letting $x_1 = R_1/R_0$ and $x_2 = R_2/R_0$, the ratio of the energy efficiencies is

$$\frac{E(R_1)}{E(R_2)} = \frac{1 + x_2^6}{1 + x_1^6} = \frac{1 + 3^6}{1 + 1^6} = \frac{1 + 729}{1 + 1} = 365$$

Thus the energy transfer efficiency decreases by a factor of 365 when the distance between the fluorophores increases from $R_1 = R_0$ to $R_2 = 3R_0$.

Table 9.2 lists some commonly used FRET pairs, the corresponding Förster distance, the donor excitation wavelength, and the acceptor emission wavelength.

9.2.2 Fluorescence Lifetime Imaging Microscopy

In a FRET process, the decay times of fluorescent emissions from an acceptor molecule can be used at the cellular level to study protein interactions, conformational changes, and parameters such as viscosity, temperature, pH, refractive index, and ion and oxygen concentrations. An imaging procedure that makes use of FRET is *fluorescence lifetime imaging microscopy* (FLIM), which is an advanced version of the fluorescence microscopy technique described in Sect. 8.4. By generating fluorescence images based on the differences in the exponential decay times

Table 9.2 Commonly used FRET pairs and their spectral parameters

Donor (excitation wavelength)	Acceptor (emission wavelength)	Förster distance (nm)
Fluorescein (512 nm)	QSY-7 dye (560 nm)	6.1
Fluorescein (512 nm)	Tetramethylrhodamine (TRITC: 550 nm)	5.5
Cyan fluorescent protein (CFP: 477 nm)	Yellow fluorescent protein (YFP: 514 nm)	5.0
IAEDANS (336 nm)	Fluorescein (494 nm)	4.6
Blue fluorescent protein (BFP: 380 nm)	Green fluorescent protein (GFP: 510 nm)	3.5
EDANS (340 nm)	DABCYL (393 nm)	3.3

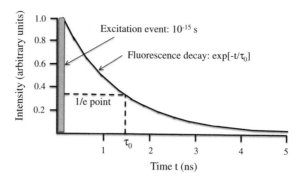

Fig. 9.6 Example of fluorescence decay for a fluorophore with lifetime τ_0

of fluorescence from excited molecules, FLIM can detect fluorophore-fluorophore interactions and environmental effects on fluorophores [13–17].

As is noted in Fig. 9.4, following a transition to an excited energy state a molecule will transition to a lower ground-state level in the order of 10^{-9} s. As Fig. 9.6 shows, in an ensemble of excited molecules, the fluorescence intensity from spontaneously emitted photons will decay with time t according to the exponential function

$$I(t) = I_0 \exp(-t/\tau_0) \tag{9.2}$$

Here τ_0 is the normally observed fluorescence lifetime and I_0 is the initial fluorescence intensity at time t = 0. The time interval needed for the fluorescence intensity to decay to 1/e of its initial intensity is called the *fluorescence lifetime* τ_0. By examining Eq. (9.2) it can be seen that the fluorescence lifetime can be found from the slope of a plot of the natural logarithm ln $[I(t)/I_0]$ versus time t.

Example 9.3 Consider the fluorescence decay for the particular fluorophore shown in Fig. 9.6. Using the illustrated intensity decay curve, what is the fluorescence lifetime τ_0?

Solution: The fluorescence lifetime is the time interval needed for the fluorescence intensity to decay to 1/e of its initial value I_0. In the curve in Fig. 9.6 the fluorescence lifetime occurs at the time where $I(t) = I_0/e = 0.368$ I_0. This occurs at t = τ_0 = 1.45 ns.

A primary application of FLIM is to identify different fractional amounts of the same fluorophore when it is in different states of interaction with its environment. This is done by using the basic property that *fluorescence lifetimes are independent of fluorophore concentration and laser excitation intensity*. Because the fluorescence lifetime of a fluorophore is sensitive to the local environment (for example, pH, molecular charge, the presence of fluorescence quenchers, refractive index, and

Fig. 9.7 Examples of fluorescence decay for fluorophores with lifetimes τ_0 and τ_Q

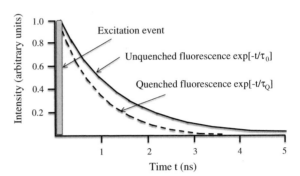

temperature), lifetime measurements under a microscope allow the observation of molecular effects through the spatial variations of the fluorescence lifetimes.

After an acceptor molecule is raised to an excited state, it can dissipate part of the absorbed energy through interactions with other molecules. This process is called *fluorescence quenching*. In this case the fluorescence lifetime τ_Q becomes shorter than τ_0, as is shown in Fig. 9.7. Quenching molecules include oxygen, halogens (bromine, chlorine, iodine), heavy atoms (iodides, bromides), heavy metal ions (cesium, copper, lead, nickel, silver), and a variety of organic molecules. For the accurate use of quenchers in FLIM it is important that the fluorescent quenching rate depends linearly on the concentration of the quenchers. Thereby the quencher concentration can be determined directly from the decrease in the fluorescence lifetime [13].

When a quencher with a fluorescence lifetime τ_Q is added, the intensity decay becomes a double exponential. This can be expressed as

$$I(t) = \alpha_1 \exp(-t/\tau_0) + \alpha_2 \exp(-t/\tau_Q) \tag{9.3}$$

Here the parameters α_1 and α_2 are called the *intensity factors*.

Example 9.4 Consider the case in which two fluorophores are attached to a molecule. Suppose the fluorescence lifetimes of the fluorophores are 1 and 5 ns, respectively, and let $\alpha_1 = \alpha_2 = 0.5$.

(a) What is the expression for the fluorescence intensity?
(b) Make plots of ln I(t) versus t for the two cases first when $\alpha_1 = 0$ and then when $\alpha_2 = 0$. Let t range from 0 to 8 ns.
(c) Plot the ln I(t) versus t curve for the combined fluorophores.

Solution: From Eq. (9.3) the fluorescence intensity is given by

$$I(t) = 0.5 \exp(-t/5) + 0.5 \exp(-t/1)$$

The three curves for parts (b) and (c) are given in Fig. 9.8.

Fig. 9.8 Logarithmic plots of
single and double exponential
fluorescence decays

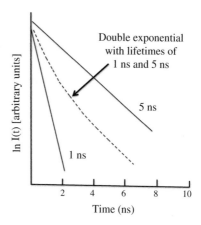

9.3 Fluorescence Correlation Spectroscopy

In addition to its use for identifying molecules and examining their external
behaviors, fluorescence spectroscopy has the ability to examine molecular diffusion
processes and to carry out noninvasive measurements inside of living cells. Toward
this goal, a technique called *fluorescence correlation spectroscopy* (FCS) has been
devised for precisely investigating the basic physical and chemical (referred to as
physicochemical) processes in the smallest functional units in biological systems,
such as individual proteins and nucleic acids [18–21]. In general, FCS measurements are made using fluorescently labeled biomolecules that are diffusing in an
aqueous buffer solution. A laser spot in the focus of a confocal microscope or a
multiphoton fluorescence microscope defines the detection volume.

In contrast to other fluorescent spectroscopic methods that concentrate on
examining the spectral emission intensity, FCS is a statistical method that examines
the *spontaneous fluorescent intensity fluctuations* caused by minute deviations of
the small biological system from thermal equilibrium. The intensity fluctuations can
originate from Brownian motion of dye labeled molecules, enzymatic activity,
rotational molecular motion, or protein folding. A key parameter of interest is the
lateral diffusion of fluorescent molecules in and out of the detection volume (in the
viewing plane of the microscope). The time duration in which molecules remain
within the laser spot depends on their size. For example, if a small, dye-tagged
molecule binds to a larger molecule, the tagged molecule will slow down and emit
photons for a longer time during its diffusion through the laser spot than if it were
attached to a smaller molecule. The intensity fluctuations due to the lateral diffusion
process range from milliseconds to seconds, whereas photochemical processes are
usually much faster. Thus, the study of the contributions to the intensity fluctuations
from the photochemical processes can be separated from the diffusion effects.
Experimental setups for FCS use nanomolar concentrations of molecules in sample

volumes as low as a few microliters. The measurements can be performed in a solution and in living cells.

When considering the spreading or *diffusion* of ions and molecules in solutions, often it is important to estimate the time required for the molecule to diffuse over a given distance. A key parameter for such an estimate is the *diffusion coefficient* D, which has a unique value for each type of molecule and must be determined experimentally. The diffusion coefficient typically is expressed in units of cm^2/s and is a function of factors such as the molecular weight of the diffusing species, temperature, and viscosity of the medium in which diffusion takes place. The *diffusion time* t is inversely proportional to the diffusion coefficient and can be approximated by

$$t \approx \frac{x^2}{2D} \qquad (9.4)$$

where x is the mean distance traveled by the diffusing molecule in one direction along one axis after an elapsed time t.

Example 9.5 Consider the following diffusion coefficients for CO_2, glucose, and hemoglobin for diffusion in water at 25 °C: (a) $D(CO_2) = 1.97 \times 10^{-5}$ cm^2/s, (b) D(glucose) = 5.0×10^{-6} cm^2/s, and (c) D(hemoglobin) = 6.9×10^{-7} cm^2/s What is the time required for these molecules to diffuse 50 nm in water?

Solution: From Eq. (9.4) the diffusion times are (a) 635 ns, (b) 2.5 μs, (c) 18.1 μs.

Parameters that can be examined readily with FCS include local molecular concentrations, molecular mobility coefficients (which describe the rate at which molecules diffuse), and chemical and photophysical rate constants (which describe the rate at which intermolecular or intramolecular reactions of fluorescently labeled biomolecules take place). One important application of FCS is the measurement of lipid and protein diffusions in planar lipid membranes to study the factors influencing membrane dynamics, such as membrane composition, ionic strength, the presence of membrane proteins, or frictional coupling between molecules.

As Fig. 9.9 shows, the detection volume normally is a spheroid with equatorial axis and polar axis radii of approximately $\underline{a}$ = 0.3 μm and $\underline{c}$ = 2.0 μm, respectively. This yields a volume of around 1 μm^3 or one femtoliter (fL), which is about the volume of an *E .coli* bacterial cell. The solutions have concentrations in the nanomolar range for the molecules of interest. In such a case, only a few molecules are detected simultaneously, which allows good signal-to-noise ratios for making precise measurements on small biological units.

Fig. 9.9 Example detection
volume of a confocal
microscope in FCS
procedures

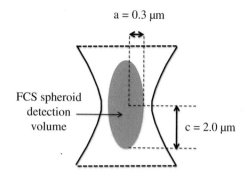

a = 0.3 μm

FCS spheroid
detection
volume

c = 2.0 μm

Example 9.6 Figure 9.9 shows a spheroid with the radii of the equatorial and
polar axes being 0.3 and 2.0 μm, respectively. What is the volume of the
spheroid?

Solution: The volume of a spheroid is found from the equation $(4\pi/3)a^2c$.
Here a is the equatorial radius and c is the polar radius. Thus the volume is

$$V = (4\pi/3)(0.3\ \mu m)^2(2.0\ \mu m) = 0.75\ \mu m^3 = 0.75\ fL$$

The analysis procedure of FCS for determining molecular processes is made
through a statistical analysis of the fluorescence signal fluctuations. This analysis
uses a Poisson distribution, which describes the statistical fluctuations of particle
occupancy. In a Poisson system the variance is proportional to the average number
of fluctuating species. This is carried out by means of an autocorrelation function G
(τ) of the fluctuations in the fluorescence intensity emission. Note that in statistics
the *autocorrelation* of a random process describes the correlation between values of
the process at different times. An example of this is given in Fig. 9.10 where
measurements are made at two different observation times t_1 and t_2 in time intervals
τ. Then the *autocorrelation function* $G(\tau)$ is defined by

Fig. 9.10 Sampling of a
time-varying signal at two
different time intervals

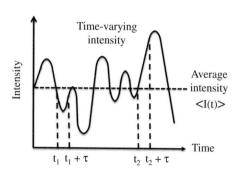

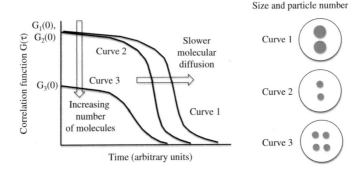

Fig. 9.11 Examples of the correlation function for different number and sizes of molecules

$$G(\tau) = \frac{\langle \delta I(t)\delta I(t+\tau)\rangle}{\langle I(t)\rangle^2} = \frac{\langle I(t)I(t+\tau)\rangle}{\langle I(t)\rangle^2} + 1 \tag{9.5}$$

where the pointed brackets signify averaging over all values of time t and where $\delta I(t) = I(t) + \langle I(t)\rangle$ is the deviation from the mean intensity.

For a two-dimensional sample with N particles the correlation function can be written as

$$G(\tau) = \frac{1}{N}\frac{1}{1+(\tau/\tau_D)} \tag{9.6}$$

where τ_D defines the *diffusion time* (the time it takes a particle to move a specific distance by diffusion). Examples of a typical correlation function are shown in Fig. 9.11. The shape of $G(\tau)$ gives information on molecular diffusion. As Fig. 9.11 indicates, for a given liquid viscosity the diffusion time becomes longer for larger particles, that is, the diffusion is slower and the curve shifts to the right. In addition, Fig. 9.11 shows that for an increasing number of particles in the observation volume the correlation function curve moves downward. Note that at $\tau = 0$ the value of $G(0)$ is inversely proportional to the average number of particles in the measurement volume. For illustration purposes on the right-hand side of Fig. 9.11, two and four different sized particles are shown to be in the measurement volume.

Example 9.7 Suppose curve 2 in Fig. 9.11 gives the result for a measurement volume that contains two medium sized particles. What do curve 1 and curve 3 represent in relation to curve 2 for a liquid with the same viscosity?

Solution: Because $G_1(0) = G_2(0)$, the measurement volume for curve 1 contains the same average number of particles as for the setup of curve 2, but the particles for test 1 are larger (so they diffuse slower). Because $G_3(0)$ is half the value of $G_2(0)$, there are twice as many particles in the measurement volume for $G_3(\tau)$ compared to that for $G_2(\tau)$.

In practice, FCS is combined with confocal microscopy to yield a good signal-to-noise ratio for examining individual molecules. In order to ensure that all of the processes that are being analyzed actually are statistical fluctuations around an equilibrium condition, the samples under investigation are maintained at thermal equilibrium. Depending on what category of fluorescent markers are used, the excitation light sources can be argon or argon-krypton multiline lasers, single-line He–Ne lasers, or laser diodes.

The flexibility of the FCS method can be increased by using two fluorescent dyes and employing separate photodetectors for the two emission spectra. This method is called *cross-correlation*. Two other common methods based on using fluorescence fluctuations to probe molecular interactions are *photon counting histograms* (PCH) and *fluorescence intensity distribution analysis* (FIDA). In addition, *confocal fluorescence coincidence analysis* is a highly sensitive and ultrafast technique for probing rare events in the femtomolar range [22–24].

9.4   Elastic Scattering Spectroscopy

Elastic scattering spectroscopy (ESS), which also is known by the names *diffuse reflectance spectroscopy* (DRS) and *light scattering spectroscopy* (LSS), has found important applications for the in vivo diagnoses of diseases in a wide range of organs, such as brain, breast, colon, esophagus, oral cavity, pancreas, and skin. ESS is based on the analysis of elastic scattering in tissue of broadband light that can range anywhere from the UV through the visible to the near-infrared regions [25–31]. In an elastic scattering process, photons that impinge on tissue components are scattered without experiencing a change in their energy. This means that the photons will change their travel direction but not their wavelength.

An example of an ESS optical fiber probe is shown in Fig. 9.12. The light is injected into a tissue sample (which is a turbid medium) through a dual-fiber optical fiber probe. Typical core diameters can be 400 and 200 μm for the illumination and collection optical fibers, respectively, with a center-to-center separation of 350 μm. The spectrum of the injected broadband light, which ranges from 330 to 760 nm, undergoes several scattering events in a typical depth of between 200 and 600 μm from the surface of the tissue.

The collected photons come from the characteristic banana-shaped region between the illumination and the collection fibers. This region has a volume of approximately 0.06 mm^3 for the probe configuration shown in Fig. 9.12. The information contained in the collected light depends on the anisotropy factor (the angular scattering probability distribution; see Sect. 6.3) and the distance between the illumination and detection fibers. The optical probe generally is placed in direct contact with tissue. The diagnosis of the backscattered signal can be done within milliseconds of the tissue illumination. Note that only light that has experienced multiple scatterings can be collected by the detection fiber.

Fig. 9.12 Example of an
ESS probe with separate
illumination and collection
fibers

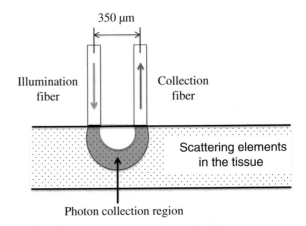

Illumination
fiber

Collection
fiber

350 μm

Scattering elements
in the tissue

Photon collection region

The relative intensity of the backscattered light that enters the collection fiber depends on the sizes and concentrations in the tissue of scattering components (e.g., nuclei, mitochondria, and connective tissue) and absorbing components (e.g., hemoglobin and oxyhemoglobin) [2]. Because the sizes and densities of these biological units change when the tissue becomes diseased, the ESS process assists pathologists in diagnosing the development of abnormal cell growth. By using appropriate mathematical models, the measured reflectance spectrum can be analyzed to yield scattering and absorption coefficients of the tissue sample. The values of these optical parameters depend on the cell morphology (the cell form and structure), the extracellular matrix (the biological constituents lying outside of the cell), and the biochemistry and vascular structure of the tissue sample. Because these tissue characteristics undergo predictable changes during the progression of a disease, the variations in the optical properties with time can be used to get information about the status of a tissue disease.

Example 9.8 Consider the photon collection volume in Fig. 9.12 to be a simple semicircular curved cylinder of tissue with a diameter $2a = 200$ μm. If the separation between the fiber cores is 350 μm and the center of the cylinder has a maximum depth of 175 μm, what is the volume of the cylinder?

Solution: The depth of 175 μm means the center of the cylinder is simply a semicircle with a radius of $r = 175$ μm. Thus the length of the cylinder is $\pi r = \pi(175$ μm$) = 550$ μm and its cross sectional area is πa^2 μm$^2 = \pi(100)^2$ μm$^2 = \pi 10^4$ μm^2. Then the volume is

$$V = (\text{cylinder cross section}) \times (\text{length}) = (\pi 10^4 \, \mu m^2) \times (550 \, \mu m)$$
$$= 0.017 \, mm^3.$$

9.5 Diffuse Correlation Spectroscopy

Diffuse correlation spectroscopy (DCS) is a continuous noninvasive technique that is used for probing the flow of blood in thick blood vessels or in vessels located in deep tissue, such as brain, muscle, and breast tissues [32–35]. DCS is based on using a time-averaged intensity autocorrelation function of a fluctuating diffusely reflected light signal. This technique also is known as *diffusing-wave spectroscopy*. DCS procedures are carried out in the 600-to-900-nm spectral range where the relatively low absorption enables deep penetration of light into tissue (see Fig. 6.8). The blood-flow measurements are made by monitoring the speckle fluctuations of photons that are diffusely scattered by blood cells as they move in the tissue (see Sect. 6.6).

The typical DCS setup shown in Fig. 9.13 uses a laser that has a long coherence length, a photodetetctor that can count single photons (such as an avalanche photodiode or a PMT), and a hardware autocorrelator. The purpose of the *autocorrelator* is to compute the degree of similarity of a signal with itself at varying time intervals. If the measurement times of the signal intensities are widely separated, then for a randomly fluctuating signal the intensities are not going to be related in any way. If the intensities are compared within a very short time interval δt, there will be a strong relationship or correlation between the intensities of two signals.

To measure the blood flow, near-infrared light from a laser that has a long coherence length is launched into the tissue through a MMF that has its probing end placed on the tissue surface [2]. A fraction of the light that is scattered by blood cells in the tissue is collected by a SMF or a few-mode fiber, which is placed a few millimeters or centimeters away from the illuminating fiber. The temporal fluctuation of the light intensity in a single speckle area is related to the movements of the red blood cells in microvasculature (the smallest blood vessels located throughout the tissue). The blood flow can be quantified by calculating the decay of the light intensity, which is derived from the autocorrelation function results.

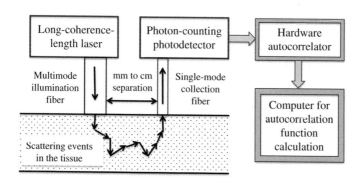

Fig. 9.13 Schematic of a representative experimental DCS setup

9.6 Raman Spectroscopy

Raman spectroscopy is a non-invasive, label-free biomedical optics tool that is used for evaluating the chemical composition of biological tissue samples [2]. Detailed compositional information is obtained by analyzing photons that have been scattered inelastically from vibrational and rotational level transitions in the chemical bonds of the tissue sample. A number of variations of the basic Raman spectroscopic method have been investigated, such as surface-enhanced Raman scattering, coherent anti-stokes Raman scattering, stimulated Raman scattering, time-resolved spectroscopy, polarization modulation, and wavelength-modulated Raman spectroscopy. The first three of these techniques are discussed in Sects. 9.7 through 9.9. Designs of miniaturized fiber-optic Raman probe can be made from a simple single optical fiber, two fibers with lenses for better sensitivity, and coherent fiber bundles that have optical filtering modules integrated on the bundle ends. The biomedical Raman spectroscopy applications include in vivo examinations of organs such as the bladder, breast, colon, esophagus, larynx, lung, oral cavity, and stomach [36–40].

The possible molecular transitions in a Raman inelastic scattering event are shown in Fig. 9.14 and are compared with Rayleigh scattering. During the Raman interaction of a photon with a molecule either a small amount of energy is transferred from the photon to the molecule, or the molecule can transfer some energy to the photon (see Sect. 6.3.4). The first transfer process (from the lowest molecular vibrational energy level to a higher vibrational state) is called *Stokes scattering*. If the incoming photon interacts with a molecule that already is in a higher vibrational state, the molecule can transfer some of its energy to the photon during the scattering event. This process is called *anti-Stokes scattering*.

In a Raman spectroscopy system, first laser light emerging from an excitation fiber is scattered by the tissue and then a fraction of the scattered light is collected by a detection fiber [2]. In Raman spectroscopy only inelastically scattered photons are of interest. Therefore the elastically scattered photons are suppressed by long-pass or bandpass optical filters, which are located in the detection channel (see Sect. 5.6). At the analysis end of the collection fibers, the inelastically scattered photons are separated according to wavelength in a spectrograph (an instrument that

Fig. 9.14 State transitions in Rayleigh elastic and Raman inelastic scattering

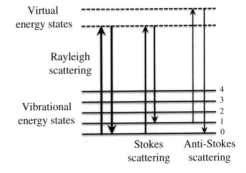

disperses light into a spectrum) and then are registered by a CCD detector array that has a large number of pixels (>1000) in the spectral axis in order to acquire a high-resolution Raman spectrum. Commonly used laser wavelengths in Raman spectroscopy include 532 nm (a green diode-pumped Nd:YAG laser) and 785 nm (an AlGaAs diode laser), which enable the achievement of a lateral resolution of better than half the wavelength (i.e., 250–350 nm). This sub-cellular resolution is similar to that attainable in fluorescence imaging.

Example 9.9 What are some optical sources that can be used for Raman spectroscopy?

Solution: Some common lasers and their selected emission wavelengths that are applicable to Raman spectroscopy include the following:

- Ar (488 nm, 514.5 nm)
- Kr (530.9 nm, 647.1 nm)
- He–Ne (623 nm)
- Diode lasers (782 nm, 830 nm)
- Nd:YAG (1064 nm)

A general illustration of a microscope system used for Raman spectroscopy is shown in Fig. 9.15. The main microscope components are the following:

- An excitation light source, for example, a 532-nm or a 785-nm laser
- Various reflecting and light-collecting and focusing optical elements; lens #1 can be a cylindrical lens to illuminate the sample with a slice of light or the lens can be removed so that only a spot of light falls on the sample
- An excitation optical filter that passes only the selected spectral band (≈40 nm wide) needed for absorption by the fluorophore being used
- A dichroic mirror that blocks light from the excitation wavelengths and passes to the detector only the spectral region in which the fluorescent emission occurs
- A spectrometer that spreads the wavelengths across the face of a detector, such as a photodiode array, CCD, or PMT

Fig. 9.15 Diagram of a microscope system used for Raman spectroscopy

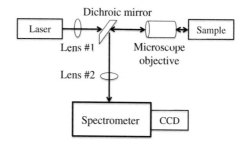

Fig. 9.16 A typical spectrum of Raman intensity as a function of wavenumber

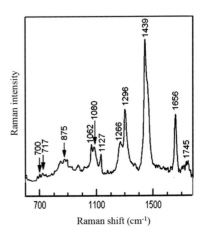

A typical spectrum of Raman intensity as a function of wavenumber is shown in Fig. 9.16. The difference between the frequency of the incident laser light and that of the Raman-shifted light is equal to the frequency of the vibrational bond that was excited. Molecules of different substances have Raman spectral peaks at specific frequencies, which are unique to the chemical bonds of that particular substance. A Raman spectrum thus serves as a type of fingerprint for specific biomolecules. Traditionally the Raman frequency shifts are recorded in wavenumbers (in units of cm^{-1}) with the spectrum of vibrations ranging from about 600–3000 cm^{-1}.

Based on Boltzmann distribution statistics, at room temperature the vibrational ground state typically is significantly more populated than the higher vibrational excited states. As a result, the intensity of the Stokes Raman spectrum tends to be much higher than the anti-Stokes Raman spectrum. An example of this is shown in Fig. 9.17 for CCl_4. Using Boltzmann statistics, the ratio of the anti-Stokes intensity

Fig. 9.17 Anti-Stokes and Stokes frequency spectra of Raman scattering in CCl_4 based on argon laser excitation (Reproduced with permission from R. Menzel, *Photonics*, Springer, 2nd ed., 2007)

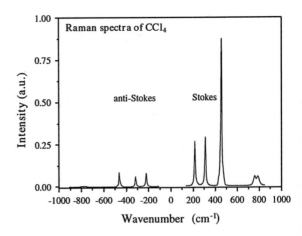

$I_{anti-Stokes}$ to the Stokes intensity I_{Stokes} for a given molecular vibration can be approximated by

$$\frac{I_{anti-Stokes}}{I_{Stokes}} \approx \exp(-h\nu_R/k_BT) \tag{9.7}$$

Here $h\nu_R$ is the Raman energy shift of the excited vibrational mode (that is, $h\nu_R$ is the energy gained or given up by a photon through Raman scattering).

Example 9.10 What is the ratio of the anti-Stokes intensity $I_{anti-Stokes}$ to the Stokes intensity I_{Stokes} for a 218-cm^{-1} Raman mode for CCl_4 at a temperature of 20 °C?

Solution: First,

$$h\nu_R = hc/\lambda = hc(218/10^{-2}\ m) = (6.626 \times 10^{-34}J\ s)(3 \times 10^8\ m/s)\,(218/10^{-2}\ m)$$
$$= 4.33 \times 10^{-21}\ J = 0.027\ eV$$

At 20 °C = 293 K, k_BT = (1.38 × 10^{-23} J/K)(293 K)/(1.6 × 10^{-19} J/eV) = 0.025 eV

Using Eq. (9.7), the ratio is

$$\frac{I_{anti-Stokes}}{I_{Stokes}} \approx \exp(-h\nu_R/k_BT) = \exp[-(0.027/0.025)] = \exp(-1.08)$$
$$= 0.340$$

This intensity difference can be seen in Fig. 9.17, which shows the major Raman spectral lines for CCl_4.

9.7   Surface Enhanced Raman Scattering Spectroscopy

Unlike Raman spectroscopy, which depends on evaluating weak spectral scattering signatures from intrinsic molecular components of the sample under investigation, *surface enhanced Raman scattering* (SERS) spectroscopy is based on using a large array of efficient scattering molecules. These collections of molecules can be conjugated to metallic substrates and then can produce distinct, optically strong spectra upon illumination by a laser. Basically SERS combines Raman scattering effects with surface plasmon resonance (see Sect. 7.8) that takes place on a noble metal surface (such as silver or gold), which has been roughened with nanoparticles [41–46]. By applying an excitation wavelength that coincides with the plasmon absorption resonance peak of the particular nanoparticle, surface plasmons are

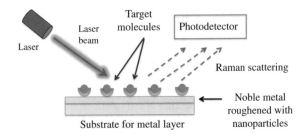

Fig. 9.18 Basic plasmonic-based setup for SERS

excited on the coated metal layer. Through this process, a strong evanescent electromagnetic field is induced on the metal surface, as shown in Fig. 9.18. Thereby the Raman modes of molecules that are close to the metallic surface are enhanced dramatically, because the Raman intensity is proportional to the square of the incident electromagnetic field amplitude.

Thus SERS now is widely used for biomedical research, because this technique provides narrow Raman spectra ($\sim$ 1 nm full width at half-maximum) that serve as unique biological fingerprints to identify different chemical species. These specific enhanced Raman spectra therefore can be used to identify a particular molecular species in a biological sample and to quantify distinct components in a mixture of different types of molecules. Because the SERS spectra have strong distinct spectral lines, this technique can be used for the simultaneous detection and analysis of multiple diseases that have different lines [46].

Example 9.11 As an alternative to attaching target molecules onto a nanoparticle roughened substrate, another technique is to immobilize biomarkers on the inside core surface of a hollow-core photonic crystal fiber (HCPCF). How would such a setup be realized?

Solution: The preparation and use of three types of Raman-reporter biomarkers immobilized inside a HCPCF is described in [46]. The biomarkers were cyanine 5 (Cy5), malachite green isothiocyanate (MGICT), and naphthalenethiol (NT). Each of these SERS nanotags have at least two distinct spectral peaks. This situation allows for the simultaneous multiplex detection of three different diseased conditions. The peaks are Raman shifts of 1120 and 1595 cm^{-1} for Cy5, 1175 and 1616 cm^{-1} for MGICT, and 1066 and 1378 cm^{-1} for NT.

9.8 Coherent Anti-stokes Raman Scattering Spectroscopy

Coherent anti-Stokes Raman scattering (CARS) spectroscopy is a nonlinear optical four-wave-mixing process [47–49]. *Nonlinear optics* deals with the interaction of light with matter in a manner in which the amplitude response of the material to the applied electromagnetic field is nonlinear with respect to this field. For example, doubling the optical input intensities does not simply result in a doubling of the output intensities. In a general *four-wave-mixing* (FWM) process, three laser fields with frequencies ω_1, ω_2, and ω_3 generate a fourth field with a frequency

$$\omega_{FWM} = \omega_1 \pm \omega_2 \pm \omega_3 \tag{9.8}$$

If the frequency difference $\omega_1 - \omega_2$ of two of the laser fields is tuned to the vibrational resonance of a Raman mode of the nonlinear medium, the FWM process

$$\omega_{FWM} = \omega_1 - \omega_2 + \omega_3 = \omega_{CARS} \tag{9.9}$$

results in coherent anti-Stokes Raman scattering.

Thus the CARS process requires the use of at least two coherent input laser pulses, which have a frequency difference equal to that of the Raman mode being investigated. Basically a CARS setup consists of two stimulated Raman scattering steps, as shown in Fig. 9.19. First two separate laser beams emit a pump photon of frequency ω_{pump} and a Stokes photon of frequency ω_{Stokes}, respectively. These two photons then resonantly excite a Raman oscillator of vibrational frequency $\omega_{vib} = \omega_{pump} - \omega_{Stokes}$, which is an excited vibrational state of the molecule. This step is known as *stimulated Stokes emission* because the pump photon is inelastically scattered into the Stokes photon along the direction of the Stokes beam. This action sets up a periodic modulation of the refractive index of the material. In the second step, which is known as *stimulated anti-Stokes emission*, the index modulation can be probed with a laser beam of photon frequency ω_{probe}. The interaction of the probe beam with the Raman oscillator produces the CARS anti-Stokes photon of frequency $\omega_{CARS} = \omega_{probe} + \omega_{vib} = \omega_{probe} + \omega_{pump} - \omega_{Stokes}$, which is the measurement photon.

The CARS method enables label-free imaging of many types of molecular assemblies by examining their resonant vibrational spectra. Similar to Raman spectroscopy, there are many variants of CARS. Although Raman spectroscopy and CARS spectroscopy are equally sensitive because they use the same molecular transitions, the CARS technique often produces imaging signals that are orders of magnitude stronger than those obtained using conventional Raman scattering spectroscopy. This factor allows the CARS signal from a single molecular transition to be collected by a factor of about 10^5 faster than a Raman signal in practical situations.

Fig. 9.19 Basic CARS setup consists of two stimulated Raman scattering steps

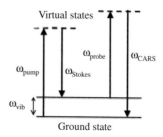

9.9 Stimulated Raman Scattering Spectroscopy

Stimulated Raman scattering (SRS) spectroscopy offers an alternative analytical tool to CARS [50–52]. In SRS spectroscopy, two laser beams are used to coherently excite the sample. One laser provides the pump beam with frequency ω_p and the other laser provides the Stokes beam with frequency ω_s. When the *beat frequency* $\Delta\omega = \omega_p - \omega_s$ (also called the *Raman shift*) is equal to a particular molecular vibration of the sample, SRS signals are produced as a result of the nonlinear interaction between the photons and the molecules. These SRS signals include both *stimulated Raman loss* (SRL) and *stimulated Raman gain* (SRG). Advantages of SRS that enable straightforward chemical analyses are that it does not have a resonant background signal, it yields an identical spectrum as Raman scattering, and it is linearly proportional to the concentration of the analyte. SRS has a major advantage in imaging speed over Raman scattering and gives an improved image contrast and spectral fidelity compared to CARS. However, a challenge to its use is the need for an expensive and sensitive ultrafast dual-wavelength tunable laser source.

9.10 Photon Correlation Spectroscopy

Photon correlation spectroscopy (PCS) is based on *dynamic light scattering* techniques for probing time variations of the density or concentration fluctuations of small particles in a highly diluted suspending fluid or polymers in a solution [53–55]. Dynamic light scattering measures the Brownian motion of the suspended particles and relates this motion to their sizes. For a liquid of a given viscosity, larger particles will have a slower Brownian motion than smaller ones. The sizes for the particles being analyzed range from 1 nm to 10 μm. The particle size is calculated from the translational diffusion coefficient D by using the Stokes-Einstein equation

$$d_H = \frac{k_B T}{3\pi\eta D} \tag{9.10}$$

Here T is the temperature (K), η is the viscosity of the suspending solution, and k_B is Boltzmann's constant. The parameter d_H is called the *hydrodynamic diameter*, which is the diameter of a sphere that has the same translational diffusion speed as the particle being analyzed.

A basic setup for PCS is illustrated in Fig. 9.20a. First a light pulse from a laser is injected into a solution and as it hits the molecules in the liquid sample the light travels in all directions due to Rayleigh scattering. The scattered light is collected by a photodetector, such as a photomultiplier tube, and then is directed to a signal analysis unit. The pulsing process is repeated at short time intervals. The scattering

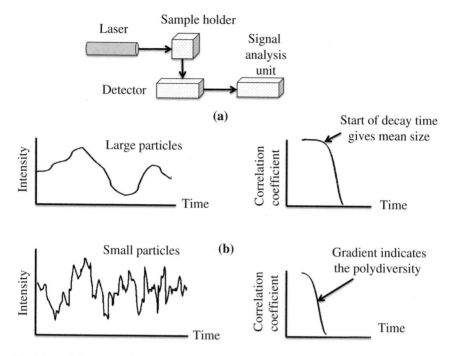

Fig. 9.20 a A basic setup for PCS; **b** Signal fluctuations and the related correlation coefficient for large and small particles

intensity will fluctuate over time because the distances between the molecules in the solutions are constantly changing due Brownian motion. As a result of this motion, the scattered light then interferes either constructively or destructively, which produces a dynamically changing speckle pattern. These sets of speckle patterns then are analyzed by an autocorrelator that compares the intensity of light at each spot over time.

Information for evaluating the sizes and the movements of the scattering particles can be obtained from the time correlations of the intensity fluctuations. As Fig. 9.20b shows, larger particles move slower than smaller ones. Thus the intensity fluctuations are slower for large particles compared to faster moving small particles for which the intensity varies more rapidly. In the correlation functions for the two different sized particles, the onset of the correlation decay (loss of correlation) indicates the mean size of the particles. The gradient of the curves are related to the *polydispersity* of the particles, which is a measure of the distribution of molecular mass in a given polymer sample.

For a given temperature T, the PCS method only requires knowing the viscosity of the suspending fluid in order to estimate the average particle size and the particle distribution function. To obtain accurate information from the PCS method, the light can only be scattered once. Thus, PCS requires highly diluted suspensions in order to avoid multiple scattering that can lead to incorrect results.

Example 9.12 Consider a PCS test for examining a hemoglobin molecule that is diffusing in water. (a) If $D = 1 \times 10^{-6}$ cm^2/s for hemoglobin in water and if $\eta = 1 \times 10^{-3}$ N s/m^2 for water, what is the hydrodynamic diameter of the hemoglobin molecule? (b) How does the volume of the equivalently diffusing sphere compare with the actual volume of 321.6 nm^3 for a hemoglobin molecule?

Solution:

(a) From Eq. (9.10), the hydrodynamic diameter is

$$d_H = \frac{(1.38 \times 10^{-23}\,\text{J/K})(293\,\text{K})}{3\pi(1.0 \times 10^{-3}\text{N s/m}^2)(0.1 \times 10^{-9}\,\text{m}^2/\text{s})} = 4.29\,\text{nm}$$

(b) The volume of the equivalent sphere is

$$V = \frac{4\pi(d_H/2)^3}{3} = \frac{4\pi(2.15\,\text{nm})^3}{3} = 41.3\,\text{nm}^3$$

9.11   Fourier Transform Infrared Spectroscopy

Fourier transform infrared (FTIR) spectroscopy is used to determine how much light a material sample emits (from luminescence or Raman scattering) or absorbs at each wavelength in a broad spectral range [56–58]. The information that can be deduced from the FTIR spectrum include an identification of the material sample, a measure of the consistency or quality of a material, and a determination of the amount of specific compounds in a mixture. Some key advantages and features of FTIR include the following:

- It is a nondestructive measurement and analysis method
- It provides a precise measurement method that requires no external calibration
- The signal-to-noise ratio of the spectral display is a factor of 100 better than that of previous generation spectrometers
- It has a high wavenumber measurement accuracy of ± 0.01 cm^{-1}
- Information from all wavelengths is collected simultaneously in scan times of less than a second
- It has a wide scan range of 1000–10 cm^{-1}

As shown in Fig. 9.21, the FTIR spectrometer consists of a collimated laser source, an interferometer component consisting of a beam splitter and two mirrors, a sample compartment, a photodetector for capturing the optical signal,

Fig. 9.21 Basic FTIR
spectrometer configuration

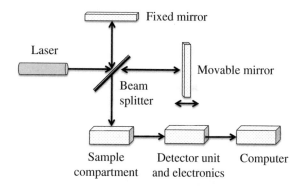

amplification and analog-to-digital conversion electronics, and a computer for calculating the Fourier transform.

The FTIR spectrometer simultaneously collects spectral data in a wide wavelength range. The resulting detected spectrum represents a molecular fingerprint of the material sample, because similar to an actual fingerprint no two unique molecular structures produce the same infrared spectrum. The designation *Fourier transform* is in the FTIR name because this mathematical procedure is used to decode the individual spectral frequencies in the measured data in order to obtain an interpretable plot of intensity versus frequency.

9.12   Brillouin Scattering Spectroscopy

Brillouin scattering spectroscopy is defined as the inelastic scattering of light by *thermally excited acoustical phonons*. Basically it is an optical technique for noninvasively determining the elastic moduli or stiffness of materials. Because diseases can alter the elasticity of body fluids or tissues, Brillouin scattering spectroscopy can be used to distinguish healthy from diseased fluids or tissues. Brillouin spectroscopy is similar to Raman spectroscopy because the physical scattering processes are identical. However, whereas Raman scattering involves high frequency molecular rotational and vibrational modes, Brillouin scattering involves the scattering of photons by low frequency phonons. The Brillouin scattering process thus provides information regarding elastic properties of the scattering medium. Optical phonons measured in Raman spectroscopy have wavenumbers on the order of $10–10{,}000 \ \text{cm}^{-1}$, whereas wavenumbers of phonons involved in Brillouin scattering are on the order of $0.1–6 \ \text{cm}^{-1}$. This biophotonics tool has been used for functions such as in vivo measurements of the rheological properties of the eye lens, screening for increased total protein in cerebrospinal fluid during bacterial meningitis, and assessing changes of the microscopic viscoelasticity associated with skin injury [59–62].

9.13   Summary

Optical spectroscopic methodologies are being used worldwide for in vitro applications in research laboratories and medical clinics for rapid, accurate, and non-invasive in vivo detection and diagnosis of various health conditions. Table 9.1 lists some major methodologies. In particular, techniques involving the observation and interpretation of fluorescence spectra are being applied for studying the characteristics of molecules, identifying infectious diseases, and performing noninvasive soft tissue biopsies. These methods include fluorescence spectroscopy, fluorescence lifetime imaging microscopy, fluorescence correlation spectroscopy, elastic scattering spectroscopy, diffuse correlation spectroscopy, a variety of methodologies based on Raman spectroscopy, and Fourier transform infrared spectroscopy.

9.14   Problems

9.1   (a) Show that the wavelength equivalents in nm to the spectroscopic wavenumbers $\upsilon = 5500$ cm^{-1}, 12,500 cm^{-1}, and 22,000 cm^{-1} are 1818 nm, 800 nm, and 455 nm, respectively. (b) Show that the wavenumber equivalents in cm^{-1} to the wavelengths 300 nm, 532 nm, and 785 nm are $\upsilon = 33,333$ cm^{-1}, 18,800 cm^{-1}, and 12,740 cm^{-1}, respectively.

9.2   Verify the plot in Fig. 9.22 of FRET efficiency as a function of the molecular separation R given in units of the Förster radius R_0.

9.3   Suppose the distance between the donor and acceptor in a FRET setup is increased by a factor of two from $R = R_1 = R_0$ to $R = R_2 = 2R_0$. Show that the decrease in FRET energy transfer efficiency is 32.5.

9.4   If the Förster distance of a FRET pair is 6.0 nm, show that the efficiency has dropped to 25 % at a separation distance of $R = 3^{1/6}R_0 = 1.20R_0 = 7.2$ nm.

Fig. 9.22 FRET efficiency as a function of molecular separation

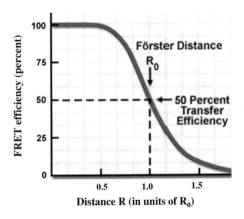

9.5   Consider the fluorescence decay for the two particular unquenched and quenched fluorophores shown in Fig. 9.7. Show that the approximate fluorescence lifetimes are $\tau_0 \approx 1.4$ ns and $\tau_Q \approx 1.0$ ns.

9.6   The diffusion coefficients for O_2 is 2.10×10^{-5} cm^2/s for diffusion in water at 25 °C. Show that the diffusion times in which an oxygen molecule travels distances of 50 nm, 100 μm, and 1 cm are 595 ns, 2.38 s, and 6.61 h, respectively. This shows that diffusion alone is not an adequate mechanism for distributing oxygen in the human body.

9.7   Suppose that for the spheroid shown in Fig. 9.9 the radii of the equatorial and polar axes are 0.3 μm and 2.4 μm, respectively. Show that the volume of the spheroid is 0.91 μm^3 = 0.91 fL.

9.8   When interpreting optical signals, various autocorrelation functions are used to analyze the characteristics of optical pulses in the femtosecond region. An autocorrelation is an electronic tool used for examining the similarity of observations of a signal at different points in time. Using Web or vendor resources describe in a few paragraphs how an autocorrelator works.

9.9   Consider the photon collection volume in Fig. 9.12 to be a simple semicircular curved cylinder of tissue with a 250-μm diameter. If the separation between the fiber cores is 350 μm, show that is the volume of the cylinder is 0.027 mm^3.

9.10 Consider the banana-shaped scattering region in Fig. 9.12 to consist of two cylindrical cones and a semicircular curved cylinder as shown in Fig. 9.23. For the conical sections, let $D_{top} = 200$ μm, $D_{bottom} = 300$ μm, and $h = 100$ μm. Thus, similar to the setup in Example 9.8, the curved cylinder has a length of 550 μm and its radius is $a = 0.5\, D_{bottom} = 150$ μm. For the cylindrical conical section shown in Fig. 9.23, the volume is given by $V = \frac{\pi h}{12}\left(D_{bottom}^2 + D_{bottom}D_{top} + D_{top}^2\right)$. Show that the total volume of the banana-shaped region is $2(0.005) + 0.039$ mm^3.

9.11 Verify that the ratio of the anti-Stokes intensity $I_{anti\text{-}Stokes}$ to the Stokes intensity I_{Stokes} is 0.11 for the 459-cm^{-1} Raman mode shown for CCl$_4$ in Fig. 9.17 at a temperature of 20 °C.

9.12 Consider a PCS test for examining a glucose molecule that is diffusing in water. If $D = 6.7 \times 10^{-6}$ cm^2/s for hemoglobin in water and if $\eta = 1 \times 10^{-3}$ N s/m^2 for water, show that the hydrodynamic diameter of the glucose molecule is 0.64 nm.

Fig. 9.23 Volume estimate of a banana-shaped scattering region

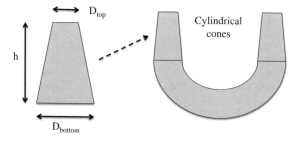

References

1. R. Richards-Kortum, E. Sevick-Muraca, Quantitative optical spectroscopy for tissue diagnosis. Annu. Rev. Phys. Chem. **47**(10), 555–606 (1996)
2. G. Keiser, F. Xiong, Y. Cui, P.P. Shum, Review of diverse optical fibers used in biomedical research and clinical practice. J. Biomed. Opt. **19**, 080902 (2014)
3. M. Olivo, R. Bhuvaneswari, I. Keogh, Advances in bio-optical imaging for the diagnosis of early oral cancer. Pharmaceutics (Special issue: Molecular Imaging) **3**(3), 354–378 (2011)
4. A. Wax, M.G. Giacomelli, T.E. Matthews, M.T. Rinehart, F.E. Robles, Y. Zhu, Optical spectroscopy of biological cells. Adv. Opt. Photonics **4**(3), 322–378 (2012)
5. J.R. Lakowicz, *Principles of Fluorescence Spectroscopy*, 3rd edn. (Springer, New York, 2006)
6. U. Kubitscheck, *Fluorescence Microscopy: From Principles to Biological Applications* (Wiley-Blackman, Weinheim, 2013)
7. M. Olivo, J.H. Ho, C.Y. Fu, Advances in fluorescence diagnosis to track footprints of cancer progression in vivo. Laser Photonics Rev. **7**(5), 646–662 (2013)
8. Y. Engelborghs, A.J.W.G. Visser (eds.), *Fluorescence Spectroscopy and Microscopy: Methods and Protocols* (Springer, New York, 2014)
9. P.P. Mondal, A. Diaspro, *Fundamentals of Fluorescence Microscopy* (Springer, Dordrecht, 2014)
10. S.C. Hovan, S. Howell, P.S.-H. Park, Förster resonance energy transfer as a tool to study photoreceptor biology. J. Biomed. Opt. **15**(6), 067001 (2010)
11. T. Förster, Energiewanderung und Fluoreszenz, *Naturwissenschaften*, **33**(6), 166–175 (1946). Translation: Energy migration and fluorescence. J. Biomed. Opt. **17**(1), 011002 (2012)
12. R.S. Knox, Förster's resonance excitation transfer theory. J. Biomed. Opt. **17**(1), 011003 (2012)
13. W. Becker, *The bh TCSPC Handbook*, 6th edn. (Becker & Hickl, Berlin, 2015)
14. J.W. Borst, A.J.W.G. Visser, Topical review: fluorescence lifetime imaging microscopy in life sciences. Meas. Sci. Technol. **21**, 102002 (2010)
15. T. Dellwig, P.Y. Lin, F.J. Kao, Long-distance fluorescence lifetime imaging using stimulated emission. J. Biomed. Opt. **17**(1), 011009 (2012)
16. F.J. Kao, G. Deka, N. Mazumder, Cellular autofluroscence detection through FLIM/FRET microscopy, in *The Current Trends in Optics and Photonics*, ed. by C.-C. Lee (Springer, Dordrecht, Netherlands, 2015), pp. 471–482
17. K. Suhling, L.M. Hirvonen, J.A. Levitt, P.H. Chung, C. Tregidgo, A. Le Marois, D.A. Rusakov, K. Zheng, S. Ameer-Beg, S. Poland, S. Coelho, R. Henderson, N. Krstajic, Fluorescence lifetime imaging (FLIM): basic concepts and some recent developments. Med. Photonics **27**, 3–40 (2015). (Review Article)
18. S.T. Hess, S.H. Huang, A.A. Heikal, W.W. Webb, Biological and chemical applications of fluorescence correlation spectroscopy: a review. Biochemistry **41**, 697–705 (2002)
19. R. Macháň, M. Hof, Recent developments in fluorescence correlation spectroscopy for diffusion measurements in planar lipid membranes. Int. J. Mol. Sci. **11**, 427–457 (2010). (Review Article)
20. P. Schwille, J. Ries, Principles and applications of fluorescence correlation spectroscopy (FCS), in *Biophotonics: Spectroscopy, Imaging, Sensing, and Manipulation*, ed. by B. Di Bartolo, J. Collins (Springer, Berlin, 2011), pp. 63–86
21. Y. Tian, M.M. Martinez, D. Pappas, Fluorescence correlation spectroscopy: a review of biochemical and microfluidic applications. Appl. Spectrosc. **65**, 115–124 (2011). (Review Article)
22. L.N. Hillesheim, J.D. Müller, The photon counting histogram in fluorescence fluctuation spectroscopy with non-ideal photodetectors. Biophys. J. **85**, 1948–1958 (2003)
23. C. Eggeling, S. Jäger, D. Winkler, P. Kask, Comparison of different fluorescence fluctuation methods for their use in FRET assays: Monitoring a protease reaction. Curr. Pharma. Biotechnol. **6**, 351–371 (2005)

24. T. Winkler, U. Kettling, A. Koltermann, M. Eigen, Confocal fluorescence coincidence analysis: an approach to ultra high-throughput screening. Proc. Natl. Acad. Sci. USA **96**, 1375–1378 (1999)
25. I.J. Bigio, J.R. Mourant, Ultraviolet and visible spectroscopies for tissue diagnostics: fluorescence spectroscopy and elastic-scattering spectroscopy. Phys. Med. Biol. **42**, 803–814 (1997)
26. X. Cheng, D.A. Boas, Diffuse optical reflection tomography with continuous-wave illumination. Opt. Express **3**(3), 118–123 (1998)
27. O.M. A'Amar, L. Liou, E. Rodriguez-Diaz, A. De las Morenas, I.J. Bigio, Comparison of elastic scattering spectroscopy with histology in ex vivo prostate glands: potential application for optically guided biopsy and directed treatment. Lasers Med. Sci. **28**(5), 1323–1329 (2013)
28. K.W. Calabro, I.J. Bigio, Influence of the phase function in generalized diffused reflectance models: review of current formalisms and novel observations. J. Biomed. Opt. **19**(7), 075005 (2014)
29. A. Douplik, S. Zanati, G. Saiko, C. Streutker, M. Loshchenov, D. Adler, S. Cho, D. Chen, M. Cirocco, N. Marcon, J. Fengler, B.C. Wilson, Diffuse reflectance spectroscopy in Barrett's esophagus: developing a large field-of-view screening method discriminating dysplasia from metaplasia. J. Biophotonics **7**(5), 304–311 (2014)
30. B. Yu, A. Shah, V.K. Nagarajan, D.G. Ferris, Diffuse reflectance spectroscopy of epithelial tissue with a smart fiber-optic probe. Biomed. Opt. Express **5**(3), 675–689 (2014)
31. K. Vishwanath, K. Chang, D. Klein, Y.F. Deng, V. Chang, J.E. Phelps, N. Ramanulam, Portable, fiber-based, diffuse reflection spectroscopy (DRS) systems for estimating tissue optical properties. Appl. Spectrosc. **62**, 206–215 (2011)
32. J. Dong, R. Bi, J.H. Ho, P.S.P. Thong, K.C. Soo, K. Lee, Diffuse correlation spectroscopy with a fast Fourier transform-based software autocorrelator. J. Biomed. Opt. **17**, 097004 (2012)
33. Y. Shang, K. Gurley, G. Yu, Diffuse correlation spectroscopy (DCS) for assessment of tissue blood flow in skeletal muscle: recent progress. Anat. Physiol. **3**(2), 128 (2013)
34. T. Durduran, A.G. Yodh, Diffuse correlation spectroscopy for noninvasive, microvascular cerebral blood flow measurement. NeuroImage **85**, 51–63 (2014). (Review Article)
35. E.M. Buckley, A.B. Parthasarathy, P.E. Grant, A.G. Yodh, M.A. Franceschini, Diffuse correlation spectroscopy for measurement of cerebral blood flow: future prospects. Neurophotonics **1**(1), 011009 (2014)
36. A. Downes, A. Elfick, Raman spectroscopy and related techniques in biomedicine. Sensors **10**, 1871–1889 (2010). (Review Article)
37. Y. Huang, P.P. Shum, F. Luan, M. Tang, Raman-assisted wavelength conversion in chalcogenide waveguides. IEEE J. Sel. Topics Quantum Electron. **18**(2), 646–653 (2012)
38. C. Krafft, B. Dietzek, M. Schmitt, J. Popp, Raman and coherent anti-Stokes Raman scattering microspectroscopy for biomedical applications. J. Biomed. Opt. **17**, 040801 (2012). (Review article)
39. P. Matousek, N. Stone, Recent advances in the development of Raman spectroscopy for deep non-invasive medical diagnosis. J. Biophotonics **6**(1), 7–19 (2013). (Review Article)
40. W. Wang, J. Zhao, M. Short, H. Zeng, Real-time in vivo cancer diagnosis using Raman spectroscopy. J. Biophotonics **8**(7), 527–545 (2015). (Review Article)
41. K.W. Kho, C.Y. Fu, U.S. Dinish, M. Olivo, Clinical SERS: are we there yet? J. Biophotonics **4**(10), 667–684 (2011). (Review Article)
42. D. Cialla, A. Maerz, R. Boehme, F. Theil, K. Weber, M. Schmitt, J. Popp, Surface-enhanced Raman spectroscopy (SERS): progress and trends. Anal. Bioanal. Chem. **403**(1), 27–54 (2012)
43. U.S. Dinish, G. Balasundaram, Y.T. Chang, M. Olivo, Sensitive multiplex detection of serological liver cancer biomarkers using SERS-active photonic crystal fiber probe. J. Biophotonics **7**(11–12), 956–965 (2014)
44. C. Yuen, Q. Liu, Towards in vivo intradermal surface enhanced Raman scattering (SERS) measurements: silver coated microneedle based SERS probe. J. Biophotonics **7**(9), 683–689 (2014)

45. A. Shiohara, Y. Wang, L.M. Liz-Marzan, Recent approaches toward creation of hot spots for SERS detection. J. Photochem. Photobiol. C: Photochem. Rev. **21**, 2–25 (2014). (Review Article)
46. U.S. Dinish, G. Balasundaram, Y.T. Chang, M. Olivo, Actively targeted in vivo multiplex detection of intrinsic cancer biomarkers using biocompatible SERS nanotags. Sci. Rep. **4**, 4075 (2014)
47. G.S. He, *Nonlinear Optics and Photonics* (Oxford University Press, Oxford, 2015)
48. H. Tu, S.A. Boppart, Coherent anti-Stokes Raman scattering microscopy: overcoming technical barriers for clinical translation. J. Biophotonics **7**(1–2), 9–22 (2014). (Review article)
49. A.F. Pegoraro, A.D. Slepkov, A. Ridsdale, D.J. Moffatt, A. Stolow, Hyperspectral multimodal CARS microscopy in the fingerprint region. J. Biophotonics **7**(1–2), 49–58 (2014)
50. R. Pecora (ed.), *Dynamic Light Scattering: Applications of Photon Correlation Spectroscopy* (Springer, New York, 1985)
51. M. Plewicki, R. Levis, Femtosecond stimulated Raman spectroscopy of methanol and acetone in a noncollinear geometry using a supercontinuum probe. J. Opt. Soc. Am. B **25**(10), 1714–1719 (2008)
52. F.-K. Lu, M. Ji, D. Fu, X. Ni, C.W. Freudiger, G. Holtom, X.S. Xie, Multicolor stimulated Raman scattering microscopy. Mol. Phys. **110**(15–16), 1927–1932 (2012)
53. C.W. Freudiger, W. Yang, G.R. Holton, N. Peyghambarian, X.S. Xie, K.Q. Kieu, Stimulated Raman scattering microscopy with a robust fibre laser source. Nat. Photonics **8**(2), 153–159 (2014)
54. M. Filella, J. Zhang, M.E. Newman, J. Buffle, Analytical applications of photon correlation spectroscopy for size distribution measurements of natural colloidal suspensions: capabilities and limitations. Aquat. Colloid Surf. Chem. **120**(1–3), 27–46 (1997)
55. W. Tscharnuter, Photon correlation spectroscopy in particle sizing, in *Encyclopedia of Analytical Chemistry*, ed. by R.A. Meyers (Wiley, New York, 2013)
56. P.R. Griffiths, J.A. de Haseth, *Fourier Transform Infrared Spectrometry*, 2nd edn. (Wiley, Hoboken, NJ, 2007)
57. C. Hughes, M. Brown, G. Clemens, A. Henderson, G. Monjardez, N.W. Clarke, P. Gardner, Assessing the challenges of Fourier transform infrared spectroscopic analysis of blood serum. J. Biophotonics **7**(3–4), 180–188 (2014)
58. J. Cao, E.S. Ng, D. McNaughton, E.G. Stanley, A.G. Elefanty, M.J. Tobin, P. Heraud, Fourier transform infrared microspectroscopy reveals unique phenotypes for human embryonic and induced pluripotent stem cell lines and their progeny. J. Biophotonics **7**(10), 767–781 (2014)
59. G. Scarcelli, S.H. Yun, Confocal Brillouin microscopy for three-dimensional mechanical imaging. Nat. Photonics **2**(1), 39–43 (2008)
60. S. Reiß, G. Burau, O. Stachs, R. Guthoff, H. Stolz, Spatially resolved Brillouin spectroscopy to determine the rheological properties of the eye lens. Biomed. Opt. Express **2**(8), 2144–2159 (2011)
61. Z. Steelman, Z. Meng, A.J. Traverso, V.V. Yakovlev, Brillouin spectroscopy as a new method of screening for increased CSF total protein during bacterial meningitis. J. Biophoton. **8**(5), 408–414 (2015)
62. Z. Meng, V.V. Yakovlev, Brillouin spectroscopy characterizes microscopic viscoelasticity associated with skin injury. In: Proceedings of SPIE 9321, paper 93210C, Photonics West, San Francisco, 5 Mar 2015

Chapter 10
Optical Imaging Procedures

Abstract Diverse optical imaging procedures have been developed and applied successfully to biophotonics in research laboratories and clinical settings during the past several decades. Technologies that have contributed to these successes include advances in lasers and photodetectors, miniaturization of optical probes and their associated instrumentation, and development of high-speed signal processing techniques such as advanced computations in image reconstructions, computer vision and computer-aided diagnosis, machine learning, and 3-D visualizations. This chapter expands on the microscopic and spectroscopic technologies described in the previous two chapters by addressing photonics-based imaging procedures such as optical coherence tomography, miniaturized endoscopic processes, laser speckle imaging, optical coherence elastography, photoacoustic tomography, and hyperspectral imaging.

The discipline of medical imaging is widely used for the non-invasive imaging and diagnoses of diseases and health impairments in humans. The image resolutions and tissue penetration depths of some medical imaging techniques are shown in the schematic in Fig. 10.1. Common non-optical techniques such as ultrasound, high-resolution computed tomography (HRCT), and magnetic resonance imaging (MRI) can penetrate deeply into the body but have large resolutions ranging from 150 to 1 mm. Imaging methods that have a lower penetration depth but which yield much finer resolutions can be obtained with optical imaging techniques such as confocal microscopy procedures, photoacoustic tomography, and various types of optical coherence tomography.

Diverse optical imaging procedures have been developed and applied successfully to biophotonics in research laboratories and clinical settings during the past several decades [1–3]. Technologies that have contributed to these successes include advances in lasers and photodetectors, miniaturization of optical probes and their associated instrumentation, and development of high-speed signal processing techniques such as advanced 3-dimensional (3D) image reconstructions, computer-aided diagnosis of imaging data, and software-based imaging data acquisition and display. In addition to the advances in microscopic and spectroscopic

G. Keiser, *Biophotonics*, Graduate Texts in Physics,
DOI 10.1007/978-981-10-0945-7_10

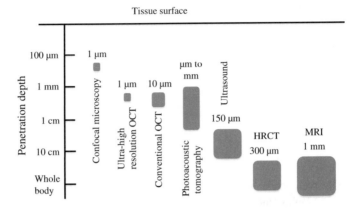

Fig. 10.1 Tissue penetration depth and resolution of some common medical imaging techniques

technologies described in the previous two chapters, photonics-based methodologies such as optical coherence tomography, miniaturized endoscopic processes, laser speckle imaging, optical coherence elastography, photoacoustic tomography, and hyperspectral imaging have emerged in biophotonics applications. These technologies are the topics of this chapter.

10.1   Optical Coherence Tomography

Optical coherence tomography (OCT) is an optical imaging modality that can capture real-time, 3D images of biological tissues and materials in vivo and non-invasively with a 1–15-μm resolution depending on the specific OCT method that is used. OCT functions by measuring the intensity and time-of-flight information collected from backscattered light coming from different points within a tissue sample, and then uses that information to reconstruct 3D depth-resolved images up to 2–3 mm deep in soft tissue and at least 24 mm deep for the more transparent eye tissue. This imaging technique has been used in a wide range of preclinical and clinical applications in fields such as ophthalmology (its initial application), cardiology, dentistry, dermatology, gastroenterology, oncology, and otolaryngology [4–11].

When reading the literature for creating 3D images with OCT and other imaging techniques, one often sees the terms A-scan, B-scan, and C-scan. These are illustrated in Fig. 10.2. The term *A-scan* or *A-line scan* refers to the use of a camera that captures one line of depth information (e.g., backscattered reflectance as a function of depth) for each readout cycle. This scan is done along an illumination axis and is known as an *axial-line scan* or an *A-line image*. In the following discussions, the illumination axis will be designated by the z-axis. Making a series of A-line scans by means of a lateral scan (for example, along a y-direction) forms a

Fig. 10.2 Concept of
A-scans (1D), B-scans (2D),
and C-scans (3D)

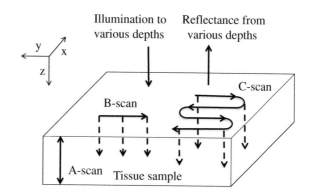

two-dimensional B-scan. A series of cross-sectional B-scans at different points along the x-direction is referred to as a *C-line scan*. That is, multiple B-scans along the x-axis form a three-dimensional C-scan image.

Example 10.1 Consider a scanning system that uses a rotating mirror that can scan at an 8 kHz rate over several mm. How much time is needed to acquire 500 A-scans (columns of data)?

Solution: The time needed to acquire 500 A-scans is

$$t_{scan} = (\text{number of scans})/(\text{scan rate}) = (500\,\text{scans})/(8000\,\text{scans/s})$$
$$= 62.5\,\text{ms}$$

As noted in Chap. 1 (Fig. 1.6), a spectral range with low optical absorption loss exists between about 600 and 1600 nm. In this range the imaging depth with light is several millimeters into nontransparent tissue. Thus OCT is used in this operating region because here light attenuation is more dependent on scattering processes than on absorption processes. The key operational characteristic of OCT is that ranging-based optical imaging can be carried out in biological tissue because a portion of the incident light is reflected by each different material interface whenever there is a change in refractive index in the biological tissue. Low-coherence light from a super-luminescent diode or other broadband optical source is needed to measure the time delay of the reflected light.

Several variations of the OCT process have been devised. The original setup is known as *time domain OCT* (TD-OCT). The TD-OCT method uses a low coherence interferometer with a scanning reference arm, which detects echo delay signals from tissue in the time domain. This method is described in Sect. 10.1.1. The second major OCT category performs detection in the Fourier domain by measuring the interferometric spectrum. The advantages of *Fourier domain OCT* (FD-OCT) are an increased sensitivity and a higher signal-to-noise ratio. This results because the Fourier domain detection principle measures all the echoes of light from different cross-sectional planes in the tissue sample simultaneously, thereby yielding an increase in imaging speed. There are two types of FD-OCT methods, as described in Sects. 10.1.2 and 10.1.3, respectively. The first is called *spectral domain OCT*, which uses a spectrometer and a line-scanning CCD camera. The other method is called *swept source OCT*, which uses a frequency-swept light source having a narrow bandwidth.

10.1.1 Time Domain OCT

An OCT system can be implemented either in free space with a dichroic beam-splitter or with optical fiber channels by using optical fiber couplers. The basic optical fiber OCT system normally is based on low-coherence interferometry using a Michelson interferometer setup as shown in Fig. 10.3. In such setups, the use of optical fiber technology enables the OCT system to be compact and portable.

In the TD-OCT procedure shown in Fig. 10.3, first light pulses from a broad-band optical source are divided in half by a 3-dB optical fiber coupler or splitter. A key feature of a broadband light source is that its low coherence length allows a well-defined imaging position in the tissue sample. In addition, it is necessary to use single-mode fibers (SMFs) in order to achieve a high image resolution. Thus the two pulse halves from the optical coupler outputs are sent separately through SMFs to the reference arm and to the sample arm. Returning reflections of the pulses from the mirror at the end of the reference arm and from the sample are recombined by means of the optical coupler and then are directed to the photodetector. Because low-coherence light from a broadband source is used, constructive interference of the light at the photodetector only occurs when the optical path lengths between the two arms of the interferometer lie within the coherence length of the light source.

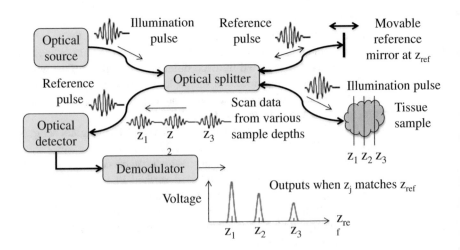

Fig. 10.3 A typical time-domain (TD) OCT imaging setup

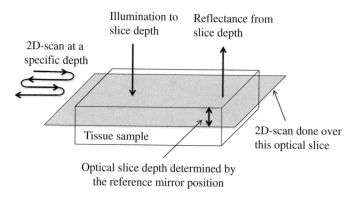

Fig. 10.4 Various positions of the OCT reference mirror yield 2D scans at different depths

As is described below, this means that the two returning pulse envelopes must overlap at the photodetector in order to have a maximum received optical signal. By scanning across the face of the reference arm mirror, a single 2D-scan from a slice of reflection data at a specific axial depth is acquired from the sample. This is illustrated in Fig. 10.4. In time domain OCT the path length of the reference arm is varied longitudinally in time. Thus once a particular two-dimensional scan has been obtained, then by slightly moving the position of the reference mirror another 2D-scan slice through the tissue sample is obtained at a different axial depth.

Example 10.2 Consider a setup for a TD-OCT A-line scan of a human eye. What are the characteristics of an A-line scan through the center of the eye?

Solution: The A-line scan will show large reflectance peaks at interfaces between materials having different refractive indices. Example boundaries are at the air-cornea, cornea-lens, lens-vitreous humor, and vitreous humor-retina interfaces, as is indicated in Fig. 10.5.

The principles of OCT operation can be understood by examining the interactions of the electromagnetic waves returning from the sample and the reference arms. For simplicity, first the approximation can be made that the source is monochromatic and emits plane waves. Thus the incident wave E_{source} from the source propagating in the z direction is written as

$$E_{source} = E_0 exp[i(kz - \omega t)] \qquad (10.1)$$

where $k = 2\pi/\lambda$ with λ being the wavelength of the source, ω is the angular frequency of the electric field, and E_0 is the amplitude of the wave with the subscript 0 indicating that a monochromatic source is used.

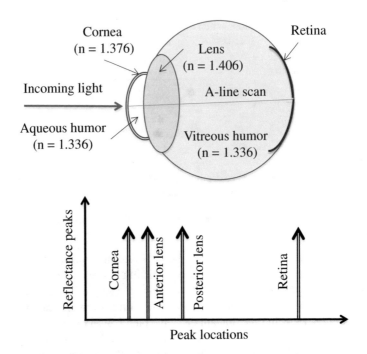

Fig. 10.5 Reflectance peaks for an A-line scan of the eye

After this wave passes through the optical coupler, the returning waves are

$$E_r = E_{r0}\exp[i(kz_r - \omega t)] \tag{10.2}$$

from the reference arm and

$$E_s = E_{s0}\exp[i(kz_s - \omega t)] \tag{10.3}$$

from the sample arm. Here z_r and z_s are the path lengths from the optical coupler to the reference mirror and to a specific depth in the sample, respectively. The monochromatic wave amplitudes E_{r0} from the mirror and E_{s0} from the sample are quite different, because only a small amount of light returns from the sample, whereas the mirror reflects most of the light incident on it. The total electric field E_{total} at the detector in Fig. 10.3 is given by the sum of the two returning waves as

$$E_{total} = E_{r0}\exp[i(kz_r - \omega t)] + E_{s0}\exp[i(kz_s - \omega t)] \tag{10.4}$$

The intensity I_D at the detector of the light beams coming from the two arms of the interferometer is given by the square of the electric field amplitude

$$I_D = \frac{n}{2\mu_0 c} E_{total} E_{total}^* = \frac{n}{2\mu_0 c} \left[E_{s0}^2 + E_{r0}^2 + 2E_{s0}E_{r0}\cos 2(kz_r - kz_s) \right] \qquad (10.5)$$

Here n is the refractive index of the transmission medium (e.g., $n \approx 1.5$ for glass), μ_0 is the magnetic permeability, and the asterisk (*) denotes the complex conjugate. Letting the intensity from the sample be

$$I_{s0} = \frac{n}{2\mu_0 c} E_{s0}^2 \qquad (10.6)$$

and letting the intensity from the reference mirror be

$$I_{r0} = \frac{n}{2\mu_0 c} E_{r0}^2 \qquad (10.7)$$

then the intensity at the detector can be expressed as

$$I_D = I_{s0} + I_{r0} + 2\sqrt{I_{s0}I_{r0}}\cos 2\left[\frac{2\pi}{\lambda}(z_r - z_s) \right] \qquad (10.8)$$

where $\lambda = 2\pi/k$ is the wavelength of the optical source. The factor $z_r - z_s$ is the path length difference between the reference and the sample arms.

Alternatively, Eq. (10.8) can be written in terms of the optical power received at the detector as

$$P_D = P_{s0} + P_{r0} + 2\sqrt{P_{s0}P_{r0}}\cos 2\left[\frac{2\pi}{\lambda}(z_r - z_s) \right] \qquad (10.9)$$

where P_{r0} and P_{s0} are the monochromatic powers from the reference and sample arms, respectively. In Eqs. (10.8) and (10.9) the first two terms are constants for particular reflection values, whereas the third term represents the wave interference. When the factor in square brackets in the cosine term is 0 or a multiple of $\pm 2\pi$ the cosine value is a maximum of 1, thereby yielding a maximum optical power at the detector.

For an actual OCT setup, the source is not monochromatic. In fact, as described below, to increase the axial resolution it is desirable to have a coherent light source with a wide spectral width. In this case the total optical power falling on the detector is found by integrating over the entire spectrum of the light source, thereby yielding

$$P_D = P_s + P_r + 2\sqrt{P_s P_r}\,\mathrm{sinc}\left[\frac{\pi(z_1 - z_s)}{l_c} \right]\cos 2\left[\frac{2\pi}{\lambda}(z_r - z_s) \right] \qquad (10.10)$$

Here P_r and P_s are the wideband spectral optical powers from the reference and sample arms, respectively, λ_0 is the center wavelength of the optical source, l_c is the source coherence length, and the *sinc function* is defined by sinc $x = (\sin x)/x$.

Analogous to the monochromatic source case, the rightmost term in square brackets in Eq. (10.10) has a peak value when the length z_r of the reference arm matches the distance z_s to the backscattering interface at a particular sample depth.

The coherence length l_c determines the broadness of the peak interference signal. In order to have a narrow width of the interference signal, which will allow a well-defined position of the backscattering interface, it is desirable for the light source to have a short coherence length or, equivalently, a broad spectral bandwidth. Thus, by examining the various peaks of the interference envelopes during a scan of the sample, the location and backscattering strength of interfaces within the sample can be determined.

> **Example 10.3** Recall from Eq. (2.36) that the coherence length for a spectral width $\Delta\lambda$ at a center wavelength λ_0 is given by $l_c = [(4 \ln 2)/\pi](\lambda_0)^2/\Delta\lambda$. Suppose an OCT setup for ophthalmology tests uses the following super-luminescent diodes: (1) an 850-nm peak wavelength with a 50-nm spectral bandwidth and (2) a 1310-nm peak wavelength with an 85-nm spectral bandwidth for the posterior segment and the anterior segment of the eye, respectively. What are the coherence lengths of these two sources?
>
> **Solution**: From Eq. (2.36) the coherence length at 850 nm is
>
> $$l_c(850) = [(4 \ln 2)/\pi](850 \, \text{nm})^2/(50 \, \text{nm}) = 12.7 \, \mu\text{m}$$
>
> At 1310 nm, $l_c(1310) = [(4 \ln 2)/\pi] (1310 \, \text{nm})^2/(85 \, \text{nm}) = 17.8 \, \mu\text{m}$

If a pulsed light source is used, an interference signal can only occur at the detector when the pulses returning from the two arms of the interferometer arrive at the detector at the same time. Therefore, as the spectral bandwidth becomes broader, the interference pulses become sharper, thereby enabling a more precise location of the backscattering interface being examined for a particular setting of the reference mirror.

Several important characteristics of an OCT system can be derived if the sample arm is treated as a reflection-mode scanning confocal microscope. In this case the SMF acts as a pinhole aperture for both the illumination and the collection of light from the sample. Then the expressions for both the lateral and the axial intensity distributions are given by the expressions for an ideal confocal microscope with a small pinhole aperture. A summary of the resulting expressions is given in Fig. 10.6 in which the OCT system is taken to be cylindrically symmetric [6].

The image resolution is an important parameter for characterizing an OCT system. The lateral and axial resolutions of OCT are decoupled from each other and thus can be examined independently. The *lateral resolution* is a function of the resolving power of the imaging optics, as is discussed in Sect. 8.2. The *axial resolution* Δz is derived from the coherence length of the source. If an OCT system uses a broadband light source with a center wavelength λ_0, the standard equation for the axial resolution is

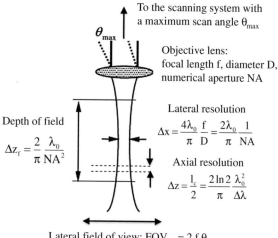

Fig. 10.6 Definitions of various resolutions for an OCT observation (Modified with permission from Izatt and Choma [6])

To the scanning system with a maximum scan angle θ_{max}

θ_{max}

Objective lens: focal length f, diameter D, numerical aperture NA

Depth of field

$$\Delta z_f = \frac{2}{\pi} \frac{\lambda_0}{NA^2}$$

Lateral resolution

$$\Delta x = \frac{4\lambda_0}{\pi} \frac{f}{D} = \frac{2\lambda_0}{\pi} \frac{1}{NA}$$

Axial resolution

$$\Delta z = \frac{l_c}{2} = \frac{2 \ln 2}{\pi} \frac{\lambda_0^2}{\Delta \lambda}$$

Lateral field of view: $FOV_{lat} = 2 f \theta_{max}$

$$\Delta z = \frac{l_c}{2} = \frac{2 \ln 2}{\pi} \frac{\lambda_0^2}{\Delta \lambda} \qquad (10.11)$$

where $\Delta \lambda$ is the FWHM spectral width of the probing source. Thus, the axial resolution is directly proportional to the square of the source central wavelength λ_0 and inversely proportional to the source spectral bandwidth $\Delta \lambda$. Note that *the axial resolution in air is equal to half of the coherence length of the light source* because of the round-trip propagation of the reference and sample beams. From Eq. (10.11) it can be seen that optical sources with broad spectral bandwidths are desired in order to achieve a high axial resolution. Higher resolutions also could be achieved by decreasing the center wavelength from the standard 1300-nm value. A drawback to this idea is that shorter wavelengths are more highly scattered in biological tissue, which results in less penetration of imaging light into the tissue.

Example 10.4 Consider the two super-luminescent diodes described in Example 10.3. What is the axial resolution for each of these sources?

Solution: Using Eq. (10.11) the axial resolution for the 850-nm source is

$$\Delta z = \frac{2 \ln 2}{\pi} \frac{(850)^2}{50} = 6.38 \ \mu m$$

For the 1310-nm source, the axial resolution is

$$\Delta z = \frac{2 \ln 2}{\pi} \frac{(1310)^2}{85} = 8.91 \ \mu m$$

The lateral point-spread function at the full-width half-maximum power in the focal plane of an OCT system defines the standard *confocal lateral resolution* or *transverse resolution* Δx. This parameter commonly is specified as the focal diameter of the incident sample beam and depends only on the wavelength and the characteristics of the objective lens optics. Thus the transverse resolution is given as

$$\Delta x = \frac{4\lambda_0}{\pi} \frac{f}{D} = \frac{2\lambda_0}{\pi} \frac{1}{NA} \tag{10.12}$$

where f is the focal length of the objective lens, D is either the diameter of the beam on the lens or the diameter of the lens (whichever is smaller), and NA is the numerical aperture of the objective defined by Eq. (8.1).

The full width at half-maximum power of the confocal axial response function gives the axial range (depth range) within which the transverse resolution is fairly constant. This range is defined as the *depth of focus* Δz_f of the OCT system and is given as

$$\Delta z_f = \frac{2}{\pi} \frac{\lambda_0}{NA^2} \tag{10.13}$$

Equations (10.13) and (10.12) show that as the numerical aperture of a conventional OCT system increases, the transverse resolution improves as it becomes smaller but the depth of focus deteriorates. Recent developments in the use of an adaptive optics method called depth-encoded synthetic aperture OCT have resulted in improvements in the depth of focus [11].

The *lateral field of view* FOV_{lat} is given by

$$FOV_{lat} = 2 f \theta_{max} \tag{10.14}$$

where f designates the focal length of the objective lens and θ_{max} is the maximum one-sided scan angle of a rotating scanning mirror.

Example 10.5 Consider an OCT system in which the objective lens has a numerical aperture NA = 0.26. (a) What is the lateral resolution at a wavelength of 1310 nm? (b) What is the depth of focus for this setup?

Solution: (a) Using Eq. (10.12) yields the following lateral resolution

$$\Delta x = \frac{2}{\pi} \frac{\lambda_0}{NA} = 0.64 \frac{1310\,nm}{0.26} = 3.22\,\mu m$$

(b) Using Eq. (10.13) yields the following depth of focus

$$\Delta z_f = \frac{2}{\pi} \frac{\lambda_0}{NA^2} = \frac{2\,(1.31\,\mu m)}{\pi\,(0.26)^2} = 12.3\,\mu m$$

10.1.2   Spectral Domain OCT

Whereas time domain OCT uses a low-coherence interferometer with a scanning reference delay arm to detect echo delay signals from tissue layers in the time domain, *spectral domain OCT* (SD-OCT) performs detection in the Fourier domain by measuring the interferometer spectrum [12, 13]. The advantage of the SD-OCT method is the speed of obtaining an image is increased greatly, because data from all reflection points along a given axial measurement are obtained simultaneously.

The basic setup for SD-OCT is shown in Fig. 10.7. In this procedure the optical source is similar to that used in TD-OCT, but the mirror in the reference arm remains stationary. In SD-OCT the spectral components of the recombined beams for a specific sample illumination point are dispersed onto an optical detector array by a spectrometer. This process forms a spectral interferogram (interference pattern), which simultaneously yields reflectance information from all backscattering points as a function of depth along a given axial pass through the sample. The reflection signals from two or more interfaces may be collected simultaneously because these signals result from different spectral oscillation components. Similar to TD-OCT, transverse scanning across the sample then produces 2D or 3D images.

10.1.3   Swept Source OCT

Swept source OCT (SS-OCT) uses a narrowband light source that can be tuned rapidly to measure spectral oscillations at evenly spaced wavenumbers [14, 15]. This method sometimes is called *optical frequency domain reflectometry, wavelength tuning interferometry,* or *optical frequency domain imaging.* An example setup for SS-OCT is shown in Fig. 10.8.

Here the light source is a tunable laser that can be swept at a nominally 20 kHz sweep rate across a spectral bandwidth of 120–140 nm. Suppose the SS-OCT system acquires signals from M evenly spaced wavenumbers with a separation $\delta k = \Delta k/M$, where Δk is the total optical bandwidth over which the source is swept. Typical values of M range from 100 to 1000. Then the SS-OCT system will have a *scan depth range* (an A-line range) z_{depth} given by

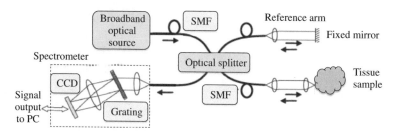

Fig. 10.7 Basic setup for a spectral domain OCT system

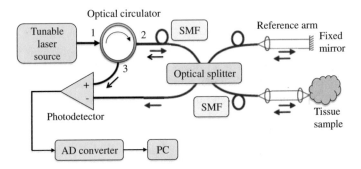

Fig. 10.8 Schematic of the setup for a swept source OCT system

$$z_{\text{depth}} = \frac{\pi}{2\delta k} = \frac{M\pi}{2\Delta k} = \frac{M}{4}\frac{\lambda^2}{\Delta\lambda} = \frac{M}{2}l_c \qquad (10.15)$$

where l_c is the coherence length of the source and $\Delta k = 2\pi\Delta\lambda/\lambda^2$.
The A-scan *axial resolution* Δz is

$$\Delta z_{\text{gaussian}} = \frac{4\ln 2}{\Delta k_{\text{FWHM}}} = \frac{4\ln 2}{2\pi\Delta\lambda/\lambda^2} \qquad (10.16)$$

when using a Gaussian source that has a full-width half-maximum bandwidth k_{FWHM}. For a rectangular source the axial resolution is

$$z_{\text{rect}} = \frac{\pi}{\Delta k} \qquad (10.17)$$

Example 10.6 Consider an SS-OCT system for which M = 220, λ = 1308 nm, and $\Delta\lambda$ = 87 nm. (a) What is the scan depth range? (b) What is the axial resolution when using a Gaussian source?

Solution: (a) Using Eq. (10.15) yields the following scan depth range

$$z_{\text{depth}} = \frac{M}{4}\frac{\lambda^2}{\Delta\lambda} = \frac{220}{4}\frac{(1308\,\text{nm})^2}{87\,\text{nm}} = 1.08\,\text{mm}$$

(b) Using Eq. (10.16) yields the following axial resolution for a Gaussian source

$$\Delta z_{\text{gaussian}} = \frac{4\ln 2}{2\pi\Delta\lambda/\lambda^2} = 4\ln 2\frac{(1308\,\text{nm})^2}{2\pi(87\,\text{nm})} = 8.7\,\mu\text{m}$$

10.2 Endoscopy

The field of *endoscopy* uses a photonics-based instrument containing one or more optical fibers that allow a clinician to examine the inside of a hollow organ or body cavity. The word *endoscope* is a generic name for this instrument. The endoscope type and the endoscopic procedure generally are named based on the body region being examined or treated, as shown in Table 10.1. For example, laparoscopy deals with operations within the abdominal or pelvic cavities, rhinoscopy is used for examinations of the nasal cavity, arthroscopy refers to imaging procedures for joints, and bronchoscopy visualizes the inside of the respiratory tract.

The designs of different endoscope types are continually improving to yield more information from biological tissues, such as acquiring images in colors, achieving finer image resolution, or observing molecular or metabolic functions. New optical fiber types and miniaturized cameras have helped to enable these capabilities and have resulted in smaller, faster-performing devices. This section first describes some endoscopic applications using different optical fiber designs, then gives examples of endoscopic or minimally invasive surgery techniques, and concludes with a description of a procedure called tethered capsule endomicroscopy.

Example 10.7 The laparoscope is a widely used endoscope for examining the peritoneal cavity. (a) What is the peritoneal cavity? (b) What is a generic structure of a laparoscope?

Table 10.1 Endoscope type examples and the endoscopic procedures

Endoscope type	Endoscope application
Angioscope	The angioscope is used to observe blood vessels
Arthroscope	The arthroscope is used for examining the interior of a joint
Bronchoscope	The bronchoscope is used to examine the respiratory tract
Colonoscope	The colonoscope is used to examine the lower part of the bowel
Cystoscope	The cystoscope is used for examining the urinary tract
Encephaloscope	The encephaloscope is used for examining cavities in the brain
Esophagoscope	The esophagoscope is used to examine the inside of the esophagus
Gastroscope	The gastroscope is used for inspecting the interior of the stomach
Laparoscope	The laparoscope is used to examine the peritoneal cavity
Laryngoscope	The laryngoscope is used to examine the larynx
Mediastinoscope	The mediastinoscope examines the area separating the two lungs
Nephroscope	The nephroscope is used to examine the kidneys
Proctoscope	The proctoscope is used for inspecting the rectum
Rhinoscope	The rhinoscope is used in nasal examinations
Thoracoscope	The thoracoscope is used for examining the chest cavity

Solution: (a) The peritoneal cavity contains the pelvis, stomach, spleen, gall bladder, liver, and the intestines. The fluid in the peritoneal cavity lubricates the tissues and organs in the abdomen during digestion. Excess fluid in the peritoneal cavity (for example, from a liver disease) can cause the abdomen to swell, which can lead to breathing difficulties. (b) The most common laparoscope is a rigid tube with a diameter of 5 or 10 mm that has a working insertion depth of approximately 300 mm. The distal end contains imaging lenses and the viewing end has a simple eyepiece.

10.2.1 Basic Endoscopy

The following units make up a basic endoscope system [16–21]:

- A flexible tube that encapsulates one or more optical fibers for illumination and viewing functions. As Fig. 10.9 illustrates, the encapsulating endoscope tube also can contain miniature air, water, and suction or biopsy tubes, plus wires for tip control.
- An external light source that is coupled to the optical fibers in the encapsulating tube for illuminating the organ, tissue area, or body cavity being diagnosed
- A lens system which collects reflected or fluorescing light from the diagnostic site for transmission via optical fibers to a viewer
- A viewing mechanism such as a simple eyepiece, a monitor, or a camera

An endoscope also can contain capillaries for collecting tissue specimens, miniature manipulative tools for microsurgery, or more optical fibers for therapeutic purposes. Traditionally endoscopes have used a small flexible optical fiber bundle to transmit light to a diagnostic area and a coherent fiber bundle for image retrieval.

Although the use of fiber bundles results in a noninvasive exploratory instrument, the number of individual fibers in the bundle and their characteristics limit the image resolution and can increase the outer diameter of the instrument [21]. The

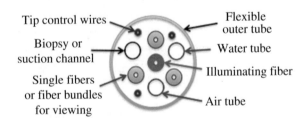

Fig. 10.9 Cross-sectional view of a generic endoscope tube showing various possible optical fibers, miniature tubes, and wires for different functions (*J. Biomed. Opt.* 19(8), 080902 (Aug 28, 2014). doi:10.1117/1.JBO.19.8.080902)

larger size can result in limited flexibility of the endoscope tube, which may require patient sedation during use of the endoscope. Consequently, during the evolution to modern endoscopes, a replacement for the fiber bundle was sought using single optical fibers, such as SMF, MMF, DCF, or hollow-core PCF, to reduce the size and to increase the flexibility of endoscopes, thereby enabling safer, faster, and less expensive clinical procedures. In some endoscope designs a single fiber is used for both delivery of the incident light and image transport. Other endoscope designs use a central fiber for illuminations and either another individual fiber or a circular array of fibers for image retrieval.

A major limitation when using MMFs is the temporal variation of the modal distributions in the fiber, which can lead to image flickering [21]. Actually this situation occurs in all imaging systems that employ MMFs both for the delivery and the collection of light. SMFs offer better resolution and a decrease in speckle noise compared with MMFs. Compared to conventional SMF and MMF, the DCF types described in Sect. 4.3 can support single-mode light transmission through the core for illumination of a target area and multimode image transfer (consisting of partially incoherent light reflected from the sample) through the inner cladding back to an imaging instrument. A DCF thus achieves high-resolution imaging, reduced speckle effects, and a large depth of field resulting from the larger collection diameter of the cladding.

In conjunction with evaluating the potentials of using single fibers are innovative optical designs, miniature image scanning mechanisms, and specific molecular probes [21]. The aims of the combined fiber plus the technology advances are to provide images with various colors, more depth, and multiple dimensions in order to acquire greater biological details from tissue.

10.2.2 Minimally Invasive Surgery

The traditional use of endoscopes has been to insert them through the natural openings in a body. Another branch of endoscopy is known as endoscopic surgery or *minimally invasive surgery*. This procedure also is referred to as *laparoscopic surgery*. In this surgical discipline, a small incision (nominally 0.5–1.5 cm) is made in the body through which a small plastic tube containing miniaturized surgical tools then is inserted in order to treat a medical condition [21–25]. In addition, as shown in Fig. 10.10 for orthopedic surgery, a separate optical fiber endoscopic probe containing a light source and a miniature video camera normally is inserted through another small incision in the body to monitor and evaluate the treatment procedure. This probe, which is known as an *arthroscope*, can magnify images from inside a joint up to 30 times. The images obtained by the arthroscope typically are transmitted to a TV monitor. This setup gives the orthopedic surgeon a clear and enlarged view of the inside of a joint in order to operate using small surgical tools inserted through separate minor incisions. Table 10.2 lists some of many other kinds of minimally invasive surgery.

Fig. 10.10 Generic setup for a minimally invasive surgery procedure

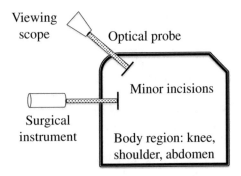

Table 10.2 A selection of minimally invasive surgical procedures

Minimally invasive surgical procedure
Adrenalectomy to remove one or both adrenal glands
Colectomy to remove parts of a diseased colon
Colon and rectal surgery
Ear, nose and throat surgery
Endovascular surgery procedures (e.g., aneurysm repair)
Gallbladder surgery (cholecystectomy) to remove gallstones
Gastroenterologic and general surgery
Heart surgery, for example, to treat heart valve diseases
Hiatal hernia repair to relieve gastroesophageal reflux
Nephrectomy to remove a kidney or for donation
Neurosurgery to treat fluid buildup in the brain
Orthopedic surgery for a wide range of joint injuries and disorders
Splenectomy to remove the spleen
Thoracic surgery: lung and esophageal tumors, cysts, or diseases
Urologic surgery, such as prostate treatment

   Optical fibers used for minimally invasive surgery must be able to withstand tight bends in order to have a light path with a very low bending-induced optical power loss [21]. This characteristic allows the catheter to follow sinuous paths through veins and arteries and around organs and bones as it moves toward its destination. An important parameter is the NA of the fiber. This characteristic influences how tightly the light is confined within the optical fiber core and dictates how much the fiber can be bent before optical bending loss becomes noticeable. A number of high-NA single-mode and multimode fibers are commercially available, some of which can be bent to diameters as small as 3–4 mm with minimum optical loss.

   The overall optical fiber size is another important characteristic [21]. Because the fibers are inside of space-constrained catheters, the outer fiber diameter should be as small as possible without sacrificing light transmission or mechanical durability.

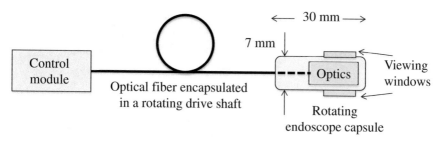

Fig. 10.11 Example schematic of a tethered capsule endoscope system

For this application, 80-μm cladding-diameter SMFs and reduced-cladding MMFs are available from various suppliers.

10.2.3   Tethered Capsule Endomicroscopy

A recent development in endoscopy for the digestive tract involves using a swallowable capsule that is connected with an optical fiber link to a measurement instrument outside the body. This setup enables the use of high-resolution confocal microscopic techniques inside the body to diagnose and monitor diseases in the digestive tract. The method is known as *tethered capsule endomicroscopy*. The clinical procedure involves swallowing a nominally 7-mm-diameter and 30-mm long rotating viewing capsule that is tethered to an optical fiber link and captures cross-sectional microscopic images as it travels through the digestive tract [26–28]. Images that are obtained have a nominal 30-μm lateral resolution and a 7-μm axial resolution. As illustrated in Fig. 10.11, typically the tethered capsule endoscope system consists of (a) an external control module that contains a light source, a photodetector, capsule manipulation controls, and signal-processing elements, (b) a tether consisting of a flexible drive shaft for rotating the capsule, (c) a high-NA SMF contained within the drive shaft, and (d) the rotating capsule that contains the viewing optics.

10.3   Laser Speckle Imaging

Whenever coherent laser light illuminates a diffuse or random scattering medium, the light that is scattered from different places on the illuminated surface will travel through a distribution of distances. This distance distribution will give rise to granular constructive and destructive interference patterns in the scattered light distribution [29–34]. If a camera or photodetector array is used to observe the scattered light, a randomly varying intensity pattern known as *speckle* will be seen. As shown in Fig. 10.12, the granular speckle patterns look like sheets of moving

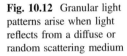

Fig. 10.12 Granular light patterns arise when light reflects from a diffuse or random scattering medium

sand grains. Fluctuations in the interference patterns will be observed if the scattering particles are moving, which will cause intensity variations at the photodetector. These intensity variations result in a blurring of the speckle pattern.

Note that the formation of the laser speckle pattern is a random process, so the speckles need to be described statistically. The goal of laser speckle imaging is to determine the relations between the statistical characteristics of the detected intensity fluctuations and the time-varying and structural properties of the probed tissue. Because the temporal and spatial statistics of this speckle pattern yield information about the motion of the scattering particles, speckle-imaging techniques have been used for procedures such as monitoring blood flow in capillaries, analyzing blood samples in vitro, and imaging of blood coagulation. Two-dimensional maps of blood flow can be obtained by examining the degree of speckle pattern blurring that results from blood flow. In those regions where the blood flow is faster, the intensity fluctuations of the speckle pattern are more rapid, which results in a more blurred speckle pattern. The exposure time when using a CCD camera for these measurements is typically 1–10 ms.

Whereas speckle patterns can be used advantageously for measurement procedures such as blood flow, in other cases speckles can lead to an unwanted blurring of an image. For example, because OCT is a coherent imaging modality, speckles can corrupt its images by giving them a grainy or irregularly spotted appearance. On the other hand, speckle is fundamental to OCT image formation. Thus, attempts to suppress or eliminate speckles results in a decrease in the OCT image resolution. A discussion of the effects of speckle on OCT is beyond the scope of this chapter, but can be found in the literature [29–33].

An important parameter for quantifying the blurring of the speckles is the *speckle contrast*. This parameter is designated by K and can be expressed as a function of the exposure time T of the camera and the *speckle correlation time* τ_c, which is inversely proportional to the local velocity of the scattering particles. The speckle contrast is defined as the ratio of the standard deviation σ of the speckle intensity to the mean intensity $\langle I \rangle$ and in terms of the ratio T/τ_c is given by

$$K = \frac{\sigma}{\langle I \rangle} = \left[\beta \frac{\exp(-2T/\tau_c) - 1 + 2T/\tau_c}{2(T/\tau_c)^2} \right]^{1/2} \tag{10.18}$$

Here the coefficient β ($0 \le \beta \le 1$) is determined by the ratio of speckle size and pixel size at the detector and by the polarization characteristics and coherence properties of the incident light. The theoretical limits of K range from 0 to 1. A spatial speckle contrast value of 1 means that there is no blurring of the speckle pattern, which indicates that there is no motion. For particle velocities corresponding to values of τ_c less than about 0.04 T, the speckle contrast is very low because the scattering particles are moving at such a high speed that all the speckles have become blurred.

Example 10.8 Consider a laser speckle imaging systems operating at 532 nm. (a) Suppose the exposure time of the camera is T = 10 ms. If the speckle coefficient β = 0.5, what is the speckle contrast factor K if the speckle correlation time is τ_c = 100 μs? (b) What is the value of K if τ_c = 20 ms?

Solution: (a) Using Eq. (10.18) yields the following value for K when the ratio T/τ_c = 10 ms/100 μs = 100

$$K = \left[\beta \frac{\exp(-2T/\tau_c) - 1 + 2T/\tau_c}{2(T/\tau_c)^2} \right]^{1/2} = \left[0.5 \frac{\exp(-200) - 1 + 200}{2(100)^2} \right]^{1/2}$$
$$= 0.071$$

(b) Using Eq. (10.18) yields the following value for K when T/τ_c = 10 ms/20 ms = 0.5

$$K = \left[0.5 \frac{\exp(-1) - 1 + 1}{2(0.5)^2} \right]^{1/2} = 0.61$$

A number of factors must be considered carefully when making measurements with speckle patterns. These issues are described in detail in the literature [29–34]. One factor deals with proper spatial sampling of the speckle pattern. A key point involves the speckle size $\rho_{speckle}$ relative to the camera pixel size ρ_{pixel}. When the speckle pattern is viewed by a camera, the minimum speckle size will be given by

$$\rho_{speckle} = 1.22(1 + M)(f/\#)\lambda \tag{10.19}$$

where λ is the wavelength of the detected light, M is the image magnification, and f/# is the f number of the system (e.g., f/5.6), which is the ratio of the focal length of a camera lens to the diameter of the aperture being used. To satisfy the Nyquist

Fig. 10.13 The minimum speckle size that can be resolved must be two times larger than the camera pixel size

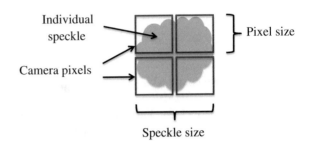

sampling criteria, the minimum speckle size that can be resolved must be two times larger than the camera pixel size ρ_{pixel}, that is, $\rho_{speckle} \geq 2\rho_{pixel}$. This is illustrated in Fig. 10.13.

> **Example 10.9** Consider a speckle imaging system operating at 532 nm in which the magnification of the optics is M = 10. Suppose the camera has f/# = 4.0. (a) What is the minimum speckle size that can be resolved by the camera? (b) What are the requirements for the pixel size?
>
> **Solution:** (a) Using Eq. (10.19) yields the following speckle size
>
> $$\rho_{speckle} = 1.22(1+M)(f/\#)\lambda = 1.22(1+10)(4.0)(532\,nm) = 28.6\,\mu m$$
>
> (b) The maximum pixel size of the camera must be 28.6/2 µm = 14.3 µm.

10.4   Optical Coherence Elastography

The field of *elastography* evolved from the observation that diseases such as edema or dropsy (an excess of watery fluid collecting in the cavities or tissues of the body), fibrosis, cancer, and calcification alter the elastic properties of tissues [35]. The basis of elastography is to measure the mechanical properties of tissue, such as the Young's modulus, by means of optical medical imaging, ultrasound imaging, or magnetic resonance imaging when a mechanical loading is applied. An example of an optical medical imaging procedure is shown in Fig. 10.14. Here a circular piezoelectric transducer puts a mechanical loading on a tissue sample, and the response to the pressure variation is monitored by means of a light beam impinging on the tissue through the center of the transducer. This process gives rise to images called *elastograms*, which are used to examine and monitor the onset and progression of a disease through the temporal variations in mechanical properties of the tissue. The first elastographic methodologies were based on using ultrasound and magnetic resonance imaging. Subsequently, optical elastographic techniques

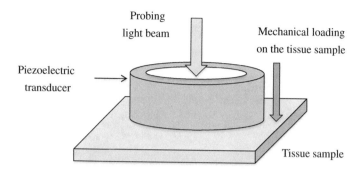

Fig. 10.14 Example of an OCE setup using a circular piezoelectric transducer and a central light beam probe

combined with optical coherence tomography, which is known as *optical coherence elastography* (OCE), were investigated to improve the image resolution [36–41].

The goal of OCE is to find the Young's modulus of a tissue, which will indicate the tissue elasticity (or stiffness). Deviations in the Young's modulus from that of healthy tissue can help identify diseased cells. A challenge in applying OCE is that biological tissues are anisotropic and their response to an external mechanical loading has both viscous and elastic components. Thus elaborate mathematical techniques are needed for elastic modulus imaging.

However, a basic idea of the phenomenological process can be obtained by assuming that a tissue is a uniform linear elastic solid with isotropic mechanical properties. The linearity assumption is usually valid because the strain level applied in elastography is typically less than 10 % of the elastic limit. Thus when a uniform stress σ is applied in the axial direction, a local stress results that is constant throughout the sample tissue and is equal to the applied stress. For a uniaxial load (compression along one axis), the local stress can be used to find the Young's modulus through the simple relationship

$$E = \frac{\sigma}{\varepsilon} \qquad (10.20)$$

where E is the Young's modulus, σ is the axial stress, and ε is the axial strain. This equation shows that stiffer materials, for example, tissues of higher density such as bones, require a higher stress to induce identical deformations and thus have a larger Young's modulus than less dense materials such as skin, arterial walls, or red blood cells. In addition, diseases can modify both the constituent elements and the structure of tissue, which results in variations in mechanical properties between healthy and diseased tissue. The value of the Young's modulus varies over more than eight orders of magnitude ranging from tens to billions of Pascals. This illustrates that a high contrast between different tissue constituents, structures, and whole organs can be obtained from elastograms.

Example 10.10 What is the measurement procedure needed for the case of phase-sensitive OCE?

Solution: Only two OCT scans need to be made in this setup [40]. One scan is made when there is no mechanical load on the tissue sample, whereas the other scan is performed when there is a compressive load resulting from a known applied axial stress σ. The local axial strain ε is then determined from the relationship ε = Δd/Δz, where

$$\Delta d = \Delta \Phi \lambda_0 / (4\pi n)$$

is the change in axial displacement over the depth range Δz. Here ΔΦ is the change in the measured OCT phase between the loaded and unloaded scans, λ_0 is the mean free-space wavelength of the OCT system, and n is the refractive index of the tissue.

A wide variety of mechanical loading methods have been used in various OCE techniques. These methods include static, quasi-static, continuous wave dynamic, or pulsed dynamic loadings. The loadings can be applied to the tissue either internally or externally. In each case, the displacement resulting from the loading is measured and a mathematical model of mechanical deformation is used to estimate a mechanical property of the tissue.

10.5 Photoacoustic Tomography

Photoacoustic tomography (PAT) is a hybrid biophotonics imaging modality that combines the processes of optical absorption in tissue with a resulting sound wave generation [42–46]. The procedure in PAT is to irradiate a localized tissue volume of interest with a short-pulsed laser beam. Some of the light is partially converted into heat in regions where enhanced optical absorption takes place. For example, some forms of diseased tissues and certain endogenous or exogenous chromophores exhibit strong light absorption properties. Strongly absorbing endogenous chromophores are molecules such as melanin, water, oxygenated hemoglobin, and deoxygenated hemoglobin that originate or are produced within an organism, tissue, or cell. As a result of their absorption properties, such endogenous chromophores can be used for examining features such as vascular structures, hemoglobin oxygen saturation, and blood flow. Exogenous chromophores are externally introduced substances such as dyes, nanoparticles, and reporter genes.

The strong light absorption by molecules from the short laser pulse results in a transient localized pressure rise due to a thermoelastic expansion effect. The sudden pressure rise creates an ultrasonic wave, which is known as a *photoacoustic wave*. An ultrasonic probe then measures this resultant ultrasonic wave and a computer is used to analyze the data and form an image of the targeted region. The amount of

Fig. 10.15 Example setup for a photoacoustic measurement method

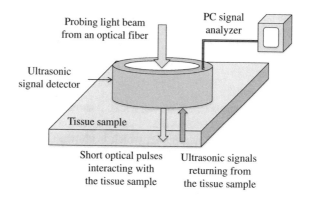

detected pressure depends on the degree of optical absorption in the target object, the thermal diffusion properties of the tissue, and the thermoelastic properties of the tissue. A number of different optical irradiation and ultrasonic probe architectures have been proposed for these measurements. The example scheme shown in Fig. 10.15 uses an annular acoustic probe with the laser pulse passing through the center opening of the probe.

The image resolution is limited by the light pulse duration, the properties of the tissue, and the response characteristics of the ultrasonic probe. *Increased resolution results from higher detected acoustic frequencies, shorter light pulses, and smaller measurement depths.* Typical resolutions are in the micrometer range for a measuring depth of millimeters and in the millimeter range for a measuring depth of centimeters. A point to note is that whereas the resolution is determined by the ultrasound parameters, the image contrast is determined by the optical absorption.

Two key parameters that need to be considered for the implementation of PAT are the thermal relaxation time and the stress relaxation time [42]. The *thermal relaxation time* (TRT) τ_{th} is a commonly used parameter for estimating the time required for thermal energy to conduct away from a heated tissue region. The TRT is defined as the time necessary for an object to cool down to 50 % of its original temperature. If a laser pulse that impinges on a tissue sample is shorter than the TRT, then mainly the target material will be heated. This condition is known as *thermal confinement.* When the pulse is longer than the TRT, the heat will travel into the surrounding tissue structure, which could result in tissue damage because the heat does not dissipate fast enough. The TRT can be estimated by

$$\tau_{th} = \frac{d^2}{\alpha_{th}} \tag{10.21}$$

where d is the size of the heated region and α_{th} is the *thermal diffusivity*, which is about 0.13 mm^2/s for soft tissue.

When a laser pulse heats a tissue, the heated sample will undergo a thermoelastic expansion, which results in the emission of acoustic waves. After the pulse stops, the stress induced by the expansion will decrease and can be characterized in terms of the *stress relaxation time* τ_s. This time is defined by

$$\tau_s = \frac{d}{v_s} \qquad (10.22)$$

where v_s is the speed of sound, which is about 1.48 mm/μs in water and soft tissue. For proper PAT procedures, the volume expansion of the absorber during the illumination period should be negligible, which is known as *stress confinement*. This condition holds when the laser pulse duration is shorter than the stress relaxation time.

Example 10.11 Suppose that in a PAT procedure the heated region in a sample of soft tissue has a dimension d = 1 mm. What are the thermal relaxation time and the stress relaxation time?

Solution: Using Eq. (10.21) with α_{th} = 0.13 mm²/s yields a thermal relaxation time of

$$\tau_{th} = \frac{d^2}{\alpha_{th}} = \frac{(1\,\text{mm})^2}{0.13\,\text{mm}^2/\text{s}} = 7.7\,\text{s}$$

Using Eq. (10.22) with v_s = 1.48 mm/μs yields a stress relaxation time of

$$\tau_s = \frac{d}{v_s} = \frac{1\,\text{mm}}{1.48\,\text{mm}/\mu\text{s}} = 0.68\,\mu\text{s}$$

The basic operation of PAT can be illustrated through a simple mathematical model [42]. When a short laser pulse excites a tissue sample, the local fractional volume expansion dV/V can be expressed in terms of the pressure change p (given in Pa) and the temperature change T (given in K) as

$$\frac{dV}{V} = -\kappa p + \beta T \qquad (10.23)$$

where κ is the isothermal compressibility and β is the coefficient of thermal expansion. Typical values of these parameters are $\kappa \approx 5 \times 10^{-10}$ Pa^{-1} for water and soft tissue and $\beta \approx 4 \times 10^{-4}$ K^{-1} for muscle tissue.

When a laser pulse impinges on a tissue sample, the temperature rise can be expressed as

$$T = \frac{A_e}{\rho C_V} = \frac{\mu_a H}{\rho C_V} \qquad (10.24)$$

Here ρ is the mass density (1 gm/cm³ for water and soft tissue), C_V is the specific heat capacity at constant volume [≈4 J/(gm K)], and $A_e = \mu_a H$ is the absorbed energy density where μ_a is the absorption coefficient and H is the local optical fluence.

Typical PAT laser pulses have durations of about 10 ns. In this case the stress confinement condition holds so that the fractional volume expansion dV/V is negligible. Thus from Eqs. (10.23) and (10.24) the local photoacoustic pressure rise p_0 immediately after the laser pulse can be written as

$$p_0 = \frac{\beta T}{\kappa} = \frac{\beta A_e}{\kappa \rho C_V} = \Gamma \mu_a H \tag{10.25}$$

Here the dimensionless factor Γ is the Grueneisen parameter, which is defined as

$$\Gamma = \frac{\beta}{\kappa \rho C_V} = \frac{\beta v_s^2}{C_P} \tag{10.26}$$

where the isothermal compressibility is expressed as

$$\kappa = \frac{C_P}{\rho v_s^2 C_V} \tag{10.27}$$

with C_P being the specific heat capacity at constant pressure.

Example 10.12 Consider a PAT procedure that uses short laser pulses with a fluence of 12 mJ/cm^2 to irradiate a soft tissue for which the absorption coefficient is $\mu_a = 0.1$ cm^{-1}. What are the temperature rise and pressure rise for this setup?

Solution: Using Eq. (10.24) yields a temperature rise of

$$T = \frac{\mu_a H}{\rho C_V} = \frac{(0.1\,\text{cm}^{-1})(12\,\text{mJ/cm}^2)}{(1\,\text{gm/cm}^3)[4000\,\text{mJ/(gm} \cdot \text{K)}]} = 0.30\,\text{mK}$$

Using Eq. (10.25) with $\beta \approx 4 \times 10^{-4}$ K^{-1} and $\kappa \approx 5 \times 10^{-10}$ Pa^{-1} yields a pressure rise of (1 bar = 10^5 Pa)

$$p_0 = \frac{\beta T}{\kappa} = \frac{(4 \times 10^{-4}\,\text{K}^{-1})(3 \times 10^{-4}\,\text{K})}{5 \times 10^{-10}\,\text{Pa}^{-1}} = 2.4 \times 10^2\,\text{Pa} = 2.4\,\text{mbar}$$

After the photoacoustic pressure wave is generated, it travels through the tissue sample and is detected by an ultrasonic probe or detector array. The shape of the wave depends on the geometry of the object. For example, a spherical object generates two spherical waves, one of which travels outward and the other travels inward. This results in a bipolar-shaped photoacoustic signal wherein the distance between the two peaks is proportional to the size of the object. Thus, a smaller object produces a photoacoustic signal with higher frequency components.

Once data on the distribution of time-resolved pressure variations have been obtained, then a high-resolution tomographic image of optical absorption can be created. The mathematical analyses and image formation methodologies are beyond the discussion of this chapter but are detailed in the literature [42–45].

10.6 Hyperspectral Imaging

Hyperspectral imaging (HSI) is a non-invasive multimodal medical imaging modality that combines reflectance and fluorescence techniques for applications such as disease diagnosis and image-guided surgery [47–49]. The advantages of HSI are related to the observation that tissue characteristics such as absorption, fluorescence, and scattering change during the development and progression of many diseases. When cells are in different disease states their fluorescent spectra or diffuse reflectance spectra change. These spectral variations result because diseased cells typically have modified structures from normal cells or they undergo different rates of metabolism. Thus as a disease progresses, further changes in the emission or reflectance spectra will take place. Because a HSI system captures data for such tissue characteristics, it can assist in identification and diagnosis of diverse tissue abnormalities. The applications include identification of various types of cancer, assessment of diabetic foot ulcers, assessment of tissue wound oxygenation, mapping of eye diseases, identification of abnormalities in gastrointestinal tissue, and identification and monitoring of changes in coronary arteries that may be associated with atherosclerosis.

The HSI system operates by collecting sets of spectral information at continuous wavelengths at each pixel of a 2D detector array. These data are then analyzed to determine the absorption spectroscopic characteristics of multiple tissue components in a tissue sample. The result is a 3D set of spatially distributed spectral information that shows where each spectral point is located in the sample. The 3D dataset is known as a *hypercube* and is illustrated on the left-hand side in Fig. 10.16. In this figure, the curve shows the spectral characteristic of an example pixel in the 2D slice through the tissue sample.

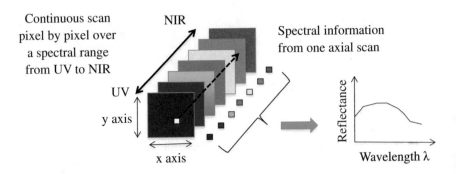

Fig. 10.16 Example of hyperspectral imaging

Two commonly used HSI techniques are spatial scanning and spectral scanning. In a *spatial scanning* method, first a complete continuous spectrum is obtained for an individual pixel in a 2D tissue sample and then the sample is scanned pixel by pixel to create a hypercube dataset, which consists of two spatial dimensions and one spectral dimension. In the *spectral scanning* method, first the entire 2D sample is scanned for a specific wavelength and then the system steps through a wavelength range to obtain the hypercube.

10.7   Summary

Light-based biomedical imaging technologies have greatly extended the ability for the non-invasive imaging and diagnoses of diseases in humans. In addition to the advances in microscopic and spectroscopic technologies described in the previous two chapters, photonics-based methodologies such as optical coherence tomography, miniaturized endoscopic processes, laser speckle imaging, optical coherence elastography, photoacoustic tomography, and hyperspectral imaging have emerged in biophotonics applications.

10.8   Problems

10.1   Consider an imaging system that records 1000 points in each 3D direction. If the line scans are carried out at rates of 100 kHz, show that the two-dimensional B-scans are recorded at rates of 100 frames per second and that the volume C-scans are done at rates of 0.1 volume per second, so that a volume C-scan takes 10 s.

10.2   Consider an imaging system that records 1000 points in each 3D direction. If the line scans are carried out at rates of 1 MHz, show that a volume C-scan is done in 1 s.

10.3   From Eq. (10.4), derive the intensity expression I_D at the detector of the light beams coming from the two arms of the interferometer as given by Eq. (10.5).

10.4   Suppose an OCT setup for ophthalmology tests uses the following super-luminescent diodes: an 800-nm peak wavelength with a 25-nm spectral bandwidth and a 1300-nm peak wavelength with a 70-nm spectral bandwidth for the posterior segment and the anterior segment of the eye, respectively. Show that the coherence lengths of these two sources are 25.6 and 24.1 μm, respectively.

10.5   Consider the following super-luminescent diodes: an 800-nm peak wavelength with a 25-nm spectral bandwidth and a 1300-nm peak wavelength with a 70-nm spectral bandwidth. Show that the axial resolution for each of these sources is 11.3 and 10.6 μm, respectively.

10.6   Consider an OCT system in which the objective lens has a numerical aperture NA = 0.22. (a) Show that the lateral resolution at a wavelength of 850 nm is 1.43 μm. (b) Show that the axial field of view for this setup is 39.2 μm.

10.7   Consider an SS-OCT system for which M = 190, λ = 1310 nm, and $\Delta\lambda$ = 85 nm. (a) Show that the scan depth range is 0.96 mm. (b) Show that the axial resolution is 8.9 μm when using a Gaussian source.

10.8   (a) Using Web resources or vendor data sheets, describe the detailed structure of a typical laparoscope. Include characteristics such as probe material, the diameter and length of the probe, optical components at the inserted end, and eyepiece structure. (b) What are some applications of a laparoscope?

10.9   (a) Using Web resources or vendor data sheets, describe the detailed structure of a typical commercially available arthroscope. Include characteristics such as the diameter and length of the probe, optical components at the inserted end, and eyepiece structure. (b) What are some applications of an arthroscope?

10.10 Fig. 10.10 shows the use of miniaturized instruments used for minimally invasive surgery. Describe some instruments for this use, such as a bone shaver, ligament repair tool, scissors, and grasping tools.

10.11 Optically guided surgical tools are being developed to address the challenges of precise control during microsurgery. Using Web resources, describe the characteristics of an optically guided forceps tool. [For example, see: C. Song, D. Y. Park, P. L. Gehlbach, S. J. Park, and J. U. Kang, "Fiber-optic OCT sensor guided "SMART" micro-forceps for microsurgery," *Biomed. Opt. Express*, vol. 4, issue 7, pp. 1045–1050, July 2013.]

10.12 Consider a laser speckle imaging systems operating at 532 nm. (a) Suppose the exposure time of the camera is T = 10 ms. If the speckle coefficient β = 0.5, show that if the speckle correlation time is τ_c = 200 μs then the speckle contrast factor K = 0.14. (b) Show that if τ_c = 10 ms then the value of K is 0.28.

10.13 Consider a speckle imaging system operating at 650 nm in which the magnification of the optics is M = 10. Suppose the camera has f/# = 2.8. Show that the minimum speckle size that can be resolved by the camera is 24.4 μm.

10.14 Suppose that in a PAT procedure the heated region in a sample of soft tissue has a dimension d = 15 μm. Show that the thermal relaxation time and the stress relaxation time are 17×10^{-4} s and 0.01 μs, respectively.

10.15 Consider a PAT procedure that uses short laser pulses with a fluence of 10 mJ/cm^2 to irradiate a soft tissue for which the absorption coefficient is μ_a = 0.1 cm^{-1}. Show that the temperature rise and pressure rise for this setup are 0.25 mK and 200 Pa, respectively.

References

1. A.R. Kherlopian, T. Song, Q. Duan, M.A. Neimark, M.J. Po, J.K. Gohagan, A.F. Laine, A review of imaging techniques for systems biology. BMC Syst. Biol. **2**, 74–91 (2008)
2. A.P. Dhawan, B.D. Alessandro, X. Fu, Optical imaging modalities for biomedical applications. IEEE Rev. Biomed. Eng. **3**, 69–92 (2010)
3. C.T. Xu, Q. Zhan, H. Liu, G. Somesfalean, J. Qian, S. He, S. Andersson-Engels, Upconverting nanoparticles for pre-clinical diffuse optical imaging, microscopy and sensing: current trends and future challenges. Laser Photonics Rev. **7**(5), 663–697 (2013)
4. J.G. Fujimoto, Optical coherence tomography for ultrahigh resolution in vivo imaging. Nat. Biotechnol. **21**, 1361–1367 (2003)
5. M. Wojtkowski, High-speed optical coherence tomography: basics and applications. Appl. Opt. **49**(16), D30–D61 (2010)
6. J.A. Izatt, M.A. Choma, Theory of optical coherence tomography, chap. 2, ed. by W. Drexler, J.G. Fujimoto, eds., *Optical Coherence Tomography Technology and Applications* (Springer, 2008)
7. R.L. Shelton, W. Jung, S.I. Sayegh, D.T. McCormick, J. Kim, S.A. Boppart, Optical coherence tomography for advanced screening in the primary care office. J. Biophotonics **7**, 525–533 (2014)
8. W. Drexler, M. Liu, A. Kumar, T. Kamali, A. Unterhuber, R.A. Leitgeb, Optical coherence tomography today: speed, contrast, and multimodality. J. Biomed. Opt. **19**(7), 071412 (2014)
9. Z. Hubler, N.D. Shemonski, R.L. Shelton, G.L. Monroy, R.M. Nolan, S.A. Boppart, Real time automated thickness measurement of the in vivo human TM using optical coherence tomography. Quant. Imaging Med. Surg. **5**(1), 69–77 (2015)
10. M.E. Brezinski, *Optical Coherence Tomography: Principles and Applications*, 2nd edn. (Academic, New York, 2016)
11. J. Mo, M. de Groot, J.F. de Boer, Depth-encoded synthetic aperture optical coherence tomography of biological tissues with extended focal depth. Opt. Express **23**(4), 4935–4945 (2015)
12. R. Leitgeb, C.K. Hitzenberger, A.F. Fercher, Performance of Fourier domain vs. time domain optical coherence tomography. Opt. Express **11**(8), 889–894 (2003)
13. Y. Zhao, H. Tu, Y. Liu, A.J. Bower, S.A. Boppart, Enhancement of optical coherence microscopy in turbid media by an optical parametric amplifier. J. Biophotonics **8**(6), 512–521 (2015)
14. I. Grulkowski, J.J. Liu, B. Potsaid, V. Jayaraman, J. Jiang, J.G. Fujimoto, A.E. Cable, High-precision, high-accuracy ultralong-range swept-source optical coherence tomography using vertical cavity surface emitting laser light source. Opt. Lett. **38**, 673–675 (2013)
15. W.J. Choi, R.K. Wang, "Swept-source optical coherence tomography powered by a 1.3-μm vertical cavity surface emitting laser enables 2.3-mm-deep brain imaging in mice in vivo. J. Biomed. Opt., **20**, article 106004 (Oct 2015)
16. R. Kiesslich, M. Goetz, A. Hoffman, P.R. Galle, Review paper: new imaging techniques and opportunities in endoscopy. Nat. Rev. Gastroenterol. Hepatol. **8**, 547–553 (2011)
17. S.F. Elahi, T.D. Wang, Future and advances in endoscopy. J. Biophotonics **4**(7–8), 471–481 (2011)
18. P.S. Thong, S.S. Tandjung, M.M. Movania, W.M. Chiew, M. Olivo, R. Bhuvaneswari, H.S. Seah, F. Lin, K. Qian, K.C. Soo, Toward real-time virtual biopsy of oral lesions using confocal laser endomicroscopy interfaced with embedded computing. J. Biomed. Opt. **17**(5), article 0560 (May 2012)
19. V. Subramanian, K. Ragunath, Advanced endoscopic imaging: a review of commercially available technologies. Clin. Gastroenterol. Hepatol. **12**, 368–376 (2014)
20. M. Gu, H. Bao, H. Kang, Fibre-optical microendoscopy. J. Microsc., **254**(1), 13–18 (Apr 2014)

21. G. Keiser, F. Xiong, Y. Cui, P.P. Shum, Review of diverse optical fibers used in biomedical research and clinical practice. J. Biomed. Optics, **19**, art. 080902 (Aug 2014)
22. F. Lucà, L. van Garsse, C.M. Rao, O. Parise, M. La Meir, C. Puntrello, G. Rubino, R. Carella, R. Lorusso, G.F. Gensini, J.G. Maessen, S. Gelsomino, Minimally invasive mitral valve surgery: a systematic review. Minim. Invasive Surg., **2013**, Article ID 179569 (Mar 2013)
23. T. Blinman, T. Ponsky, Pediatric minimally invasive surgery: laparoscopy and thoracoscopy in infants and children. Pediatrics **130**(3), 539–549 (Sept 2012) (Review article)
24. F.M. Phillips, I. Lieberman, D. Polly (eds.), *Minimally Invasive Spine Surgery* (Springer, New York, 2014)
25. P. Banczerowski, G. Czigléczki, Z. Papp, R. Veres, H.Z. Rappaport, J. Vajda, Minimally invasive spine surgery: systematic review. Neurosurg. Rev. **38**, 11–36 (2015)
26. M.J. Gora, J.S. Sauk, R.W. Carruth, K.A. Gallagher, M.J. Suter, N.S. Nishioka, L.E. Kava, M. Rosenberg, B.E. Bouma, G.J. Tearney, Tethered capsule endomicroscopy enables less invasive imaging of gastrointestinal tract microstructure. Nat. Med. **19**, 238–240 (2013)
27. G.J. Ughi, M.J. Gora, A.-F. Swager, A. Soomro, C. Grant, A. Tiernan, M. Rosenberg, J.S. Sauk, N.S. Nishioka, G.J. Tearney, Automated segmentation and characterization of esophageal wall in vivo by tethered capsule optical coherence tomography endomicroscopy. Biomed. Opt. Express **7**(2), 409–419 (2016)
28. D.K. Iakovidis, A. Koulaouzidis, Software for enhanced video capsule endoscopy: challenges for essential progress. Nat. Rev. Gastroenterol. Hepatol. **12**, 172–186 (Feb 2015). (Review article)
29. J.W. Goodman, *Speckle Phenomena in Optics* (Roberts and Company, Englewood, Colorado, 2007)
30. D.A. Boas, A.K. Dunn, "Laser speckle contrast imaging in biomedical optics. J. Biomed. Opt., **15**(1), article 011109 (Jan/Feb 2010)
31. D. Briers, D.D. Duncan, E. Hirst, S.J. Kirkpatrick, M. Larsson, W. Steenbergen, T. Stromberg, O.B. Thompson, Laser speckle contrast imaging: theoretical and practical limitations. J. Biomed. Opt. **18**(6), article 066018 (June 2013)
32. A. Curatolo, B.F. Kennedy, D.D. Sampson, T.R. Hillman, Speckle in optical coherence tomography", in *Advanced Biophotonics: Tissue Optical Sectioning*, ed. by V.V. Tuchin, R.K. Wang (Taylor & Francis, London, 2013) Chapter 6, pp. 212–277
33. J.C. Ramirez-San-Juan, E. Mendez- Aguilar, N. Salazar-Hermenegildo, A. Fuentes-Garcia, R. Ramos-Garcia, B. Choi, Effects of speckle/pixel size ratio on temporal and spatial speckle-contrast analysis of dynamic scattering systems: Implications for measurements of blood-flow dynamics. Biomed. Opt. Express, **4**(10), 1883–1889 (Oct 2013)
34. S. Ragol, I. Remer, Y. Shoham, S. Hazan, U. Willenz, I. Sinelnikov, V. Dronov, L. Rosenberg, A. Bilenca, In vivo burn diagnosis by camera-phone diffuse reflectance laser speckle detection. Biomed. Opt. Express **7**(1), 225–237 (2016)
35. S.L. Jacques, S.J. Kirkpatrick, Acoustically modulated speckle imaging of biological tissues. Opt. Lett. **23**(11), 879–881 (1998)
36. J.M. Schmitt, OCT elastography: imaging microscopic deformation and strain of tissue. Opt. Exp. **3**(6), 199–211 (1998)
37. X. Liang, V. Crecea, S.A. Boppart, Dynamic optical coherence elastography: a review. J. Innov. Opt. Health Sci. **3**(4), 221–233 (2010)
38. C. Sun, B. Standish, V.X.D. Yang, Optical coherence elastography: current status and future applications. J. Biomed. Opt. **16** article 043001 (Apr 2011)
39. K.J. Parker, M.M. Doyley, D.J. Rubens, Imaging the elastic properties of tissue: the 20 year perspective. Phys. Med. Biol. **56**(1), R1–R29 (2011)
40. B.F. Kennedy, K.M. Kennedy, D.D. Sampson, A review of optical coherence elastography: fundamentals, techniques and prospects. IEEE J. Sel. Top. Quantum Electron., **20**(2), article 7101217 (Mar/Apr 2014)
41. L. Chin, A. Curatolo, B.F. Kennedy, B.J. Doyle, P.R.T. Munro, R.A. McLaughlin, D.D. Sampson, Analysis of image formation in optical coherence elastography using a multiphysics approach. Biomed. Opt. Express **5**, 2913–2930 (2014)

42. L.V. Wang, H.I. Wu, in Biomedical optics: principles and imaging, chap. 12, in *Photoacoustic Tomography* (Wiley, Hoboken, NJ, 2007)
43. L.V. Wang, S. Wu, Photoacoustic tomography: in vivo imaging from organelles to organs. Science **335**(6075), 1458–1462 (March 23, 2012)
44. Y. Zhou, J. Yao, L.V. Wang, Tutorial on photoacoustic tomography. J. Biomed. Opt. **21**(6), 061007 (June 2016)
45. B. Zabihian, J. Weingast, M. Liu, E. Zhang, P. Beard, H. Pehamberger, W. Drexler, B. Hermann, In vivo dual-modality photoacoustic and optical coherence tomography imaging of human dermatological pathologies. Biomed. Opt. Express **9**(9), 3163–3178 (2015)
46. D. Wang, Y. Wu, J. Xia, Review on photoacoustic imaging of the brain using nanoprobes. Neurophotonics **3**(1), art. 010901 (Jan-Mar 2016)
47. R.X. Xu, D.W. Allen, J. Huang, S. Gnyawali, J. Melvin, H. Elgharably, G. Gordillo, K. Huang, V. Bergdall, M. Litorja, J.P. Rice, J. Hwang, C.K. Sen, Developing digital tissue phantoms for hyperspectral imaging of ischemic wounds. Biomed. Opt. Express **3**(6), 1433–1445 (1 June 2012)
48. G. Lu, B. Fei, Medical hyperspectral imaging: a review. J. Biomed. Opt. **19**(1), article 010901 (Jan 2014)
49. J.M. Amigo, H. Babamoradi, S. Elcoroaristizabal, Hyperspectral image analysis: a tutorial. Anal. Chim. Acta **86**, 34–51 (2015)

Chapter 11
Biophotonics Technology Applications

Abstract Biophotonics technologies are widely used in biomedical research, in the detection and treatment of diseases and health conditions, and in point-of-care healthcare clinics. This chapter describes advanced tools and implementations such as optical tweezers and optical trapping techniques that enable microscopic manipulation of cells and molecules for exploring biological materials and functions in the micrometer and nanometer regime, miniaturized photonics-based instrumentation functions and devices such as the lab-on-a-chip and lab-on-fiber technologies, microscope-in-a-needle concepts to enable 3-dimensional scanning of malignant tissue within the body, and optogenetics procedures which attempt to explore and understand the mechanisms of neuronal activity in organs such as the brain and the heart.

Biophotonics technologies are being used in a wide variety of biochemical and biomedical research disciplines, in the detection and treatment of various types of diseases and health conditions, and in point-of-care healthcare clinics. This chapter addresses several diverse technologies for such applications.

First Sect. 11.1 describes optical trapping schemes and optical tweezers that enable microscopic manipulation of cells and molecules for exploring biological materials and functions in the micrometer and nanometer regime. The basic concept uses two light-induced opposing forces on a microscopic particle. One force originates from the gradient radiation pressure of a focused light beam, and the other force arises from photon scattering, which pushes objects along the propagation direction of the light beam. By balancing or controlling these two forces, a particle can be held stationary or it can be micro-manipulated.

Next, Sect. 11.2 addresses the extension to biophotonics of the dramatic miniaturization of devices and circuits in the electronics world that has resulted in products such as compact computers, smartphones, and handheld test equipment. In the biophotonics world, greatly miniaturized photonics-based functions and devices are appearing in the form of *lab-on-a-chip technology* and *lab-on-fiber technology*. A key use of lab-on-a-chip technology is in microfluidic devices, which nominally are built on a substrate the size of a microscope slide. Such compact integrated biosensors can manipulate and analyze fluid samples in volumes on the order of microliters as the liquids pass through micrometer-sized channels. Lab-on-fiber technology envisions the creation of multiple sensing and analysis functions through the integration onto a single fiber segment of miniaturized photonic devices such as nanoparticle SPR sensors deposited on a fiber tip, fiber Bragg gratings combined with high-index surface coatings, and various types of optical filters.

Further miniaturizations have resulted in concepts such as the so-called microscope-in-a-needle. *Microscope-in-a-needle* implementations involve the use of a highly sensitive, ultrathin, side-viewing 30-gauge OCT needle probe. As is described in Sect. 11.3, the basic goal of this technology is to incorporate microscope capabilities inside a standard hypodermic needle to enable 3D scanning of malignant tissue deep within the body.

Clinical techniques that exhibit a high sensitivity to specific allergens, a large dynamic range to detect various biomarker concentrations, and the ability to assess multiple biomarkers are in high demand for disease diagnostics. Of particular interest is the ability to detect and analyze natural nanoparticles such as viruses and pollutants. A method for detecting single nanoscale particles named interferometric reflectance imaging sensor is described in Sect. 11.4.

An innovative application of biophotonics techniques in the area of neuroscience is given in Sect. 11.5. These pursuits are referred to as *neurophotonics*, which attempts to explore and understand the mechanisms of neuronal activity in organs such as the brain and the heart. A major goal is to develop photonics-based methods to control specific classes of excitable neural cells in order to determine the causes and effects of various diseases caused by malfunctions in these cells.

11.1 Optical Manipulation

The advent of the laser in 1960 led to renewed interest in the centuries old concept that light can exert forces. Experimental studies on radiation pressure in the 1970s opened up a new field of exploring biological materials and functions at the microscopic level. This activity led to the concepts of *optical tweezers* and *optical trapping* for the controlled grasping, holding, and manipulation of microscopic particles. These techniques use a non-contact force for manipulation of cells and single molecules down to the nanoscale level with forces ranging between 0.1–200 pN and can be used in a liquid environment [1–7].

Example 11.1 (a) What is a pN force? (b) What are force levels on a molecular scale?

Solution: (a) Newton's second law of motion states that $F = ma$, where F is the force in units of newtons (designated by N) applied on an object of mass $\underline{m}$ (in kilograms) and $\underline{a}$ is the acceleration of the object (in m/s^2). Thus, $1\ N = 1\ kg \cdot m/s^2$. As an example, if $a = g = 9.8\ m/s^2$ is the acceleration of

Table 11.1 Forces involved at the biological level

Type of force	Bond strength (pN)
Covalent bond such as C-C	1600
Noncovalent bond between biotin and streptavidin	160
Bond between two actin monomers	100
Force required to extract a tether out of an erythrocyte	50
Weak bond such as hydrogen	4
50 % stretching of double-stranded DNA	0.1

gravity, then the force needed to pick up a medium-sized 100-gm apple is 0.98 N. On a smaller scale, a force of $F = 3.4 \times 10^3$ pN is needed to pick up a very fine grain of sand that has a mass $m = 3.5 \times 10^{-10}$ kg. (b) Forces at the molecular scale are on the order of piconewtons (pN), which is 10^{-12} N. Table 11.1 gives some examples of forces related to breaking molecular bonds. For example, 1600 pN are needed to break a covalent bond such as C–C and 4 pN are required to break a hydrogen bond.

Among the numerous applications of optical trapping techniques are cell nanosurgery, manipulation and assembly of carbon nanotubes and nanoparticles, and studies of protein-protein binding processes, DNA-protein interactions, DNA mechanical properties, mechanical-chemical processes in the cell, elastic properties of cells, cell-cell interactions, and cell transport, positioning, sorting, assembling, and patterning [8–11].

A simple geometric ray optics picture can be used for an elementary description of the operational principle of optical tweezers and optical trapping. This operation is based on the exertion of extremely small forces on nanometer and micrometer-sized dielectric particles by means of a highly focused laser beam. First consider a spherical, transparent dielectric particle located in a light field that has an inhomogeneous intensity distribution in a plane transverse to the optical axis, as is shown in Fig. 11.1. Here two light rays (shown in red) with different intensities are incident symmetrically on a sphere. The two rays will be refracted as they enter and exit the sphere, and thus will travel in different directions from those of the original paths. Because *light has a momentum associated with it*, the changes in direction mean that the momentum has changed for the photons contained in each ray. Thus, according to Newton's third law, there is an equal and opposite momentum change on the particle. Because the force on a dielectric object is given by the change in momentum of light induced by the refraction of the light by the object, the incident rays 1 and 2 will produce corresponding forces F_1 and F_2, respectively. As a result of ray 1 being less intense than ray 2, the force F_1 is smaller than F_2. Consequently, a transverse intensity gradient will result in a net *gradient force* F_{grad}, which points towards the region that has the highest intensity.

Fig. 11.1 Basic
geometric-optics principle of
optical tweezers **a** A
transverse intensity variation
gives a gradient force pointing
towards the region of highest
intensity; **b** Scattering of
photons yields a forward
scattering force

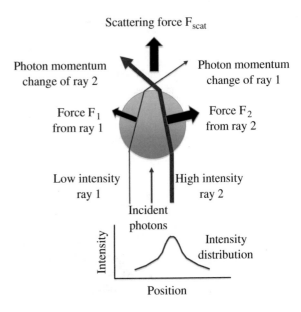

In addition to the gradient force, the scattering of incident photons from the
sphere will exert an axial force on the sphere. This force points downstream in the
direction of travel of the light beam and is known as the *scattering force* F_{scat}.

For an optical tweezers or optical trap implementation, typically a microscope
objective is used to focus a collimated input light beam into an exit beam with a
tightly focused axially symmetric Gaussian intensity profile, as is shown in
Fig. 11.2. A very strong electric field gradient exists within the narrowest portion of

Fig. 11.2 Exit beam from a
microscope objective with a
tightly focused axially
symmetric Gaussian intensity
profile

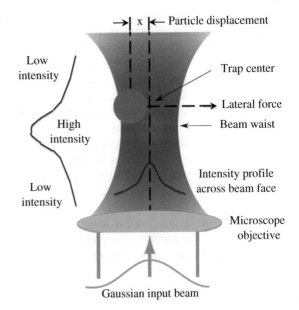

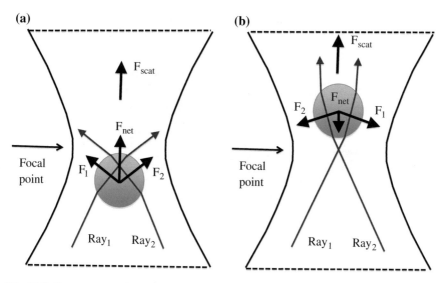

Fig. 11.3 Forces on an axially positioned particle located **a** ahead of the focal point and **b** after the focal point

this focused beam, which is called the *beam waist*. Within this region, the electric field is strongest in the center of the beam. Thus if a particle is displaced from the center of the beam by a distance x, the gradient force will push the particle back to the beam center.

Now consider the situation when the particle is located on the beam axis slightly ahead of the focal point, as is shown in Fig. 11.3a. In this case the forces due to the refraction induced momentum changes are radially symmetric on the sphere and the net gradient force points downstream in the direction of the photon flow. This causes the particle to be displaced slightly downstream from the exact position of the beam waist. In addition, the scattering force also points downstream. Once the particle is pushed past the focal point, as is shown in Fig. 11.3b, the gradient force now points upstream and eventually counterbalances the downstream scattering force. Thereby, the net result of these forces is that a stable three-dimensional trapping position is achieved for the confinement of dielectric particles.

In general, if the particle is displaced laterally from the focus of the laser beam, it will experience a restoring force that is proportional to the distance x between the center of the sphere and the focus of the laser (see Fig. 11.2). This force can be described by the equation

$$F = -kx \tag{11.1}$$

This equation is analogous to the behavior of a small spring obeying Hooke's law. Here k is a constant, which is referred to as the *trap stiffness*. The trap stiffness can vary widely depending on the particular design of the optical tweezers and the sphere size, but a nominal value is around 50 pN/μm. After the trap stiffness has been determined, then the force on a particle in the trap can be determined by measuring the position of the bead in the trap.

Example 11.2 Consider an optical trap that has a stiffness k = 50 pN/μm. If an image analysis instrument can determine the position of a micrometer-sized sphere to within 10 nm, what is the force resolution?

Solution: From Eq. (11.1) the force resolution is

$$F = kx = (50 \, pN/\mu m)(10 \, nm) = 0.5 \, pN.$$

For small particles the gradient force can be expressed in terms of the gradient of the intensity $\nabla \mathbf{I}$ as

$$\overrightarrow{F}_{grad} = \frac{2\pi n_{med} r^3}{c} \left(\frac{m^2 - 1}{m^2 + 2} \right) \nabla \overrightarrow{I} \tag{11.2}$$

where r is the radius of the particle, c is the speed of light, n_{med} is the refractive index of the surrounding medium, n_{part} is the refractive index of the particle, and $m = n_{part}/n_{med}$ is the ratio of the refractive indices of the particle and the surrounding medium, respectively. The magnitude of the scattering force is

$$F_{scat} = \frac{8\pi n_{med} r^6}{3c} \left(\frac{2\pi}{\lambda} \right)^4 \left(\frac{m^2 - 1}{m^2 + 2} \right) I \tag{11.3}$$

For a power P incident on the surface of the sphere, the total force F_{total} exerted on a particle is the sum of the force contributions due to the scattered reflected rays and the forces resulting from the emerging refracted rays. This total force can be expressed as

$$F_{total} = Q \frac{n_{med} P}{c} \tag{11.4}$$

Here, n_{med} is the refractive index of the medium in which the sphere is located. The *trapping efficiency* or *quality factor* Q depends on the Fresnel reflection and transmission coefficients R and T, respectively, and gives a measure of the momentum that the light beam transferred to the sphere. The detailed calculations of Q can be found in the literature [1, 2]. The value of Q depends on the type of optical trap, the NA of the microscope objective, and the size of the trapped particle. Optical tweezers can have trapping efficiencies up to Q = 0.3 perpendicular to the direction of the laser beam propagation.

Example 11.3 Consider an optical trap that has a trapping efficiency of $Q = 0.25$. What is the total force F_{total} exerted on a particle if a power level of 10 mW is used in the trap and the refractive index in the medium is $n_{med} = 1.36$?

Solution: From Eq. (11.4) the total force is found to be

$$F_{total} = Q \frac{n_{med}P}{c} = 0.25 \frac{(1.36)(10\,mW)}{3 \times 10^8 m/s} = 11.3\,pN$$

11.2   Miniaturized Analyses Tools

In the biophotonics world, greatly miniaturized photonics-based functions and devices are appearing in the form of *lab-on-a-chip technology* and *lab-on-fiber technology*. These compact integrated biosensors, which are based on biochip and optical fiber technology, are used to manipulate and analyze fluid samples in volumes on the order of microliters. A key use of lab-on-a-chip technology is in microfluidic devices, which nominally are built on a substrate that is the size of a microscope slide. Lab-on-fiber technology is aimed at the creation of multiple sensing and analysis functions through the integration of miniaturized photonic devices onto a single fiber segment.

11.2.1   Lab-on-a-Chip Technology

Lab-on-a-chip (LOC) technology involves the use of different configurations of micrometer-sized or smaller fluid channels that are built on glass, thermoplastic polymer, or paper-based substrates, with polymers being the preferred material [12–19]. The substrates typically have a standard laboratory format, such as a microscope slide. Such devices are compact integrated biosensors that manipulate and analyze fluid samples in volumes on the order of microliters as the liquids pass through the device channels. This technology has enabled the creation of portable and easy-to-use test and measurement tools for rapidly characterizing fluids in applications such as low-cost point-of-care diagnostics, analysis of the characteristics of biomolecules, evaluations of food conditions, and drug development. Depending on the device structure, LOCs can perform functions such as mixing of liquids, amplification of biomolecules (duplicating certain gene sets multiple times), synthesis of new materials, optical or electrical detection of specific substances, and hybridization of DNA molecules.

The two basic categories of fluid flow are laminar flow and turbulent flow. Fluids in a *laminar flow* exhibit a smooth and constant flow, whereas *turbulent flow* is characterized by vortices and flow fluctuations. For microfluidic systems the flow is essentially laminar, whereas macroscopic flows are characterized by turbulent flows that are random in space and time. The two basic factors that characterize fluid flow are viscous forces and inertial forces. The flow types are measured by the dimensionless *Reynolds number* Re, which is given by

$$\text{Re} = \frac{\text{Inertial force}}{\text{Viscous force}} = \frac{\rho u D_h}{\mu} = \frac{u D_h}{v} \qquad (11.5)$$

Here ρ is the fluid density, μ is the dynamic viscosity (a measure of the internal resistance of the channel), $v = \mu/\rho$ is the kinematic viscosity (the resistance to flow under the influence of some force acting on the fluid), and u is the velocity of the fluid. The *hydraulic diameter* of the channel D_h depends on the geometry of the channel and is given by

$$D_h = \frac{4A}{P_{wet}} \qquad (11.6)$$

where A and P_{wet} are the cross-sectional area and the wetted perimeter (the surface of the microfluidic channel that is in direct contact with the liquid) of the channel, respectively. The value of Re depends on factors such as the shape, surface roughness and aspect ratio (ratio of width to height) of the channel. For large fluid channels the value of Re ranges from about 1500 to 2500. Its value is less than 100 for microfluidic channels, which represents a laminar flow.

Example 11.4 Compare the hydraulic diameter of two square channels that have channel dimensions of 50 μm and 80 μm, respectively.

Solution: (a) The cross-sectional area for the 50-μm channel is A(50 μm) = $(50\ \mu m)^2$ = 2500 μm^2 and the perimeter is P_{wet} = 4(50 μm) = 200 μm. Thus from Eq. (11.6) it follows that D_h(50 μm) = 4(2500 μm^2)/(200 μm) = 50 μm.
(b) Similarly, D_h(80 μm) = 4(6400 μm^2)/(320 μm) = 80 μm.

Several examples from many types of microfluidic chips and associated interface devices are listed in Table 11.2. The *Lab-on-a-Chip Catalogue* (from the company microfluidic ChipShop) is an example of an online resource for a wide variety of microfluidic chips and implementation accessories [15].

Table 11.2 Selected examples of microfluidic chips and associated devices [15]

Chip or device	Key elements	Chip/device function
Sample preparation chip	Reaction chambers of various volumes	Extracts target molecules out of a given sample in preparative quantities
Plasma/serum generation chip	4 membranes for plasma/serum generation out of full blood	Each membrane can generate about 12 µl of plasma/serum out of 25 µl of full blood
Particle and cell sorting chip	Spiraled channels used to separate particles and cells according to their size	Can separate cells, analyze them, and optionally sort and collect desired cells; particles of different sizes can be received at different outlet ports.
Droplet generator chip	Contains 8 droplet channels with different channel dimensions	Creates 50–80 µm sized droplets at various droplet generation frequencies, to generate droplets with different volumes
Micro mixer chip	Applies either passive or active mixing principles	Passive elongated channels enable diffusion mixing; active integrated stir bars enable a wide range of mixing ratios
Fluidic interface	Uses lab-on-a-chip device technology	Miniaturized connectors and tubing enable connection to pumps, valves, or waste reservoirs
Liquid storage device	Uses miniature tanks with a 500 µl volume	Enables off-chip storage of liquids
Syringe pumps	Uses high-end syringe pump technology	Enable extremely precise dosing and pumping of fluids

11.2.2 Lab-on-Fiber Concept

The *lab-on-fiber* concept arose from the joining of nanotechnology and optical fiber disciplines [20–22]. One activity in this evolving technology is the deposition of patterned layers of nanoparticles either on the tip of a fiber or as an external coating over a Bragg grating written inside of a photonic crystal optical fiber. By depositing a pattern of nanoparticles on the tip of a standard optical fiber, localized surface plasmon resonances (LSPRs) can be excited on the tip surface when an illuminating optical wave exits the fiber. This condition arises because of a phase matching between the light waves scattered at the surface and the resonance modes supported by the nanostructure. The LSPRs are very sensitive to the refractive index value of the analyte material that surrounds the fiber tip. For example, the blue curve in Fig. 11.4 shows a surface plasmon resonance wave for the condition when no analyte is covering the nanoparticle pattern on the end of a fiber. Analyses of particles in a fluid then can be made by inserting the fiber tip into an analyte liquid, so that now the liquid covers the nanoparticle pattern. This action changes the refractive index of the interface between the nanoparticle layer and the fluid. The

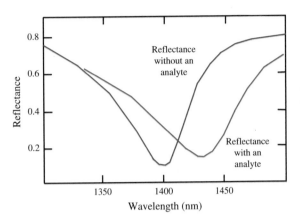

Fig. 11.4 Example shift in the surface plasmon resonant peak due to a relative index change at a fiber tip covered with a nanoarray pattern [*J. Biomed. Opt.* 19(8), 080902 (Aug 28, 2014]. doi:10.1117/1.JBO.19.8.080902)

result is a wavelength shift in the LSPR peak due to a change in the phase matching condition, as shown by the red curve in Fig. 11.4. This lab-on-fiber setup thus enables highly sensitive biological, chemical, and physical sensing functions.

11.3 Microscope in a Needle

The desire to analyze abnormal growths located deep within tissue has resulted in a number of minimally invasive needle-based optical imaging techniques. One method is to place an optical fiber inside of small gauge needles for optical imaging within soft tissues or lesions [23–26]. This method enables confocal fluorescence microendoscopy or OCT to be performed inside the body to locate and diagnose very small tumor elements. The basic technology goal is to have microscope imaging capabilities inside a standard hypodermic needle to do 3D scanning. Such techniques avoid the need for more invasive surgery and allow a quicker patient recovery from the examination process compared to a conventional biopsy.

One example is an ultrathin 30-gauge needle (310 μm in diameter), which encapsulates a SMF [23–25]. The optical elements in the needle consist of a 260-μm section of no-core fiber (NCF) and a 120-μm section of graded-index (GRIN) fiber, which is spliced to a single-mode fiber. Another piece of NCF, which is polished at a 45° angle, follows the GRIN section. This angled NCF allows side illumination and viewing through a miniature window in the side of the needle. The output beam from the window is elliptical with a transverse resolution that is smaller than 20 μm at the full-width half-maximum level over a distance of about 740 μm in tissue. The peak resolutions typically are 8.8 and 16.2 μm in the x- and y-directions, respectively.

11.4 Single Nanoparticle Detection

Clinical techniques that exhibit a high sensitivity to specific allergens, a large dynamic range to detect all clinically relevant concentrations of various biomarkers, and the ability to simultaneously assess multiple biomarkers are in high demand for disease diagnostics. Of particular interest is the ability to detect and analyze natural nanoparticles such as viruses and pollutants, which can have a major impact on human health. The detection of single nanoscale particles has been demonstrated using a method named *interferometric reflectance imaging sensor* (IRIS) [27–29].

The IRIS technique does not require complex micro-fabricated detector surfaces but instead employs a simple and inexpensive silicon-silicon dioxide (Si-SiO$_2$) substrate, commercially available LEDs, and a CCD detector. This method can detect single nanoparticles that are captured across a sensor surface by means of interference between an incident optical field that is scattered from the captured nanoparticles and a reference reflection from the sensor surface.

The basic concept of single-particle IRIS (SP-IRIS) is illustrated in Fig. 11.5. First a natural nanoparticle, such as a virus, is captured on a specially prepared sensor surface consisting of a silicon dioxide layer deposited on a silicon substrate. To detect these viral nanoparticles and to determine their size, visible LED light is used to illuminate the surface and a CCD camera captures a bright-field reflection image. In this image the particles of interest appear as diffraction-limited dots. The contrast of the dots can be used to determine their sizes and shapes.

Fig. 11.5 Illustration of the single-particle IRIS imaging technique

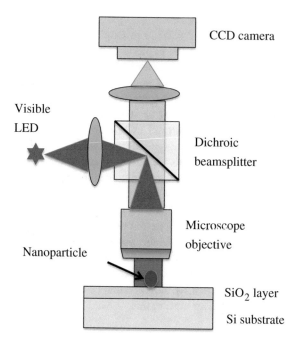

11.5 Neurophotonics

The basic activities of neuroscientists are to explore and understand the mechanisms of neuronal activity in the brain [30–32]. A major goal is to develop methods to control specific classes of excitable neural cells in order to determine the causes and effects of various diseases caused by malfunctions in these cells. Having a precise control of specific neural cells would lead to developing treatments for both neuropsychiatric diseases (e.g., Parkinson's disease, Alzheimer's disease, pain syndromes, epilepsy, depression and schizophrenia) and non-neural diseases (e.g., heart failure, muscular dystrophy and diabetes). Deep-brain stimulation (DBS) with microelectrodes has been one methodology used in this search. In this technique, microelectrodes are placed in well-defined anatomical locations for modulation of neurons in specific regions of the brain. Although these methods alleviated symptoms in some neuropsychiatric diseases, DBS electrode stimulation causes a mixture of excitation and inhibition both in targeted cells and in unrelated peripheral or transient cells.

Subsequently, the ability to precisely control either excitation or inhibition of neurons was achieved through techniques from neurophotonics [30–35], which involves the use of photonics in neuroscience research. In particular, the neurophotonics discipline of *optogenetics* uses optical fiber links to send light signals to the brain to control neurons that have been genetically sensitized to light. The basis of optogenetics is the use of microbial opsins (light-sensitive proteins found in receptor cells of the retina) that enable neurons to have both the capability to detect light and to regulate the biological activity in specific targetable molecules. Among the various opsins used in optogenetics, the two commonly used ones are channelrhodopsin-2 (ChR2) and halorhodopsin (NpHR). The opsin protein ChR2 activates neurons when exposed to 473-nm blue light and the opsin protein NpHR inhibits neural activity when it is exposed to 593-nm yellow light. Because the response spectra of ChR2 and NdHR are sufficiently separated so that they do not interfere, both can be expressed simultaneously in the same neurons to enable bidirectional optical control of neural activity.

The optical power used to activate the opsin proteins can come from either a laser diode or LED and typically is less than 1 mW. If the light has to pass through a certain thickness of tissue before reaching the opsins then the light attenuation in the tissue needs to be taken into account. In addition, when light emerges from an optical fiber the divergence of the beam depends on the core size and numerical aperture of the optical fiber. Optical fiber core sizes used in optogenetics nominally range from 100 to 400 μm. An online application that can be used to calculate light transmission through brain tissue can be found on the laboratory website of Kenneth Diesseroth (http://www.stanford.edu/group/dlab/cgi-bin/graph/chart.php). This program calculates the spread of the light beam and the depth of penetration using factors such as the wavelength of light, the fiber numerical aperture, light intensity and fiber core radius.

11.6 Summary

Biophotonics technologies are widely used in biomedical research, in the detection and treatment of diseases and health conditions, and in point-of-care healthcare clinics. In addition to a broad selection of microscopic and spectroscopic methods described in previous chapters, this chapter describes other advanced tools and implementations. These include the following:

- Optical tweezers and optical trapping techniques that enable microscopic manipulation of cells and molecules for exploring biological materials and functions in the micrometer and nanometer regime
- Miniaturized photonics-based instrumentation functions and devices such as the lab-on-a-chip and lab-on-fiber technologies
- Microscope-in-a-needle concepts to enable 3-dimensional scanning of malignant tissue within the body
- The detection of single nanoscale particles (e.g., viruses) using a method named interferometric reflectance imaging sensor (IRIS)
- Optogenetics procedures, which attempt to explore and understand the mechanisms of neuronal activity in organs such as the brain and the heart.

11.7 Problems

11.1 Show that a force of $F = 2.5 \times 10^3$ pN is needed to pick up a grain of maize pollen that has a mass $m = 2.5 \times 10^{-10}$ kg.

11.2 Consider an optical trap that has a stiffness $k = 45$ pN/μm. If an image analysis instrument can determine the position of a micrometer-sized sphere to within 12 nm, show that the force resolution is 0.54 pN.

11.3 Consider an optical trap that has a trapping efficiency of $Q = 0.30$. Show that if a power level of 150 mW is used in the trap and if the refractive index of the medium is $n_{med} = 1.36$, then the total force F_{total} exerted on a particle is 203 pN.

11.4 Consider an optical trap that has a trapping efficiency of $Q = 0.25$. Let the refractive index of the medium be $n_{med} = 1.36$. Show that if the total force F_{total} exerted on a particle is 80 pN, then a power level of 70.8 mW is used in the trap.

11.5 Using Web resources (for example, www.microfluidic-ChipShop.com), describe the operation of a microfluidic device such as a droplet-generator chip, a micro-mixer chip, or a particle and cell-sorting chip.

11.6 The applications of lab-on-a-chip devices routinely require interfaces between a microfluidic chip and the macroscopic world. Using Web resources (for example, www.microfluidic-ChipShop.com), describe how an interface chip enables such interconnections through the use of components such as miniaturized connectors and tubing, syringe pumps, valves, or waste reservoirs.

11.7 Using Web resources or journal articles, describe some configurations and uses of a lab-on-fiber device.

11.8 Using Web resources or journal articles, describe a typical setup for an optogenetics procedure.

References

1. A. Ashkin, History of optical trapping and manipulation of small-neutral particel, atoms, and molecules. IEEE J. Sel. Topics Quantum Electron. **6**(6), 841–856 (2000)
2. A. Chiou, M.T. Wei, Y.Q. Chen, T.Y. Tseng, S.L. Liu, A. Karmenyan, C.H. Lin, Optical trapping and manipulation for biomedical applications, chap. 14, in *Biophotonics*, ed. by L. Pavesi, P.M. Fauchet (Springer, New York, 2008)
3. D.J. Stevenson, F. Gunn-Moore, K. Dholakia, Light forces the pace: optical manipulation for biophotonics J. Biomed. Opt. 15(4), 041503 (2010)
4. I. Verdeny, A. Farré, J. Mas, C. López-Quesada, E. Martín-Badosa, M. Montes-Usategui, Optical trapping: a review of essential concepts. Opt. Pura Apl. **44**(3), 527–551 (2011)
5. P.M. Bendix, L. Jauffred, K. Norregaard, L. B. Oddershede, Optical trapping of nanoparticles and quantum dots. IEEE J. Sel. Topics Quantum Electron. 20(3), article 4800112 (2014)
6. J.-B. Decombe, S.K. Mondal, D. Kumbhakar, S.S. Pal, and J. Fick, Single and multiple particle trapping using non-Gaussian beams from optical fiber nanoantennas. IEEE J. Sel. Topics Quantum Electron, 21(4), article 4500106 (2015)
7. C. Pacoret S. Régnier, Invited Article: a review of haptic optical tweezers for an interactive microworld exploration. Rev. Sci. Instrum, 84, article 081301 (2013)
8. I. Heller, T P. Hoekstra, G.A. King, E.J.G. Peterman, G.J.L. Wuite, Optical tweezers analysis of DNA–protein complexes. Chem. Rev, 114(6), 3087–3119 (2014)
9. D. Wolfson, M. Steck, M. Persson, G. McNerney, A. Popovich, M. Goksör, T. Huser, Rapid 3D fluorescence imaging of individual optically trapped living immune cells. J. Biophotonics **8**(3), 208–216 (2015)
10. A.J. Crick, M. Theron, T. Tiffert, V.L. Lew, P. Cicuta, J.C. Rayner, Quantitation of malaria parasite-erythrocyte cell-cell interactions using optical tweezers. Biophys. J. 107(4), 846–853 (2014)
11. J. Mas, A. Farré, J. Cuadros, I. Juvells, A. Carnicer, Understanding optical trapping phenomena: a simulation for undergraduates. IEEE T. Educ. **54**, 133–140 (2011)
12. F.A. Gomez, *Biological Applications of Microfluidics* (Wiley, Hoboken, NJ, 2008)
13. C. Lu, S.S. Verbridge, *Microfluidic Methods for Molecular Biology* (Springer, 2016)
14. B. Lin (ed.), *Microfluidics* (Springer, Berlin, 2011)
15. Microfluidic ChipShop, *Lab-on-a-Chip Catalogue* (Jena, Germany, July 2015) www.microfluidic-ChipShop.com
16. M.W. Collins, C.S. König (eds.), *Micro and Nano Flow Systems for Bioanalysis* (Springer, New York, 2013)
17. S. Unterkofler, M K. Garbos, T.G. Euser, P.St.J. Russell, Long-distance laser propulsion and deformation monitoring of cells in optofluidic photonic crystal fiber, J. Biophotonics, 6(9), 743–752 (2013)
18. G. Testa, G. Persichetti, P.M. Sarro, R. Bernini, A hybrid silicon-PDMS optofluidic platform for sensing applications. Biomed. Opt. Express, 5(2), 417–426 (2014)
19. F.F. Tao, F. Xiao, K.F. Lei, I.-C. Lee, Paper-based cell culture microfluidic system. BioChip J, 9(2), 97–104 (2015)

20. A. Ricciardi, M. Consales, G. Quero, A. Crescitelli, E. Esposito, A. Cusano, Lab-on-fiber devices as an all around platform for sensing. Opt. Fiber Technol., 19(6), 772–784 (2013)
21. M. Consales, M. Pisco, A. Cusano, Review: lab-on-fiber technology: a new avenue for optical nanosensors. Photonic Sensors 2(4), 289–314 (2012)
22. J. Albert, A lab on fiber. IEEE Spectr. 51, 48–53 (2014)
23. R.S. Pillai, D. Lorenser, D.D. Sampson, Deep-tissue access with confocal fluorescence microendoscopy through hypodermic needles. Opt. Express 19(8), 7213–7221 (2011)
24. X. Yang, D. Lorenser, R.A. McLaughlin, R.W. Kirk, M. Edmond, M.C. Simpson, M.D. Grounds, D.D. Sampson, Imaging deep skeletal muscle structure using a high-sensitivity ultrathin side-viewing optical coherence tomography needle probe. Biomed. Opt. Express 5 (1), 136–148 (2014)
25. W.C. Kuo, J. Kim, N.D. Shemonski, E.J. Chaney, D.R. Spillman Jr., S.A. Boppart, Real-time three-dimensional optical coherence tomography image-guided core-needle biopsy system. Biomed. Opt. Express 3(6), 1149–1161 (2012)
26. C. Song, D.Y. Park, P.L. Gehlbach, S.J. Park, J.U. Kang, Fiber-optic OCT sensor guided SMART micro-forceps for microsurgery. Biomed. Opt. Express, 4(7), 1045–1050 (2013)
27. M.R. Monroe, G.G. Daaboul, A. Tuysuzoglu, C.A. Lopez, F.F. Little, M.S. Ünlü, Single nanoparticle detection for multiplexed protein diagnosis with attomolar sensitivity in serum and unprocessed whole blood. Anal. Chem. 85, 3698–3706 (2013)
28. M.S. Ünlü, Digital detection of nanoparticles: viral diagnostics and multiplexed protein and nucleic acid assays. MRS Proc, 1720 article mrsf14-1720-d01-01 (2015)
29. D. Sevenler, N.D. Ünlü, M.S. Ünlü, Nanoparticle biosensing with interferometric reflectance imaging, ed. by M.C. Vestergaard, K. Kerman, I.M. Hsing, and E. Tamiya, Nanobiosensors and Nanobioanalyses, (Springer, 2015) 81–95
30. F. Zhang, V. Gradinaru, A.R. Adamantidis, R. Durand, R.D. Airan, L. de Lecea, K. Deisseroth, Optogenetic interrogation of neural circuits: technology for probing mammalian brain structures. Nat. Protocols 5(3), 439–456 (2010)
31. E. Ferenczi, K. Deisseroth, Illuminating next-generation brain therapies. Nat. Neurosci. 19, 414–416 (2016)
32. T.N. Lerner, L. Ye, K. Deisseroth, Communication in neural circuits: Tools, opportunities, and challenges. Cell 164, 1136–1165 (2016)
33. M. Hashimoto, A. Hata, T. Miyata, H. Hirase, Programmable wireless light-emitting diode stimulator for chronic stimulation of optogenetic molecules in freely moving mice. Neurophotonics, 1(1), article 011002 (2014)
34. J.M. Cayce, J.D. Wells, J.D. Malphrus, C. Kao, S. Thomsen, N.B. Tulipan, P.E. Konrad, E.D. Jansen, A. Mahadevan-Jansen, Infrared neural stimulation of human spinal nerve roots in vivo. Neurophotonics, 2(1), article 015007 (2015)
35. C.M. Aasted, M.A. Yücel, R.J. Cooper, J. Dubb, D. Tsuzuki, L. Becerra, M.P. Petkov, D. Borsook, I. Dan, D.A. Boas, Anatomical guidance for functional near-infrared spectroscopy: atlasViewer tutorial. Neurophotonics, 2(2), 020801 (2015)

Erratum to: Biophotonics

Erratum to:
G. Keiser, *Biophotonics*, Graduate Texts in Physics,
DOI 10.1007/978-981-10-0945-7

The subjected book was inadvertently published without including the following typo errors in some of the chapters. The erratum book and the chapters are updated.

The typo errors are:

Page 86: The last part of Problem 3.7 should be "...V = 3.59."
Page 86: In Problem 3.11 the first line should read: "n_{eff} = 1.479 and Λ = 523 nm at 20 °C."
Page 115: Problem 4.1 should read: "...show that the irradiance is 3.18 W/cm^2."
Page 116: In Problem 4.4 "Table 5.1" should be "Table 4.1."
Page 143: The last part of Problem 5.3 should be "...1678 nm."
Page 166: Eq. (6.17) should be

$$Q_s = \frac{8x^4}{3} \left(\frac{n_{rel}^2 - 1}{n_{rel}^2 + 2} \right)^2$$

Page 192: In Problem 6.10 the parameter α_a should be μ_a.
Page 192: Problem 6.11: The last part should read "show that μ_s = 9 μ_a."

The updated original online version for this book can be found at
DOI 10.1007/978-981-10-0945-7

G. Keiser (✉)
Department of Electrical and Computer Engineering, Boston University,
Newton, MA, USA
e-mail: gkeiser@photonicscomm.com

© Springer Science+Business Media Singapore 2016
G. Keiser, *Biophotonics*, Graduate Texts in Physics,
DOI 10.1007/978-981-10-0945-7_12

Page 227: Problem 7.1: The last part of the second sentence should read "reflection loss of 1.1 dB at a wavelength λ_1, and a transmission loss of 1.1 dB at a wavelength λ_2."

Page 228: Problem 7.5 should read: "(a) Show that the size of the active area in a bundle with one ring is 0.22 mm^2. (b) Show that the ratio of the active area to the total cross sectional area of the bundle is 54 %."

Page 228: Problem 7.6 should read: "(a) ... the active area in a bundle with two rings is 1.13 mm^2. (b) ... the active area to the total cross sectional area of the bundle is 53 %."

Page 228: Problem 7.8 should read: "the variation in the insertion loss when the longitudinal separation changes from 0.020 mm to 0.025 mm is 0.21 dB."

Page 256: Problem 8.5 should read: "Show that the depth of field at a wavelength of 650 nm is 24.5 μm and 3.24 μm..."

Page 287: Problem 9.7 should read: "Figure 9.9 shows a spheroid with the radii of the equatorial and polar axes being 0.3 μm and 2.4 μm, respectively."

Page 287: Problem 9.10 should read: "...the curved cylinder has a length of 550 μm and its radius is a = 0.5 D_{bottom} = 150 μm."

Index

© Springer Science+Business Media Singapore 2016
G. Keiser, *Biophotonics*, Graduate Texts in Physics,
DOI 10.1007/978-981-10-0945-7

Printed in the United States
By Bookmasters